DEAD RECKONING

DEAD
RECKONING

THE STORY OF HOW JOHNNY MITCHELL
AND HIS FIGHTER PILOTS TOOK ON ADMIRAL
YAMAMOTO AND AVENGED PEARL HARBOR

DICK LEHR

HARPER

An Imprint of HarperCollins*Publishers*

HarperCollins books may be purchased for educational, business, or sales promotional use. For information, please email the Special Markets Department at SPsales@harpercollins.com.

FIRST EDITION

Designed by Leah Carlson-Stanisic
Photograph by Baranov E/Shutterstock, Inc.

Library of Congress Cataloging-in-Publication Data has been applied for.

ISBN 978-0-06-244851-4

20 21 22 23 24   LSC   10 9 8 7 6 5 4 3 2 1

Dedicated to the memory of John F. Lehr,

Staff Sergeant, USMC, Guam, 1945–1947

CONTENTS

Cast of Characters . . . . *ix*

PROLOGUE: The Day 1

PART I | THE MAKING OF WARRIORS

CHAPTER 1 Johnnie Bill and the Moon 13

CHAPTER 2 Isoroku, Aeroplanes, and a Geisha Girl 35

CHAPTER 3 The Flyboy from Enid 53

CHAPTER 4 No Ordinary Strategy 67

CHAPTER 5 A Taste of War 89

CHAPTER 6 When the Rose Petals Fell 109

PART II | THE SOUTH PACIFIC

CHAPTER 7 Wedding Bells and Pacific Blues 131

CHAPTER 8 Unfinished Business 147

CHAPTER 9 Midway: Yamamoto's Lament 161

CHAPTER 10 Mitchell on the Move 179

PART III | GUADALCANAL

CHAPTER 11 The First Kill 197

CHAPTER 12 Nights to Remember 209

CHAPTER 13 Moon over Guadalcanal 233

PART IV | VENGEANCE

CHAPTER 14 Five Days and Counting 255

CHAPTER 15 One Day More 271

CHAPTER 16 Dead Reckoning 291

EPILOGUE 311

Acknowledgments 333

Notes 339

Selected Bibliography 385

Index 389

CAST OF CHARACTERS

(in alphabetical order)

The Yamamoto Mission Pilots

ROGER J. AMES: First Lieutenant, Army Air Force, Laramie, Wyoming; cover flight

EVERETT H. ANGLIN: First Lieutenant, Army Air Force, Arlington, Texas; cover flight

REX T. BARBER: First Lieutenant, Army Air Force, Culver, Oregon; attack flight

DOUGLAS S. CANNING: First Lieutenant, Army Air Force, Wayne, Nebraska; cover flight

DELTON C. GOERKE: First Lieutenant, Army Air Force, Syracuse, New York; cover flight

LAWRENCE A. GRAEBNER: First Lieutenant, Army Air Force, St. Paul, Minnesota; cover flight

RAYMOND K. HINE: First Lieutenant, Army Air Force, Harrison, Ohio; attack flight

BESBY F. HOLMES: First Lieutenant, Army Air Force, San Francisco, California; attack flight

JULIUS "JACK" JACOBSON: First Lieutenant, Army Air Force, San Diego, California; cover flight

LOUIS R. KITTEL: Major, Army Air Force, Fargo, North Dakota; cover flight

THOMAS G. LANPHIER, JR.: Captain, Army Air Force, Detroit, Michigan; attack flight

ALBERT R. LONG: First Lieutenant, Army Air Force, Taft, Texas; cover flight

JOHN W. MITCHELL: Major, Army Air Force, Enid, Mississippi; ace pilot, mission planner, cover flight leader

WILLIAM E. SMITH: First Lieutenant, Army Air Force, Glendale, California; cover flight

ELDON E. STRATTON: First Lieutenant, Army Air Force, Anderson, Missouri; cover flight

GORDON WHITAKER: First Lieutenant, Army Air Force, Goldsboro, North Carolina; cover flight

Other Notable Characters

WALLACE L. DINN: Lieutenant, Army Air Force; friend of John Mitchell, killed in action

JOSEPH FINNEGAN: Lieutenant Commander, US Navy; code breaker, Station Hypo, Pearl Harbor

ELLERY GROSS: Lieutenant, Army Air Force; friend of John Mitchell, killed in P-38 Lightning test flight

WILLIAM J. "BULL" HALSEY, JR.: Admiral, US Navy; commander, US forces, Solomon Islands

HIROSHI HAYASHI: Chief Pilot, Betty bomber No. 326, April 18, 1943

WILFRED J. HOLMES: Lieutenant Commander, US Navy; code breaker, Station Hypo, Pearl Harbor

CHIYOKO KAWAI: geisha and Yamamoto's mistress

ALVA "RED" LASSWELL: Major, US Marine Corps; code breaker, Station Hypo, Pearl Harbor

EDWIN T. LAYTON: Commander, US Navy; intelligence officer, US Pacific Fleet

JAMES MCLANAHAN: First Lieutenant, Army Air Force; originally assigned to the attack flight in the Yamamoto mission but replaced when engine trouble forced him to turn back

ANNIE LEE MILLER MITCHELL: wife of John Mitchell

EUNICE MASSEY MITCHELL: stepmother of John Mitchell

LILLIAN FLORENCE DICKINSON MITCHELL: mother of John Mitchell

NOAH BOOTHE MITCHELL: father of John Mitchell

MARC "PETE" MITSCHER: Rear Admiral, US Navy; commander, Air Forces, Solomon Islands

JOSEPH MOORE: First Lieutenant, Army Air Force; originally assigned to the attack flight in the Yamamoto mission but replaced when engine trouble forced him to turn back

CHUICHI NAGUMO: Vice Admiral, Imperial Japanese Navy; oversaw Pearl Harbor attack

CHESTER W. NIMITZ: Admiral, US Navy; commander in chief, US Pacific Fleet

JOSEPH J. ROCHEFORT: Commander, US Navy; officer in charge, Station Hypo, Pearl Harbor

SHOICHI SUGITA: Flight Petty Officer; pilot of one of six Zeros escorting Yamamoto from Rabaul to Bougainville on April 18, 1943

SADAYOSHI TAKANO: father of Isoroku Yamamoto

MATOME UGAKI: Vice Admiral; chief of staff, Combined Fleet, Imperial Japanese Navy; passenger aboard Betty bomber No. 326, April 18, 1943

HENRY "VIC" VICCELLIO: Colonel, Army Air Force; John Mitchell's commander

YASUJI WATANABE: Captain, Imperial Japanese Navy; longtime aide to Yamamoto

ISOROKU YAMAMOTO: Admiral; commander in chief, Combined Fleet, Imperial Japanese Navy; passenger aboard Betty bomber No. 323, April 18, 1943

REIKO MIHASHI YAMAMOTO: wife of Isoroku Yamamoto

KENJI YANAGIYA: Flight Petty Officer; pilot of one of six Zeros escorting Yamamoto from Rabaul to Bougainville on April 18, 1943

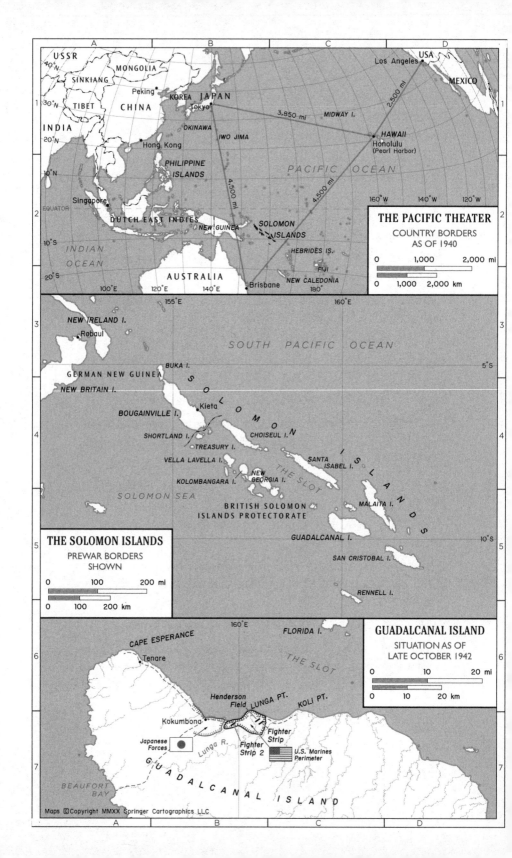

THE PACIFIC THEATER

COUNTRY BORDERS
AS OF 1940

0	1,000	2,000 mi

0	1,000	2,000 km

THE SOLOMON ISLANDS

PREWAR BORDERS
SHOWN

0	100	200 mi

0	100	200 km

GUADALCANAL ISLAND

SITUATION AS OF
LATE OCTOBER 1942

0	10	20 mi

0	10	20 km

Maps ©Copyright MMXX Springer Cartographics LLC

DEAD RECKONING

THE DAY

THE FIRST ONE APPEARED FROM THE CLEAR BLUE SKY EARLY ON A Sunday morning, out of nowhere, it seemed, as sailors, soldiers, and civilians aboard ships, at airfields, in bunkhouses and in bungalows were just waking up, some shaving, others showering, still others already sipping coffee in the mess halls where they gathered for breakfast, many nursing hangovers after a night spent socializing on shore leave or at the base itself. Indeed, at Wheeler Field, some were still going strong, wearing tuxedoes and stumbling out of the officers' club at 7:51 a.m. into the bright light of day, laughing and carrying on after their all-night party. These men dismissed the hum of engines in the sky, assuming that Navy planes were performing maneuvers, and they likely felt sorry for the pilots having to fly at the crack of dawn, on God's day, no less, while most of the thousands of men and women assigned to Pearl Harbor were off duty. But as the hum grew louder, some of the men in their tuxes realized the engine noise did not sound quite right, different from their Navy planes, and in that moment the casual mood turned dark. As the planes closed in, the rising sun insignias on their wings were revealed, and in the next seconds a wave of Japanese dive-bombers opened fire, strafing and dropping their payload. The officers in tuxedoes, festive seconds before, ran for their lives, just as Americans stationed all around the Hawaiian island of Oahu did, as the shock and awe at Wheeler Field were repeated in the next nine minutes at other airfields and naval stations and then extended to the sprawling US Pacific Fleet moored at Pearl Harbor.

It was a surprise naval aerial assault involving 183 Japanese attack planes and bombers loaded with armor-piercing bombs, shallow-water torpedoes, and machine guns, all of which, in a flash, found targets on the ground and in the harbor. Bombs whistled to earth, sending up columns of oily black smoke from the hangars they flattened and the rows and rows of US fighter planes they wrecked, while men raced outside, dazed and wild eyed, pulling on pants, shirts, or a bath towel, dodging ear-shattering explosions, bullets, and shrapnel, some reaching safety, others not, their bodies bludgeoned by the blasts and bullets. At the mess hall at Hickham Field, a bomb plummeting through the roof exploded, killing some thirty-five men who only had thoughts about what to eat for breakfast on their minds. At 7:55 a.m., the *Nevada* ship's band was readying to play "The Star-Spangled Banner" when two Japanese planes strafed the deck and shredded an American flag, just as an officer at the Ford Island Command Center at 7:58 a.m. frantically typed what became one of the most famous radio dispatches ever: AIR RAID, PEARL HARBOR. THIS IS NOT DRILL. At the same moment, an ensign using binoculars aboard the battleship *Arizona* stared at Japanese planes, unimpededly dropping one bomb after another onto Ford Island, before he broke to sound the ship's three-blast alarm signaling an air raid, a howling siren that made its way throughout the six-hundred-foot-long destroyer. Men on other ship decks, too, gaped at the aerial savaging of airfields until something else caught their eyes: slender black masses in the waters below racing toward them, which the seamen instantly identified as shallow-water torpedoes, bombs that had splashed into the water from low-flying Japanese dive-bombers and were now speeding toward their battleships, the *Oklahoma, West Virginia, Nevada, California,* and *Arizona.*

The *Oklahoma's* hull was struck, sending everything inside toppling, from tables and dishes to sailors and weaponry, including massive shells that broke free and crushed men in their path, the torpedoes creating breaches through which ocean waters rushed and continued into hatches, down ventilator shafts, and through any other open passageway, flooding the ship so that it listed and began to sink. At 8:06 a.m., a high-flying Japanese bomber hit a bull's-eye more than nine thousand feet below, smack on the *Arizona's* forward deck, dropping a 1,765-pound bomb that cut through the armored skin to detonate the more than million pounds of explosives and ammunition stored below, an explosion so powerful it caused the 31,400-ton battleship to lurch upward, buckle, and then collapse into itself; an explosion so huge that the fireball soared five hundred feet into the air along with burning flesh, body parts, and sailors split in two; an explosion so fatal that more than a thousand men on board were killed. By 8:50 a.m., the catastrophic first wave of attack planes was followed by a second, as 167 additional Japanese fighters and bombers continued the relentless horror in the harbor and increased the count of American dead and wounded, until the surprise assault finally ended a full two hours and fifteen minutes after it had begun.

The very next day in Washington, at 12:30 p.m. local time—less than twenty-four hours after the devastating raid—the president of the United States appeared before a joint session of Congress. Looking drawn, Franklin Delano Roosevelt stood at a lectern packed with microphones. His voice crackling as it was broadcast on radio to all of America, the president listed other locations that Japan had attacked the preceding day—Hong Kong, Singapore, Guam, Wake Island, Malaysia, the Philippines—and as for the deadly assault on Pearl Harbor, he said, "Yesterday, December 7, 1941—a day which will live in infamy—the United States was suddenly and deliberately attacked by naval and air forces of the Empire of Japan."

The president then asked Congress for a declaration of war, and he got it.

SIXTEEN MONTHS LATER

Army Air Forces pilot Major John W. Mitchell—Mitch, as he was called—tried as best he could to get comfortable on the hill overlooking Fighter Strip Two. It wasn't easy. He and his comrades lived in perpetual discomfort. The island of Guadalcanal could seem a tropical paradise one moment—cloudless skies, aquamarine surf, and reasonably warm and dry air. But most days it was a tropical hell on the US base bordered by thick, fetid jungle, marshy beaches, and mud. Mud everywhere. Mosquitoes, too. Mitchell's home atop the airfield was a canvas tent with a cot and mesh netting that hardly served as a solid line of defense against the malaria-carrying insects.

Secretary of the Navy Frank Knox, after a recent trip to Guadalcanal, had told reporters, "One of our greatest enemies is malaria." In one month alone, nearly two thousand men had been hospitalized with it. And the flu-like illness wasn't the only invisible jungle disease lying in wait. Ground troops, especially, coped with intestinal infections and diarrhea, skin rashes, dengue with its joint pain, the yucky crud of fungus, and a foot rot that wrapped an affected man's ankles in pus-filled blisters. Then there was the dank foxhole. Everyone had a foxhole, either inside or right outside his tent, for quick refuge when the Japanese bombs began falling, which happened regularly. The coastal plain where the US troops were stationed was pockmarked with craters left by those bombs, while the shrapnel that ripped through the canvas tents seemed as common as seashells.

Mitchell pushed all that out of his mind as he took his pen and began a letter to "My Darling." It was his latest installment in a long-distance marriage by mail. "Do you realize," he told Annie Lee, "that I haven't seen you in over 15 months and that we have been married 16 months and of that 16 months I haven't been with you even a month?" He felt cheated in love by the war and missed his wife sorely. "I love you more now than ever before," he said. "'Tis true, my love."

He wrote the words in the neat cursive script he'd been taught while growing up in the red clay hills of north-central Mississippi. His hometown, Enid, a tiny hamlet alongside a railroad stop, was some forty miles south of the college town of Oxford, where William Faulkner, the future bard of the South, had been a teenager when Mitchell was born on June 14, 1914. Strong and supple, of medium height, with dark, wavy hair, John William Mitchell was now two months shy of twenty-nine.

Mitchell's letter was largely upbeat, full of chitchat and with only a bit of war news. That was because he'd just returned to Cactus—the military's code name for Guadalcanal—from a brief leave. It had been a scheduled layoff, part of the system worked out for him and the other US Army Air Forces pilots, given that there weren't enough of the new, speedy P-38 Lightning fighters to go around. Mitchell and his "flight"—the term used for a group of pilots—would fly missions for a few weeks and then take a break, replaced by a second flight.

Each flight featured a lineup of top-notch pilots. Mitchell, an ace pilot, had recently been promoted to unit commander of the 339th Fighter Squadron. The other flight had standouts, too, namely Rex Barber, a bullnecked kid from Oregon farm country; Tom Lanphier, a military brat who had grown up mostly outside Detroit; and Besby Holmes, a mix of brains and brawn from the streets of San Francisco.

Mitchell didn't tell Annie Lee much about the respite in New Zealand, which had included plenty of drinking and gambling and some rest. He did brag about winning at dominoes, where he had turned an initial $18 stake into a $435 haul—enough to pay his way home to San Antonio, Texas, he said, and "for a couple of champagne parties." But that assumed he would be coming home soon, and the question of when that might be hung over the five pages, something he looped back to time and again, like a refrain in a song, when he told his wife he wanted to return as soon as possible to be with "the sweetest woman on earth." He'd overheard talk that his name was on the list to go home—maybe, just maybe, by mid-June, he wrote longingly. But each comment like that was met with the equal and opposite slap of wartime reality. He cautioned against her getting her hopes up, citing a fact of life for every soldier: "In these days and times one hardly knows where he will be the next hour, much less the next day."

Truer words were never written. Because for all his aching desire to reunite with Annie Lee, to celebrate with champagne paid for from his gambling winnings, Mitchell wasn't going home anytime soon. As he finished the letter that Friday, April 16, 1943, unbeknownst to him, preparations were under way. In the command dugout not far from his tent, mission plans were being hashed out—plans that would have cataclysmic and far-reaching consequences for the pilot from a tiny rural town in Mississippi and, bigger still, for a world at war.

SIX HUNDRED THIRTY MILES NORTH FROM MAJOR MITCHELL, ON the island of New Britain, the largest island in New Guinea, another soldier had a woman on his mind as he made plans for a troop inspection at bases in the southern Solomon Islands. He was Isoroku Yamamoto, commander in chief of Japan's Combined Fleet. Diminutive in stature at five feet, three inches, the stoic-looking admiral none-theless cast the long shadow of a giant. He was the most powerful naval officer in Japan and the most famous of all of Japan's military leaders—a symbol of the nation's navy, present and past. As one US commander put it, "Yamamoto represented more than one person, or individual—he represented the Japanese Navy, its strength, its morale and its early victories."

The decorated admiral had arrived by seaplane two weeks ear-lier, on April 3, to set up a temporary command at Rabaul. He was staying by himself in a cottage located some 1,000 feet above the base on what was called "Residency Hill," a quieter and, more im-portant, cooler spot, where without fanfare he had turned fifty-nine the day after his arrival. As did all of Rabaul, the cottage sat beneath towering volcanoes on the island's northeastern tip. And never far from his thoughts was the woman who'd captured his heart, Chiyoko Kawai. Chiyoko was not Yamamoto's wife but an attractive geisha nineteen years younger than he. The summer upcoming would mark the tenth anniversary of their first meet-ing at a restaurant in Tokyo. Even after a decade the admiral was still enchanted with her, savoring romantic feelings that had never been part of his arranged marriage to his wife of twenty-five years, Reiko Mihashi. In her favor, Reiko was from a town not far from his birthplace, Nagaoka, a remote village with harsh winters. But the best Yamamoto could say about Reiko after they'd met was that she seemed "strong" and "sturdy," and, as he told an older brother, "I'm thinking of taking a look at her, and if there's nothing wrong, settling for her." Though often apart, the couple had two sons and two daughters.

Yamamoto had certainly come a long way from his modest start in Nagaoka in northwest Japan. Born Isoroku Takano on April 4, 1884, it was only later that his identity changed to Isoroku Yama-moto. His parents died when he was in his late twenties, and he was adopted by the Yamamoto family, a common practice in Japan, even at his age. Isoroku began his military career early; he was just six-teen when he was accepted into the Naval Academy. He rose steadily over the years, being promoted in 1940 to the highest naval rank of admiral, an ascendency that included vast international experience in England and the United States and earned him a reputation as a tactical genius.

Indeed, 1941 saw him achieve singular renown. Isoroku Yamamoto was the architect of the massive Japanese assault on Pearl Harbor on December 7. It was his idea—in the making for nearly a year—to initiate the surprise air and sea attack that killed more than 2,400 US sailors, soldiers, pilots, marines, and civilians and left another 1,178 wounded, while twenty-one US ships, including eight battleships, were sunk or disabled. Upon his return to Japan, Yamamoto was showered with praise from the nation's political and military leaders. Emperor Hirohito sent a messenger with his personal congratulations to the newly minted hero. The polar opposite was the case in the United States, however: to Americans, Yamamoto was a monster on a par with Adolf Hitler.

Now, sixteen months later, on Friday, April 16, 1943, the admiral made ready for an inspection tour, overriding objections by aides who worried that traveling to the front lines would put him too close to the flame. Yamamoto insisted that he wanted to thank his men in person for their courage during the bombing raids the previous week, first at Guadalcanal and then at other US-held islands. The itinerary set, a radio cable was sent out from Rabaul detailing his travel plans. Yamamoto then stayed busy reviewing combat reports and attending meetings.

But as the hours passed and his departure drew closer, his aides continued to fret about their admiral traveling so close to live combat. The risks were too great, they argued; war could be so unpredictable. They urged him to cancel. But Yamamoto would have none of it. He was not traveling alone, he reassured them. Six of Japan's best fighter aircraft, Mitsubishi Zeros, were to escort him.

Besides, his men were expecting him. "I have to go," he said.

IN FORTY-EIGHT HOURS—ON PALM SUNDAY IN THE UNITED States—the two men, the US Army Air Forces fighter pilot and the Japanese Imperial Navy admiral, would meet for the first time. One of them would survive. But more than a fatal face-off, theirs was an epic moment involving two nations at war and a clash of cultures, a moment loaded with xenophobia, spycraft, special military operations, wartime sacrifice, and broken hearts. On Sunday, April 18, 1943, Major John Mitchell led fifteen pilots from Guadalcanal to intercept Admiral Isoroku Yamamoto en route to Ballale. Expecting to face dozens of enemy fighters, few on the Mitchell flight expected to return.

It was a historic, high-stakes US attempt at a wartime "targeted kill"—but hardly the last. The circumstances in many ways resembled those of a future surprise attack against the United States, one that killed thousands of civilians and again plunged the nation into war. The chief planner, like Yamamoto, instantly became the hated face of the enemy. His name, as strange to the American tongue as Yamamoto's, became synonymous with evil. Then came a day when spies pinpointed the precise location of the planner and US leaders authorized a secret "decapitation operation." They ordered an elite team of soldiers to nullify the man by death or capture, preferably the former. In a new century the horrific event was the attacks of September 11, 2001, and the subsequent secret decap op was called Operation Neptune Spear, the May 2011 raid by SEAL Team Six commandos that killed Osama bin Laden.

The bin Laden raid was a case of history echoing itself—of John Mitchell and his men receiving orders to get Admiral Yamamoto in an earlier war. Sixteen months had passed since the Pearl Harbor attack had staggered Americans, like a heavyweight boxer rocked by a sucker punch. And the pummeling kept coming, with Japanese army and navy forces sweeping through the western Pacific before the month of December 1941 even ended. The United States rushed to get onto a war footing. Men raced to enlist as volunteer coastal watchers took up positions along the oceanfront, looking for enemy ships. In intelligence circles, code breakers in Hawaii and Washington, DC, unfairly suffered from a sense of guilt. Having previously broken parts of Japan's key naval code, called JN-25, the cryptanalysts despaired that they hadn't detected Yamamoto's huge armada assembling within striking distance of Hawaii in time. They knew that in the future, they would have to do better.

From coast to coast the fear of a "yellow peril" gripped the country. The FBI rounded up 1,379 Japanese immigrants within four days of the attack—men and women, teachers and community leaders—and detained them as "dangerous enemy aliens." Mistrust and animosity toward "Japs" and "Nips," the latter slur based on an abbreviation for Nippon, the Japanese name for Japan, erupted into the open. In Washington, DC, one Chinese newspaper reporter began wearing a large badge pinned on his lapel that read "Chinese Reporter—NOT Japanese—Please."

Ethnic animosity raged throughout the Pacific battlefronts—in the air, on the high seas, and especially on the ground. Fighting between Americans and Japanese took on what one marine called a "brutish, primitive hatred" that instigated "ferocious killing with no holds barred." Wartime atrocities abounded in the "existential struggle for annihilation." And Admiral Isoroku Yamamoto was the marquee face for the loathing that Americans felt toward their Asian enemy.

Only late in 1942 did the picture begin to change, starting with the grueling, bloody takeover of Guadalcanal. John Mitchell was there, and by year's end he was riding in the newly arrived P-38 Lightning, learning on the job how to fly the twin-engine fighter that became a game changer in the skies against the famed Japanese Zero and helped to turn the tide in the Pacific. Then, on Friday, April 16, 1943, when the ace John Mitchell pined to go home, came the extraordinary—and dangerous—opportunity to make history. Yamamoto's departure from Rabaul was set for dawn on Sunday, April 18. It was a chance to mark the end of the beginning—a period during which Japan had dictated the terms of engagement—and pivot to a new beginning of US dominance in the Pacific. Immediately after Pearl Harbor, Yamamoto had cited surprise as the crucial factor: "That we could defeat the enemy at the outbreak of the war was because they were unguarded," he'd written a friend.

Mitchell and his men would now be the ones banking on surprise as they raced across the ocean hoping to meet up with the unsuspecting Japanese icon.

PART I

THE MAKING
OF WARRIORS

CHAPTER 1

JOHNNIE BILL AND THE MOON

John Mitchell's mother, Lillian Mitchell, and father,
Noah Mitchell *Courtesy of the Mitchell family*

IN THE SPRING OF 1943, JOHN MITCHELL'S WIFE, ANNIE LEE, WASN'T
the only one aching for his return from the Pacific War. His father,
Noah Boothe Mitchell, wanted him stateside, too, a feeling expressed
in a poem he composed at the family homestead in the Deep South.
The verse, typed on a Royal typewriter and mailed to the island of Guadalcanal half a world away, was titled "Rainbow in My Heart." It began:

> *There is a rainbow in my heart,*
> *But no pot of gold lies at the end.*
> *The pot contains my cherished hopes*
> *Of my dear one's return again.*

The poem's rhymed quatrains, with their tender love along with a fervent hope for a safe return, also displayed a pacifist streak. The closing stanza went:

And when this grim war is over
And he comes home at last,
Let us keep that hope eternal
That ALL such wars have passed.

JOHN WILLIAM MITCHELL, THE FUTURE WORLD WAR II PILOT, grew up in the rural village of Enid in Mississippi's Tallahatchie County—Tallahatchie being a Choctaw name meaning "rock of waters." He was a third-generation Mississippian. It was his great-great-grandfather Washington Mitchell who, in the first half of the 1800s, had moved from North Carolina to an area east of the Tallahatchie River. Enid did not yet exist officially, and the early 1840s was when a survey crew started a settlement while working on the Mississippi and Tennessee Railroad. One of the patriarch's sons, William Washington Mitchell, nearly a man when the family arrived, married a woman of Irish descent named Jane Carson, and family lore has it that she was related to Kit Carson, the famous frontiersman and scout.

The Mitchell-Carson marriage produced eight children. The first son, William Carson Mitchell, arrived on October 1, 1845; he was John Mitchell's grandfather. William Carson's eventual marriage to Josephine Wilson of nearby Sardis, Mississippi, drew more frontier glamour into the family sphere; Josephine was said to be a distant relative of an earlier pioneer of even greater fame: Daniel Boone. If all that is true, John Mitchell's ancestral lines intertwined with those of two of America's best-known folk heroes.

William was a teenager when he enlisted in the Confederate Army, joining the nearly 80,000 other white men from Mississippi, the second state to secede from the Union, to defend slavery in the "War for Southern Independence," as it was known. No surviving family records indicate that the Mitchells—most of whom were merchants and storekeepers—were slaveholders, but slavery was vital to their state's cotton economy. The state's slave population exceeded its white population—437,000 to 354,000. William fought in northern Mississippi and Tennessee under the command of Confederate general Nathan Bedford Forrest, who built a reputation as "The Wizard of the Saddle" for his tactical use of the cavalry in mobile, quick strikes. Forrest later became an early leader of the Ku Klux Klan.

William returned home to Mississippi after the war. He and Josephine married two years later, in 1867. They raised three sons. The youngest, Noah, born in the summer of 1881, became John Mitchell's father. It was during Noah's teen years that the settlement where a handful of Mitchell families had lived for decades, which had been called various names over the years, officially became the township of Enid. With its railroad depot, grist mill, bank, barber shop, several stores, school, post office, and even a few saloons, Enid had grown into what one resident later called "a thriving place." It was home to Civil War veterans, doctors, and merchants, such as the Mitchells, who built a brick store in the center of town that various members of the family were involved in operating for decades to come. With its train depot, Enid had also become a shipping point for white and red clay mined from the hills eight miles north of town. But Enid was "thriving" in the context of a region still largely rural and undeveloped—the wilderness. The village's population in 1900, according to the US Census, was only 180.

Noah went to high school more than two hundred miles north of Enid at "Bell Buckle," the nickname for the prestigious Webb School in the Tennessee town of Bell Buckle, a stop on the Nashville and Chattanooga Railroad. The school had been founded by another Confederate Army veteran, William R. "Sawney" Webb, a disciplinarian who had graduated from the University of North Carolina and whose own classical education had included instruction in Greek and Latin. Webb and his wife had first started the school in Culleoka, Tennessee, in 1870, but when the town had legalized the sale of liquor in 1886, the couple, ardent prohibitionists, would have none of it; they had relocated thirty-five miles west to Bell Buckle.

Young Noah Mitchell arrived on campus the next fall, in 1887, paying the semester's tuition of $39 in cash his father had given him. The Webbs had built a new schoolhouse on six acres of beech forest a short walk from the train station. Noah joined a sophomore class of forty-seven students that included other boarders, day students, and, unusual for the time, a handful of local girls. When he graduated in 1900, he was one of twenty-six classmates still standing. The schoolwork proved classical and demanding, and Sawney Webb insisted on a strong work ethic, trustworthiness, and honesty. He despised deception, admonishing students with a line that eventually became the school's motto: "Do nothing on the sly," or, in its Latin translation, "Noli res subdole facere." The schoolmaster also had little tolerance for social pretense, a view he had included in the school's original mission, which was, in part, "To turn out young people who are tireless workers . . . who are always courteous without the slightest trace of snobbery."

Compared to his peers at school, Noah was on the small side. Indeed, in the sophomore class photograph taken during his first year at the school, Noah was placed in the first row, seated on the ground with the other shorter boys. His expression is one of seriousness, almost sadness, but then, not a single student in the photograph wears a smile; there is a gravitas to them all. Noah is dressed in a jacket, white shirt, and tie. His clasped hands rest in his lap; his brown hair is parted neatly down the middle. His brown eyes look dreamily to one side. The period dress notwithstanding, the facial similarities to his future son John are unmistakable.

JUST AS HIS OWN FATHER HAD RETURNED HOME AFTER FIGHTING in the Civil War, Noah returned to Enid after graduating from the Webb School, so it was in Enid where John Mitchell would be born into the modest, unpretentious life the Mitchell clan had built. Noah soon met Lillian Florence Dickinson from Temple, Texas, and when they were married on December 23, 1904, he was already established as a salesman marketing the clay from nearby mines as well as Mississippi long-leaf yellow pine from forests farther south. With railroad construction taking off, Mississippi was experiencing a timber boom; just two other states, Washington and Louisiana, produced more lumber in 1904. It was work that took Noah—who was only in his twenties—to parts north, such as New York City, as well as overseas, to Europe and England, where he sold clay to foundries. Nonetheless, he and Lillian began a family the year after their marriage; two boys and two girls were born in the next decade before John William's birth on June 14, 1914. The baby boy with the same brown eyes and hair as his father was called "Johnnie Bill" at first, then "Johnnie" as he got older, and later still, in the military, he became "Mitch."

Johnnie Bill Mitchell grew up in an Enid that hadn't changed much over the years. "Just a wide place in the road," his father would continue to say about the village for years to come. The village's population actually decreased slightly from 1900 to 1920—from 180 to 174 residents, according to the US Census. Meanwhile, Tallahatchie County, where Enid was situated, experienced steady growth; its population increased from 19,600 in 1900 to 35,953 in 1920. The nearest "big city" was Charleston, twelve miles to the south on the east side of the Tallahatchie River. Its population was 1,834, a hub by comparison, featuring several churches and schools, a courthouse, and a jail. Getting around the county was never easy, though; most roads were dirt and gravel, or mud in winter and dust in summer. It was in 1914, the year Johnnie Bill was born, that state officials gave their approval to construction of the first paved road in all of Mississippi, an eleven-mile stretch in Lee County that eventually became the first leg of US Highway 45.

Johnnie Bill grew up in the same house his father had, a single-story, wood-shingled structure located in the village center. The front of the house looked out onto the railroad tracks less than a hundred yards away. The trains rumbling through town at night, their whistles blowing, were jarring to visitors but something Johnnie Bill and his family slept right through. The front porch running the length of the house was a popular spot on hot summer evenings, while inside a large fireplace in the living room kept the house warm during occasional winter cold spells. The bedrooms were toward the front of the house, while the dining room and large kitchen were in the back, with easy access to the well just outside the rear door, where Johnnie Bill and his siblings fetched buckets of water for cooking and cleaning. The privy was farther back; it would be thirty years after Johnnie Bill's birth before basic plumbing was installed.

Pecan and maple trees filled the large backyard. Even with his frequent travels as a salesman for the Memphis-based Gayoso Lumber Company and his own Mitchell Clay Company, Johnnie Bill's father found time to maintain a pear orchard just south of town, producing an annual yield of about a thousand bushels. "The fruit is excellent and abundant," noted a local newspaper, adding that Noah was "selling the pears cheap, only one dollar per bushel." People came from all over for his pears. In their large yard Noah kept a sprawling vegetable garden with flowers that attracted birds of all kinds. "Birds seem to know those who love them and it seems to me that a good many more birds come to our place than to some of the neighbors," Noah wrote proudly in an essay titled "Around Home." The family kept track of goldfinches, mockingbirds, orioles, bright red Kentucky cardinals, and northern flickers, a type of woodpecker. They listened to the doleful croaking of slate-colored rain crows, and their days often began and ended to the lyrical sounds of brown thrushes. "I do not believe there is a bird on earth that can surpass them in sweetness of song," Noah wrote of thrushes. "The morning notes are bright, clear and cheerful, while the evening song seems a little sad and sleepy and much like a lullaby, but both morning and evening renditions are purest melody."

These were the sights and sounds of Johnnie Bill's home. His grandfather William Carson lived next door in a smaller house, and on the other side was the redbrick, mazelike general store. Inside, a balcony with iron railings—accessible by a staircase in each of the four corners—ran the length of the interior walls, increasing the store's capacity tenfold. Upstairs was office space that at one point was used as a funeral parlor, at another time as a barbershop. Over the years various uncles and townspeople owned or operated the bustling emporium, and it was a place where Johnnie Bill and his brothers and sisters were always welcome. The main floor was lined with shelves, and cubbyholes were filled with everything imaginable from candy and treats to such foodstuffs as meal, flour, and dried lentils, as well as cookware, horse collars, shoelaces, matches, and axle grease. Separated by a window toward the rear of the store was a meat house, and the ground floor even had room for displays of women's apparel and menswear.

A big iron potbellied heater in the middle and a small fireplace on the balcony provided warmth.

Johnnie Bill's world was Enid and the surrounding countryside; it was family centered and small town. The life beyond seemed to be passing him by, even with the railroad right there. Automobiles, the "good roads movement" and new railroad lines—hallmarks of progress—were happening elsewhere, not in his backyard. But if he missed out on some of the big events of the time, he was fortunate to go largely untouched by others. In 1917, the United States entered the war that had been raging across Europe since 1914. Congress passed a law that spring requiring all men between the ages of twenty-one and thirty to register for military service, and in April, President Woodrow Wilson's call to arms promised that "the spirit of ruthless brutality will enter into every fiber of our national life, infecting Congress, the courts, the policeman on the beat, and the man on the street." Johnnie Bill's father was one of those men. Noah Mitchell went to his local draft office on September 12, 1918, joining 157,606 other Mississippians who registered to be called. He listed his occupation as "clay miner" for the Mitchell Clay Company and his nearest relative as his wife, Lillian Mitchell. But Lillian, her toddler Johnnie Bill, and the other children were spared losing their husband and father to combat. By the Great War's end in November 1918, 43,362 men from Mississippi had been inducted into the military, but Noah had not been one of them. Though willing to go, he was never called; he was too old—thirty-seven.

Then, the same year as Noah's draft registration, the influenza pandemic swept around the planet, a lethal virus that infected cells deep in the lungs and, in fifteen months, killed up to 50 million people worldwide. In the United States, the death toll was put at 675,000, and in Mississippi, which the prior year had seen 442 flu-related deaths, 6,219 residents succumbed. In cities and towns throughout the state, Sunday church services were canceled, public meetings postponed, and schools closed. The Mississippi Delta region was especially hard hit, but farther north in Enid, the Mitchell family managed to dodge the disease's deadly grasp.

Johnnie Bill's young life was not without tragedy, however. In

May 1919, a month before he turned five, a baby sister named Elizabeth was born, Noah and Lillian Mitchell's sixth child. Six months later, the baby was dead. The family soon suffered an even greater setback: in February 1922, Noah and Lillian were in Memphis when she took ill and was hospitalized. Eleven days later—and only in her late thirties—she died. The culprit was encephalitis, a rare and sometimes fatal infection that causes swelling in the brain. Johnnie Bill was seven, with four older siblings ranging from nine to sixteen years old. The funeral was a sad affair, attended by the extended Mitchell family except for Johnnie Bill; he had a bout of the flu and remained home. Lillian Dickinson Mitchell was buried in the family plot at Enid Oakhill Cemetery, about a mile north of the village.

His father faced parenting alone while juggling his diverse business interests. Johnnie Bill's sixteen-year-old sister, Florence, shouldered some of the household responsibilities, but Noah Mitchell also arranged for an elderly black woman to move in as the family's nanny. Evelyn Lott, a Mississippi native, was in her mid-sixties when she joined the family, herself a widow. She was quite possibly born a slave in 1860, but little information about her has survived. The 1930 US Census listed Evelyn as "Negro" living in the Mitchell home as "Servant." The little there is in Mitchell family letters and materials shows affection for her, albeit in the broader social context of unalloyed racism. Fifty years removed from the Civil War, a horrific backlash to Reconstruction was well under way throughout the Deep South. The movie director D. W. Griffith's racist epic, *The Birth of a Nation*, had been released to great acclaim in 1915, a year after Johnnie Bill's birth; the film had become a runaway hit and the nascent Hollywood's first blockbuster. The three-hour dramatization of the Civil War and Reconstruction portrayed the Ku Klux Klan as heroes restoring order to a South torn asunder by lawless, sexually predatory ex-slaves. Its message of white supremacy resonated powerfully—a review in the *Atlanta Constitution* gushed, "Ancient Greece had her Homer, modern America has her David W. Griffith." Lynching throughout the South during that period occurred with frightening regularity, while Klan membership soared. Mississippi politics, meanwhile, featured its own "Great White Chief." US senator James K. Vardaman, spewing bigotry while dressed in a white suit, white hat, and white boots, his flowing hair worn long, embod-ied the archetypal southern demagogue. The popular Mississippi po-litico once asserted, "If it is necessary every Negro in the state will be lynched; it will be done to maintain white supremacy."

Vardaman's keenness for lynching seemed another of the events out there, beyond Johnnie Bill, who, as was the custom, would ca-sually refer to black men and women as "niggers." But his family's brand of racism may have been more paternalistic than venomous. Tucked in the family's well-worn, 832-page Bible from those years is a yellowed newspaper clipping of a poem, "Negro Mammy." The overtly bigoted poem is mixed with caring and was perhaps saved in appreciation of Evelyn Lott, the fixture in the Mitchell home during Johnnie Bill's formative years.

"Negro Mammy" reads, in part:

> Send me a negro mammy
> From the good old Dixie land,
> Tie her wooly old head
> In a bandana red;
> Put a dishrag in her hand,
> And, oh, let her be in her cotton-checked dress.
> Tell her a song to moan,
> For I miss her sadly.
> Will welcome her gladly
> In my new-made Western home.

EVENTUALLY JOHNNIE BILL'S FATHER FELL IN LOVE AGAIN. IN DE-cember 1927, Noah hosted a holiday dinner party for twelve at his house. Several of the female guests were schoolteachers, including one named Eunice Massey. Eunice taught in the Batesville School fifteen miles north of Enid. Dinner—a feast built around two geese with seasonal vegetables—was served at 9:00 p.m. and was followed by music and dancing late into the night. The party made the social notes of the local weekly newspaper. "Everyone declared it to be a *real party* and all were delighted." Especially Noah and Eunice; they hit it off and were soon a couple. The next year Noah, or, as the wedding notice called him, "Mr. N. B. Mitchell, our handsome townsman," was married to "the beautiful Miss Eunice Massey." Noah was forty-seven years old, and Eunice was twenty years younger. "Their many friends wish that their pathway through life shall be strewn with lovely and fragrant flowers."

Johnnie (no longer Johnnie Bill) grew to like his stepmother, whom he called by her first name, but he was never as close to her as he had been to his mother or even Evelyn Lott. He was practically grown—fifteen—when his father and Eunice married on July 14, 1928, and by then his days were filled with sports, schoolwork, and the outdoors. In just about any sport he was a quick study, able to pick up baseball or tennis in short order. He loved and excelled at basketball, relying on his speed and athleticism to make up for his small size; as a teenager he weighed only about 135 pounds. He loved the woods just as much, heading out with his father and two older brothers to camp and hunt. His father had taught him to shoot as a boy, although the first time, he had been too small to hold the rifle; he had pulled the trigger as the barrel rested on his father's shoulder. When he managed to shoot his first dove out of the sky, his father carried on as excitedly as if he had swished a basketball through the hoop from midcourt. Starting when he was twelve or so years old, Johnnie began to camp out for a few nights alone, taking along bare essentials—a blanket, frying pan, salt and pepper, maybe some bacon—to live off the land. He fished and hunted squirrels, rabbits, and birds, developing a self-reliance that seemed ordinary to him and his peers but was soon being championed by President Herbert Hoover as the defining feature of American men: "rugged individualism." While camping, Johnnie found himself looking up into the sky, fascinated. He especially liked the moon and its monthly phases, and he became known in his family for his sense of direction; he was able to study the sun, moon, and stars to find his way home.

It was around that same time—when he was twelve or so—that he got his first taste of being airborne. Fliers, known as "barnstormers," came through on occasion, landing in a cow pasture in a two-seater biplane that was often a salvaged leftover from the First World War. Johnnie did not have the $1.50 fee to take a ride, so he pulled a fast one by convincing the cashier at the emporium to advance him cash and charge the family account. Up he went. He was hooked—and wanted more.

"I'd always beg my dad for a couple of bucks so I could go and take a ride in one of those open cockpits," he said years later. "I always liked airplanes."

Though Noah Mitchell never went to college after the Webb School, he always treasured the lessons he had learned there about the value of education and strong character. That meant nothing less than academic excellence was acceptable for Johnnie. Enid had a single school, covering grades one through twelve, and the boy indeed shined in the classroom, so much so that he skipped the eighth grade. But his success at school was certainly complemented by homeschooling. Noah drilled Johnnie in Latin, running him through the paces of verb conjugations. He filled the house with books, and Johnnie was required to read Charles Dickens's collected works. Johnnie was drilled in mathematics as well; he was pretty good at it, but it did not come as easily to him as other subjects did. And, as a result of extra practice at home, he developed a neat writing style.

Johnnie was seventeen in 1931 when he graduated with high honors from a high school whose senior class consisted of fewer than a dozen graduates. His father, having missed out on higher education, was likely the one who pushed the idea of college and, given Noah's extensive travels, had him looking northward, specifically at Columbia University in New York City. Johnnie sought—and won—a prized $500 scholarship, one of only twenty scholarships the university awarded annually. That fall, Johnnie Mitchell left the Deep South and, in a sense, left for good. Both his grandfather and father had returned to Enid after being away at the Civil War and boarding school, respectively, but not Johnnie. He returned for visits but never again looked homeward to Enid.

EVEN WITH THE ACADEMIC SCHOLARSHIP, JOHNNIE SOON FOUND himself under constant financial pressure. He came from a large family with a father's modest income, in which frugality and money consciousness were ingrained. He started to study economics with interest, but his need to always be working interfered. His steadiest job was in a college cafeteria as a student waiter or as a server behind the counter, slinging hash. When he could, he picked up extra odd jobs that typically paid fifty cents an hour. Every little bit helped. In the fall he worked college football games, using the pass that came with being an usher to view the game and then selling the free ticket given to all students. He tried pursuing personal interests. He began to box and was good at it, winning an intramural bantamweight championship, and he continued playing basketball in an intramural league. But the Ivy League college proved demanding in every way, as he juggled his studies, jobs, and sports.

Johnnie managed to get through nearly three years, but that's when he decided he'd had it with the financial strain. He eyed West Point, which was free—if he could get in. To gain admission, he devised a plan. First he dropped out of Columbia in the spring of his junior year. Then he enlisted in the US Army, figuring that within a year he'd have the training needed to test into West Point. He wasn't interested so much in military service as he was in a free ride to complete his college education. In the spring of 1934, just as he turned twenty years old, he was dispatched to an army base in Hawaii, farther away from home than he'd ever been.

But things did not work out as planned. Instead of a quick entry into the elite military college, Johnnie spent the next four years in Hawaii as an army grunt. "I passed the tests all right, but I couldn't pass it high enough," he said about falling short of West Point's admission standards. "I got left by the wayside." He never earned a degree.

Hawaii did work out in other ways. The island was a paradise, after all, its beaches among the most idyllic in the world, a playland of sorts for a young man in his early twenties. Moreover, the nation was staggering through the Great Depression, and for someone who had yet to decide what he wanted to do in life, the army took care of his every need. Johnnie was stationed at Fort Ruger, built in 1906 at Diamond Head on the coast of Oahu, thirty miles southeast of Pearl Harbor. He served in what was known as the Coast Artillery Corps, a branch of the army created as the first line of defense against an enemy invasion. "In order that no foreign soldier put foot upon our shores, Coast Artillery personnel must be expert in the efficient use of fixed and mobile cannon and anti-aircraft guns," said the corps's commander, Major General Andrew Hero, Jr. Johnnie was a "plotter" on a team manning a 155 mm cannon with its nearly twenty-foot-long barrel. During target practice, a ship towed a target offshore and Johnnie calculated the lead time for the cannon shot.

He earned a promotion to corporal, starting out earning $18.75 a month and, with pay raises, eventually $42 monthly. Never idle, he found ways to work off the base and always played sports. He dabbled for a while in journalism as a sportswriter for the *Honolulu Advertiser*. He honed his tennis game, taking lessons from a local pro and playing in tournaments all over the island. He was a starter on the base's basketball team, which played in a league against teams from other bases. Etched in his mind forever was Oahu's natural beauty. It was a "world of greenness and beautiful flowers on every tree and shrub." Once his three-year tour of duty expired, he decided to re-up for another hitch. Why not? He was unable to get into West Point, but Hawaii was great: he soldiered, partied, and played sports. Then, during his fourth year at Fort Ruger, his thinking changed, as if something had clicked; he realized that while the living might be easy and he would always love the central Pacific island, he needed to get on with life. Before the year was out, he had bought his way out of the corps, paying $150 to do so.

Johnnie returned to the continental United States in 1938, apparently for some soul-searching. He did not go back to Enid. Instead he headed to Atlanta, Georgia, where his oldest brother, Philip, was living. He bunked in Philip's apartment for the next several months. He spent his days working in his brother's liquor store and at night managed to knock off a couple more college credits by taking courses at the University of Georgia's extension school. He made new friends, including a young couple living in his brother's apartment house and a worker at the store with whom he stayed in touch for years, a "nigger porter," as he later described his friend, who was "clean, intelligent and courteous." But overall, Johnnie seemed to lack any plan, just going along, getting along, unsure of his next move.

Until one night he experienced an epiphany. He and Philip sometimes went to watch planes take off and land at the Atlanta airport, then called Candler Field and built on an abandoned car racetrack less than a decade before. Johnny had always liked gazing up at the big sky, the stars and moon, and he'd liked airplanes ever since flying with barnstormers back home. He and his brother hung around the airfield killing time, especially liking to watch the Douglas DC-3s, the new twin-engine propeller planes that were fueling the popularity of air travel. One time Johnnie offhandedly said, "I could learn to fly." Philip scoffed, which stopped Johnnie in his tracks. His brother did not believe he could do it. Johnnie took it as a dare, and at that moment his next move became clear. He grew serious and this time asserted—as fact—that he would learn to fly. And he meant it. Within days he headed downtown to the recruiting offices of the US Army Air Corps and applied to aviation school. He was accepted and, in October 1939, left Atlanta and his brother for San Antonio, Texas, to start basic training. Johnnie Mitchell was twenty-four years old and at last had found direction. He would become an army pilot.

LIVING IN SAN ANTONIO WHEN FLYING CADET JOHNNIE MITCHELL settled in at Kelly Field was a petite and pretty brunette named Annie Lee Miller. Just twenty-two, she was sharing an apartment on East Woodlawn Avenue with her cousin Josephine, or Josie. She worked for her uncle, Dr. Joseph Kopecky, who liked being called Dr. Joe. She was the "gal Friday" in his general medical practice.

Annie Lee was from El Campo, a rural farming town a three-hour drive to the east of San Antonio. She was a homegrown Texan through and through—had rarely, in fact, ever traveled beyond the state's borders. Her young life had so far orbited around her hometown and two small but growing cities, San Antonio and Austin. And no matter where she happened to be, she always had relatives nearby, as she was from a big extended family. Her grandparents on her father's side were German immigrants who had settled in the central Texas town of Temple to farm and raise eleven children, including her father, Armin W. Miller. Her grandparents on her mother's side were from Czechoslovakia, and they had started a farm and family just north of Fayetteville, Texas. The couple raised eleven children as well, including Annie Lee's mother, Theresa Kopecky.

In 1916, two years after Armin Miller and Theresa Kopecky married, Annie Lee was born on November 2. She was the couple's second child. They were renting a farm just east of El Campo at the time, but Armin soon eyed eighty acres of available land further out from town. With a bank loan he bought the parcel in the fall of 1918. He dug a well, built a barn and a four-room, bungalow-style house, and then put everyone to work. Cotton was a main crop. Annie Lee and her brother, Joe, began picking cotton at the ages of six and seven, respectively, working alongside seasonal migrant workers from Mexico. The kids were taught to hoe the rows of crops, chopping away the grass and thinning out the cotton. Come picking time, grown-ups strapped long canvas sacks over their shoulders—bags holding up to a hundred pounds of seed cotton—while Annie Lee and Joe used much smaller sacks made just for them. They'd load a wagon stationed at one end of the patch, and it was work that Annie Lee and Joe actually liked, because their father had them climb atop mounds of cotton and tumble around in them to pack the cotton down. Armin Miller then hitched horses to the wagon to transport the load to the cotton gin in town.

While Armin managed the fields, Theresa managed the house. She was a tiny woman, barely five feet tall, and a feisty bundle of energy who never hesitated to speak her mind and tell the kids and her husband they weren't doing enough. Theresa Miller was an excellent cook; her fried fish and frog legs, baked cheese kolaches and poppy-seed cakes were just a few of the dishes and desserts always in demand at the frequent family gatherings. Annie Lee loved a drink her mother made using clabber—milk that had curdled—shaken in a glass with sugar and cinnamon. In addition to her work in the kitchen, Theresa maintained an expansive flower garden to brighten the grounds, and she'd send Annie Lee out when need be to pick a bouquet from the beds of purple violets, white Shasta daisies, blue cornflowers, and various colored roses that circled the house.

Although Annie Lee's mother had never gone to school, she valued education; in fact, many of Annie Lee's uncles and aunts who had finished school had ended up staying in education as a career. When Annie Lee began school at age six, her teacher was none other than her Uncle John Kopecky. She and her brother, Joe, would walk the dirt road to school, joining up with kids from surrounding small farms along the way. Their destination was the Tres Palacios School, a one-room schoolhouse with between thirty and forty students. The daily round-trip walk was eight miles.

In the summers, Annie Lee spent weekends at a ranch her uncle Dr. Joe had bought, which everyone considered a "real paradise." It was about twelve miles north of San Antonio, a property heavily wooded with cedar and live oak trees, and featuring a spring-fed stream filled with bass, catfish, and perch running from the Guadalupe River. Cousins and uncles lined the banks to fish, and to cool off they'd jump into the waterway's deep pools. Dr. Joe built a cabin with a wood stove, a refrigerator and freezer, and a screened-in porch along one side for sleeping. He presided over the weekend affairs, walking the grounds wearing a pith helmet, shorts, and no shirt, a hand-rolled cigarette in one hand, a beer in the other. Annie Lee, her family, and dozens of cousins, aunts, and uncles feasted on spicy homemade chili and all kinds of barbecued meats, with beer for the grown-ups stacked in ice-filled washtubs. Eventually the accordions were unpacked and the singing began. Dr. Joe, proud of his Czech heritage, often led the Slavic tunes.

When Annie Lee began high school in El Campo at age thirteen, her extended family, once again, was all around. As it was too far for them to walk from home, the Millers and Kopeckys pitched in to rent rooms two blocks from El Campo High School. Three Kopecky boys stayed in one bedroom, Annie Lee roomed with some girls in another, and her brother, Joe, stayed with cousins in a third bedroom. They all shared the kitchen with the widow who owned the house and needed the rental income. Come the end of the school week, Annie Lee and the others walked nine miles to get home, taking the most direct route across fields and meadows rather than the winding dirt roads.

She was only sixteen when she graduated in June 1933, one of sixty seniors. In its coverage of the graduation, the high school newspaper, *The Rice Bird*, included one-liners about each member of the class, things such as "Our Prince Charming," "Our Lady Preacher," and "Basketball and Tennis Shark." The entry for Annie Lee read "God's Gift to Teachers." Though her parents needed her brother, Joe, to stay home after high school to work on the farm, they were able to afford to send Annie Lee to college for two years at the University of Texas. It was in Austin that Annie Lee tasted real independence for the first time. Yet another relative lived nearby—this time Ludmila Kopecky, her mother's sister—but Aunt "Ludma" was hardly the hovering type. Ludma, in her late twenties, worked as a nurse at the University of Texas Health Center and was considered "modern." She kept her own apartment, owned her own car, dressed stylishly, and played a steel guitar. On weekends she performed in clubs, singing and playing her guitar. Ludma lived life "in the fast lane," as one of Annie Lee's cousins once said, "with wine, men, song and dancing until dawn." While in college in Austin, Annie Lee spent lots of her free time with Aunt Ludma. They'd go to movies together or the clubs, or Ludma would make dinner in her apartment, using her secret recipe for baked chicken and herbs. Annie Lee came to regard her Aunt Ludma as a big sister of sorts, worldly and wise.

ANNIE LEE MOVED TO SAN ANTONIO IN 1935 AFTER FINISHING HER two-year college program, living with her cousin Josie and working for her uncle, Dr. Joe. That was the setup for the next four years, as she earned her own way, roomed with a favorite cousin, and enjoyed the life of a single woman barely twenty years old. She, Josie, and their girlfriends enjoyed socializing with the servicemen from the several nearby bases. On weekends, when the Officers' Clubs on the bases opened their doors to civilians, they joined the flocks of women who went. Some of Annie Lee's friends considered dating a soldier a very big deal. But Annie Lee wasn't convinced. She had fun going out, for sure, and at different times she dated one or more men but none seriously. She was wary of getting involved with someone in the service; here today, gone tomorrow kind of thing. But then came a new year, January 1940, when her cousin Josie proposed a double date. The men were aviation cadets, she said. Josie knew one, but Annie Lee's would be a blind date—at age twenty-five slightly older than the soldier boys they were used to. He was from Enid, Mississippi, Josie told her. Annie Lee hesitated, thinking "Not another flyboy." But Josie persisted, and she caved.

The first date led to a second, then a third, and soon enough Annie Lee Miller and Johnnie Mitchell were well on their way to becoming a couple. The first Sunday morning in April found them, along with cousin Josie, heading off in Johnnie's secondhand car for the eighty-mile drive to Austin to visit Annie Lee's Aunt Ludma. Annie Lee and Johnnie had gotten past the awkwardness of a blind date and discovered the kind of chemistry that's often difficult to describe but that each recognized. They liked each other right away and were comfortable in each other's company. They shared an interest in movies, an easy sense of humor, little tolerance for people who acted superior to others, and, perhaps most important, a small-town upbringing surrounded by family and regular folks.

For Annie Lee, the Sunday drive to Austin was the first chance to show off her new man to her favorite aunt. Racing to Austin was more like it, though, because Johnnie pushed the car hard along the paved parts of the road to speeds exceeding eighty miles per hour. Maybe he was just aching to move fast, as he now did in the sky during training exercises at Randolph Field, or maybe he wanted to impress his new gal, but, either way, it backfired. He got caught, pulled over by a Texas Highway Patrol officer for speeding. Johnnie tried to act nonchalant, as Annie Lee later told her aunt, uttering his version of Scarlett O'Hara's line in *Gone With the Wind*: "he'd worry about that tomorrow." But Annie Lee could tell he was embarrassed and annoyed at having to pay the stiff fine with money he was saving for a brand-new car. They told Ludma about the ticket but decided not to say anything to Annie Lee's parents. Johnnie worried that the police stop would taint their impression of him, still in the early stages. Annie Lee told Ludma, "They would think we were fools."

That glitch aside, they enjoyed their afternoon with "modern" Ludma—afterward Johnnie went on and on to Annie Lee, praising Ludma's singing voice—and enjoyed many more dates that spring. They sometimes partied with friends in Annie Lee and Josie's apartment, drinking their way through a pint of Johnnie Walker Red and chatting the night away, or went out with a crowd for dinner in San Antonio, either at Kline's, a popular restaurant and club, or at the even more popular Earl Abel's on Main Street, started during the Depression by Earl Abel, a former silent film organist and famous for its fried chicken for decades to come. The restaurant had an outdoor area known as "The Garden of Eatin'." They experienced the rough patches that new couples do, usually involving Johnnie's jealousy about two guys Annie Lee had dated and who were still around, one named Skelly and the other Clay. At one point Johnnie announced that he wasn't going "to play second fiddle any longer." But mostly there were flowers and fun and a romance that steadily gained traction throughout the first half of 1940. The clincher came on July 26, when John William Mitchell received his pilot wings and commission as a second lieutenant in the Army Air Corps. Graduation was held at the base for a class of 240 new pilots. Johnnie's father, Noah, and one of his brothers had talked about coming for the big day but in the end didn't make it. Instead, Annie Lee Miller, sporting a new blue-and-white-striped dress, was the one looking on as the presiding Army Air Corps officer, Colonel Millard Fillmore "Miff" Harmon, Jr., shook Johnnie's hand after Johnnie received his diploma and flying pin. "You pinned my first pair of wings on me," Johnnie wrote Annie Lee years later, "and I believe it was when I became thoroughly convinced that I was going to marry you, come hell or high water."

But graduation also meant separation. Johnnie was given four weeks to report to Moffett Field south of San Francisco. Before his departure, he spent as much time as he could with Annie Lee. Their last time together was lunch the day he left, during which they only "talked of little things so the conversation would not become serious," as he recalled later. Then he said farewell and embarked on a good-bye road trip of sorts. With other pilot friends, he drove all night to Jackson, Mississippi, where they dropped off one rider and then continued north. He reached Enid at daybreak. Johnnie spent a few days there, then drove nine hours with his father to Atlanta to his brother's place, where another brother and his family were also visiting. Right away Johnnie began writing letters to Annie Lee—an epistolary habit that would last for years to come. He told Annie Lee that his father and brothers and their families hadn't gotten to bed until 5:00 a.m.—a happy reunion, he said, with lots of drinking and "singing some of our old time songs." Next he told Annie Lee of a sour moment: telling his father about her. Johnnie, of course, was excited to share the news. But the first thing the wary Noah had asked him about was her religion and racial background because, explained Johnnie, "he is prejudiced against Catholics, Germans and Jews." When Johnnie had replied that Armin Miller was pure German and Theresa Miller pure Czech, his father had gasped and said, "My God." Johnnie had gotten mad. He'd told his father that he didn't give a damn what he thought and that he'd be ashamed once he met Annie Lee.

Leaving Georgia on Friday, August 16, 1940, Johnnie headed cross country, driving through the Painted Desert in Arizona, stopping to gape at the wondrous Grand Canyon, and continuing on through the Navajo Desert, where, he said, it was "117 degrees in the shade—but there was no shade." It was the same year Woody Guthrie was writing lyrics for what became one of the most famous folk songs ever, "This Land Is Your Land," and Johnnie was having his own "this land" moment, driving Guthrie's ribbon of highway and looking above at that endless skyway. To be sure, all around him were signs that the world war, which had been expanding overseas since the prior September, when Hitler's Nazis had invaded Poland, was getting closer to home. In May 1940, President Roosevelt, addressing Congress, had asked for hundreds of millions of dollars to pay for fifty thousand new airplanes, and then, addressing Americans in a fireside chat later in the year, had asserted that the nation must become "the arsenal of democracy." When Johnnie got to Reno, Nevada, he was surprised by the throngs of couples flocking to the resort not to gamble but for a quick marriage—surprised until he realized that marriage was a deferment from the first peacetime draft in US history, to be held the next month. For his part, though, Second Lieutenant John Mitchell was en route to continue his training for the rapidly expanding Army Air Corps. In Texas during basic training he'd initially flown in a Vultee BT-13 Valiant, a training plane that was tricky to fly and could stall easily. But he'd taken quickly to the task, discovering a comfort in the sky and a knack for navigation for which he'd become known—a sense of direction learned, as he later said, from a boyhood spent "in the woods, telling my position from the sun, moon and stars."

Johnnie pulled into Moffett Field seven days after leaving Georgia and, after barely getting settled, resumed flying exercises. He'd thought about flying bombers but was assigned to a pursuit squadron, which meant he was going to train as a fighter pilot. It also meant getting the chance to fly a Curtiss P-36 Hawk, a fighter plane considered the fastest at the time. When he got the chance on Wednesday, September 4, he couldn't wait to tell Annie Lee. "Yours truly put in about an hour and ten minutes soaring over this sun-kissed land at about 265 mph," he wrote excitedly before going to bed. "I took off so suddenly that I was almost thrown back through the tail assembly and by the time I had made one climbing turn, pulled my wheels up and was straightened out, why I had 10,000 feet altitude. Those little babies can climb almost straight up."

It was the early autumn of 1940, and Johnnie was finding his place—at long last. He missed Annie Lee, for sure, telling her so in his frequent letters. But he sensed, too, that he was coming into his own, that he'd found a purpose—flying—a woman he loved—Annie Lee Miller—and, of course, the moon—always Johnnie Bill and the moon. It was a feeling he tried capturing in words in late September after his first flight in the skies over California in a P-36. "This is one of the most beautiful nights," he wrote Annie Lee. "And I stepped out on our little porch to watch a plane glide down the runway and roar off into the night. The moon was shining in a cloudless sky and reflecting very brightly upon the wind-stirred bay. I stood and looked for a while and thought—it *is* good to be alive and it *is* a wonderful life."

CHAPTER 2

ISOROKU, AEROPLANES, AND A GEISHA GIRL

———

Young Isoroku Yamamoto
WW II Database/public domain

ISOROKU YAMAMOTO STUDIED THE BLACK AND WHITE STONES spread out across the grid-lined board. The ancient game of *go* was the very sort of competitive pastime he loved—a game more complex than chess and requiring great strategic foresight. The goal is to capture an opponent's stones—the term for the playing pieces—by surrounding them with your own; one or more stone at a time until victory—hopefully. Yamamoto, stones in hand, considered his next move.

The rising star of the Imperial Navy was a guest in the Tokyo home of Baron Sotokichi Uryu. The year was 1915, a year after John William Mitchell was born, and Lieutenant Yamamoto was thirty-one years old. His host was a decorated navy hero in the Russo-Japanese War, most notably at its outbreak in 1904, when his ships had escorted Japanese army troops to the Korean port of Chemulpo and driven Russian ships from the harbor. The baron, an admiral, was a mentor to Yamamoto.

Just as Yamamoto was finishing a match against a member of the admiral's family, the front door opened. Entering the house atop a summit overlooking Tokyo were the admiral and his wife, the Baroness Uryu. With them was another guest, a foreigner. Yamamoto stood to leave, not wanting to get in the way. "Don't go," Uryu said. "This is an American editor." Uryu wanted the men to meet.

The American was Willard Price, a roving foreign reporter. Price had been delighted with the access to Uryu he'd managed to finagle and was angling for an exclusive interview with the baron about his war exploits at Chemulpo harbor. But the admiral kept rebuffing him, saying he represented the past—"A goner"—and that he should talk to Japan's future, young Yamamoto. "He is a comer," Uryu said.

The admiral directed Price and Yamamoto to a separate tea pavilion, where the two talked for more than an hour. Price sulked; the story he had wanted was Uryu's, not this upstart's. He took notes but afterward put them aside and did not write anything. But that would change twenty-seven years later, when Yamamoto emerged as the mastermind of the Pearl Harbor attack. Price would dig out his old notes for a Yamamoto profile in the April 1942 issue of *Harper's Magazine*, titled "America's Enemy No. 2: Yamamoto." (Hitler was enemy number one.) In it he demonized Yamamoto, producing a piece of wartime propaganda that was at odds with most biographical sketches of Yamamoto as smart, serious, and sociable. In Price's hand, Yamamoto was surly, someone who, from the time he had been a child, had been dedicated to "the crushing of White superiority." Even so, the interview conducted that afternoon in a tea pavilion overlooking Tokyo yielded a basic outline of Isoroku Yamamoto's hard boyhood that squares with other historical accounts.

NAGAOKA, WHERE ISOROKU WAS BORN ON APRIL 4, 1884, WAS ISOlated and bleak, located in the northwest region of Japan's main island of Honshu. Winter snowstorms buried the village, including the small wooden house where his family lived. Isoroku's surname at birth was not Yamamoto. It was Takano—as he was the seventh and youngest child of Sadayoshi Takano, who at fifty-six was an old man. Takano seemed uninterested in the arrival of another son. The newborn went unnamed for a month, until Sadayoshi finally yielded to his wife's badgering and came up with Isoroku. The name is not one that could ever be mistaken as touching; "Isoroku" in Japanese means the number fifty-six, Sadayoshi's age. In English, then, the baby's name was Fifty-six.

The age range of the Takano offspring spanned decades; Isoroku's two eldest brothers were more than thirty years senior to him. In fact, the two brothers, Yuzuru and Kazuko, were old enough to have fought along with their father in the late 1860s in the country's civil war, the Boshin War—well before Isoroku was born. Sadayoshi himself was a former samurai, a member of the elite class of warriors in feudal Japan that practiced what became known as *bushido*, a chivalric code of conduct based on courage, loyalty, and a preference for death over dishonor. By the time Isoroku was born, Sadayoshi was a struggling schoolteacher barely making ends meet, which only added to the severity of Isoroku's boyhood circumstances: a remote, hardscrabble village, a strict, militaristic father, and poverty.

During the 1890s, while attending middle school, Isoroku and his classmates were dispatched from the village on long marches— the worse the weather, the better. The idea was for the boys to get tough and ready for the armed services. They were allowed to carry weapons but not live ammunition, and Isoroku actually grew to look forward to the annual military-style adventures. In school he was taught briefly by an American missionary, which meant exposure not just to an American but to the English language as well. He read the Bible, too, though he never became a Christian, and began to show interest in the workings of the United States and Great Britain.

He was a strong student overall. He was also a boy who fought off several serious bouts of influenza that sidelined him for weeks. Yet he found plenty of time for fun. He and his friends and, for that matter, the rest of Japan were discovering the American game of baseball. Isoroku and his pals played and watched the games on the fields near school. His specialty, however, became gymnastics. He discovered a passion and knack for the sport, so that as he grew the upper parts of his otherwise slender physique expanded: his shoulders broadened, and his chest widened like a barrel. Years later the grown-up Isoroku would occasionally shatter his austere image and shock comrades by leaping onto a ship's railing and, using his hands, perform a headstand. Throughout his life, Isoroku always maintained a strong emotional attachment to Nagaoka, and he later would enjoy telling others occupying the highest ranks of Japanese society and the military that he was just another country boy at heart.

He wasn't long for Nagaoka, though. Just sixteen, the fresh-faced Isoroku sought admission to the Naval Cadet School more than five hundred miles away on the island of Etajima in Hiroshima harbor. It was at the turn of the twentieth century, and Japan's leaders had made expanding the navy a high priority. After filling out an application, Isoroku scored high on the entrance examination, coming in second out of the three hundred test takers, but he almost did not make the cut; only five feet, three inches tall, he just squeaked by the minimum height requirement. The conditions at the academy were harsh, but that was nothing new to the teenager from Nagaoka. Cadets were barred from smoking, drinking, socializing with girls, and even eating sweets. Isoroku, having decided on a naval career, applied himself thoroughly during the next three years of training, with a fourth spent aboard ships. In 1904, he graduated seventh in his class.

Isoroku's timing couldn't have been better. He completed the academy just as his nation was taking up arms as the underdog in the Russo-Japanese War. Tsarist Russia was a powerful, empire-building force that, expanding its reach across Siberia to the Pacific, had its eye on a port on the coast of China as a year-round base for its fleet. Japan felt threatened, but most experts thought it was overmatched. "This upstart Asian David could not possibly prevail against St. Petersburg's imperial Goliath," one historian noted. But that's exactly what Japan did, and the combat strategy the so-called David used to quickly gain the upper hand was one the rookie seaman Isoroku would never forget. Early in 1904, Japan's Combined Fleet—a mixture of battleships, cruisers, destroyers, and torpedo boats—stealthily made its way toward Port Arthur on the Yellow Sea, where Russia's Pacific Fleet had anchored. Then, in the middle of the night of February 8, 1904, and without any declaration of war, Japan launched a surprise torpedo attack against the unsuspecting Russians. By dawn, Japan's navy had set up a blockade at the harbor, daring the Russian ships to come out to fight. The war was under way, and Japan's naval leader, Vice Admiral Heihachiro Togo, became lionized.

Isoroku joined the fight right after his graduation. He was assigned to one of the smaller ships protecting Togo's flagship, *Mikasa*, which gave him a front-row seat at the war's climax the next year, on May 27–28, 1905, when Togo oversaw the virtual destruction of the Russian fleet in the Strait of Tsushima separating Korea from Japan. To inspire his men at the start, Togo ordered a Z flag hoisted atop his flagship—a symbol to the fleet that the fate of the Japanese Empire was at stake. Then, in one of history's most lopsided naval battles, the Russians lost twenty-three ships in a single day. Isoroku, having just turned twenty-one the prior month, was in the middle of it all aboard the cruiser *Nisshin*. "When the shells began to fly above me, I found I was not afraid," he wrote afterward. Isoroku suffered serious wounds. Though some historians have concluded that he was hurt when the six-inch barrel on one of his own ship's guns exploded, Isoroku at the time wrote to his parents that "a shell hit the *Nisshin* and knocked me unconscious. When I recovered I found I was wounded in the right leg and two fingers on my left hand were missing."

His legs and thighs were ripped open by steel shrapnel, leaving them scarred and disfigured; he was hospitalized for the next several months. And it was while he was in the hospital that Isoroku became witness to an unusual wartime moment between two navy commanders—one the victor, the other the defeated. The recuperating Isoroku was alongside Togo as the Japanese admiral made a bedside visit to his wounded Russian counterpart, Admiral Zinovy

Rozhestvensky, who had been taken prisoner and was receiving medical care. Isoroku listened as the unlikely conqueror paid his respects to the admiral from one of the world's most powerful navies. The 1904–05 war became a pivotal moment in Japan's growing military strength, a conflict Togo had initiated without the usual warning between aggressors, a declaration of war. The lasting takeaway for young Isoroku was that a well-executed surprise could set the stage for victory.

Following his recovery, which included a home visit to Nagaoka, Isoroku resumed his naval career. Over the next several years, he became known for his on-duty determination, punctuality, and obsessive interest in military strategy—past, present, and future—which mirrored his intense pleasure in tactical board games. He was also uncanny in his ability to flick a switch to enjoy off-duty life; though never taking to alcohol, he embraced hedonism in other ways. He became an inveterate gambler at cards, roulette, *go*, *shogi*, mahjongg, and billiards, to name a few. The particular game almost didn't matter; it was the tension, thrill, and tactical challenge that turned him on. Just as geishas did. During that time, Isoroku discovered the geisha houses of Tokyo, where he savored the attention of the young and colorful entertainers—the tea service, the singing, the dancing and, sometimes, the sex. He became involved with one girl in particular, Masako Tsurushima, who was sixteen years his junior. They had an affair that eventually evolved into a lifelong friendship. Three, four, even ten years might pass without their seeing each other, but they stayed in touch, and Isoroku sent her small gifts purchased during his travels.

By 1915, when Isoroku sat down for the interview with Willard Price, he was, as Baron Uryu had introduced him, "a comer." He'd been to gunnery school, trained at sea, and, in 1913, been appointed to the Naval Staff College, where his intellectual prowess continued to impress. The college appointment was key in his development—a career booster—but that year was also one of personal loss. Both of his parents died, and Isoroku Takano, turning twenty-nine, became Isoroku Yamamoto following his adoption by the Yamamoto family. The adoption—and the name change—were customary in Japan, with the adopting family being one that lacked a male heir.

But no matter how impressive his story, as far as the American writer was concerned Yamamoto was detestable. Price's portrait, written after the Pearl Harbor attack, oozed with loathing and bigotry. The young officer, he wrote, was "leather-faced, bullet-headed." Regarding the 1904–05 Russo-Japanese War, Yamamoto "had been a spectator of the first great triumph of the Asiatic over the Aryan." The future admiral, he continued, had learned firsthand "that the yellow man could whip the white." Price saw seeds of anti-Americanism everywhere in Yamamoto's background. There was, however, one statement Yamamoto had made that came to haunt him. When Price had asked Yamamoto to identify the crucial war vessel of the future, Yamamoto had not named the centerpieces of the then-existing naval strategy and navies—hulking battleships and destroyers. Instead, he had replied, "The most important ship of the future will be a ship to carry aeroplanes." The answer surprised Price. Years later, he wrote, "At the time, 1915, the 'aeroplane' was as clumsy as its name, and the aircraft carrier remained in the womb of imagination." Yet, fast forward from the 1915 remark to December 7, 1941, it was the aircraft carrier and fighter planes that made possible Yamamoto's attack on Pearl Harbor.

THE ANSWER REVEALED YAMAMOTO'S FORWARD-LOOKING MILI-tary thinking. In 1915, only a few nations, mainly Great Britain and the United States, had begun experimenting with launching planes from ships. But the idea—and its potential military benefits—intrigued Yamamoto and would inform his evolving naval philosophy, especially during the next decade as he was twice assigned to work in the United States for the Japanese navy. His first tour began in May 1919, and it came just nine months after he had married Reiko Mihashi. The ceremony on August 31, 1918, was held at the Navy Club in Tokyo, and, at thirty-four, Yamamoto had waited longer to marry than was typical for a Japanese officer. One reason was Masako Tsurushima and the other geishas whose company he'd enjoyed throughout his twenties. His introduction to Reiko Mihashi, twelve years younger, had been arranged by a friend, and marrying seemed the right thing to do for someone of his standing. But when he left for the United States the next spring aboard the *Suwa-maru*, Yamamoto traveled alone, and the fact that he spent the next two years in the United States without his new wife is evidence that theirs was a largely loveless marriage. Indeed, in the years to come they would be apart far more than they were together.

Yamamoto basically spent his first year abroad as a student. He was enrolled part-time at Harvard College, where he took a course in economics as well as one called "English E," intensive language instruction for foreigners to improve their English. He rented a room in nearby Brookline and from there made his way across the Charles River to the Ivy League campus in Cambridge. In the United States, Yamamoto discovered Abraham Lincoln and devoured biographies of the sixteenth president. He was captivated by the arc of Lincoln's life—from impoverished childhood to the White House—and the way in which Lincoln had become the "champion of the emancipation of the slaves," as he later explained to an underling. "He was a very human man. Without this quality, one can't lead others."

No matter where he was, Yamamoto kept tabs on his special interest, naval aviation. It had been only a year since the British Royal Navy had launched the first-ever aircraft carrier, HMS *Argus*, with its full-length flight deck, in September 1918. Then, after a year, he left Harvard for Washington, DC, to serve as the assistant naval attaché in the Japanese Embassy. He continued his "studies" in the broadest sense of the word, meaning that his classroom was not a room inside the ivy-covered walls of Harvard Yard but the country itself. He was a sponge for all things American, whether such games as poker or the latest advances in US technology.

He did not stay put but traveled widely, a firsthand tutorial in the nation's bigness and rich natural resources. With oil a must for any military and with Japan lacking fossil fuels, its leaders always fretted about having to rely on the kindness of other countries for black gold. Yamamoto went to Detroit to view the burgeoning automobile industry, and in the West he visited oil refineries, always asking questions about the petroleum industry and the nation's oil reserves. Wherever he went, he felt welcome, the early 1920s being a time when the two nations still got along. Japan had been an ally of the United States and Great Britain during the Great War that had ended in November 1918, and afterward it had reaped the benefits of having backed the victors; it had been awarded island groups in the Pacific formerly controlled by Germany as part of the Treaty of Versailles, signed in June 1919. However, Japan came up short in other postwar accords. When the United States hosted the world's naval powers to negotiate an arms limitation agreement that would hopefully avert future wars, it produced what became known as the 5:5:3 agreement, where for every five ships built by the United States and Great Britain, Japan was allowed three. The Washington Naval Treaty never sat well with Yamamoto and other Japanese officials.

Yamamoto returned to Japan in 1921. His wife gave birth to two of their eventual four children, a boy in the fall of 1922 and a girl in the spring of 1925. But Yamamoto was mostly on the move, either traveling or on assignment—without Reiko and his children. For nine months he accompanied an admiral on a tour through Europe and back to the United States, a trip that featured casino gambling in Monaco—a first for him. Yamamoto's luck at roulette became the stuff of legend; he won so much money that the casino manager banned him, it was said. Even if the story is embellished, Yamamoto himself bragged about his winnings, telling his admiral afterward, "If only you'd give me a couple of years amusing myself in Europe, I'd earn you the money for at least a couple of battleships." In the United States, he showed the admiral the oil fields of Texas, which served to deepen Yamamoto's sense of the gradually shifting energy reliance from coal to oil and gasoline to fuel all the new fighter planes in development in the United States, a trend he championed for Japan.

By now promoted to captain, Yamamoto was assigned in September 1924 to serve as director of studies at a new air training school at Lake Kasumigaura, forty miles northwest of Tokyo. The air station was becoming the main training venue for Japan's future navy pilots, and Yamamoto seized the moment to push the ideas that had yet to fully take hold in the military establishment—that the key to Japan's navy going forward was in oil and in planes taking off from ships. He insisted that instructors drill pilots in nighttime flying, despite its inherent dangers. That played into another combat lesson Yamamoto had never forgotten: the value of a surprise attack. He viewed fighter pilots trained to attack under the cover of darkness as one way to implement that option successfully. Most of all, he saw the school as the chance to practice what he'd been preaching—because even though he'd emerged as a leading advocate for naval aviation, he did not know how to fly. So he joined the cadets to learn. He was forty years old, a little late to be getting into the cockpit and learning how to manage a joystick, but he devoted several hours a day to training until he could fly solo. In the end, the way he interacted with his cadets and the intensity of his devotion to Japan, its navy, and its airpower won the hearts of his men. When Yamamoto was ordered back to the United States in early 1926—having spent just over a year with the Kasumigaura Air Corps—pilots from the corps flew low over the departing ship Yamamoto had boarded as a show of their respect and to say farewell.

FOR THE NEXT TWO YEARS, CAPTAIN ISOROKU YAMAMOTO SERVED as Japan's naval attaché in Washington, DC, based out of the embassy on Massachusetts Avenue. It was no secret in diplomatic circles that intelligence gathering—otherwise known as spying—was a primary duty for any attaché, no matter what his country of origin. To do his part, Yamamoto threw parties—stag affairs—during which he probed his guests' naval thinking over card games lasting late into the night. "Poker was his favorite, and he played it with an unreserved and unconcealed determination," one guest, US Navy Captain Ellis M. Zacharias, wrote later. Held in his quarters, the events were hosted by Yamamoto alone, as again he had left his wife behind. For him, a party was a twofer: he could indulge his passion for poker with the bob and weave of spycraft. "He was soon interspersing his bids and bluffs with slightly concealed inquiries of a distinctly naval character," said Zacharias, who was essentially Yamamoto's counterpart. Zacharias spoke Japanese and had briefly served as naval attaché in Japan. He'd even met Yamamoto while in Tokyo. But it was at the stag parties, when they sat across from each other at the poker table, that Zacharias came to know firsthand the Japanese officer who, in the secret files of the Office of Naval Intelligence, was described as "exceptionally able, forceful, a man of quick thinking."

Yamamoto's preoccupation with the future of Japan's Pacific naval operations became more evident during the uproar surrounding the publication of a new novel depicting a titanic war between an upstart

Japan and the United States. The author, Hector C. Bywater, was a naval correspondent for the London *Daily Telegraph* and, it was later learned, led a double life as a spy for the British Secret Service. He was a prodigious writer who was fascinated with Japan's navy and, by the 1920s, had become a leading voice on naval science. The novel was titled *The Great Pacific War*, and it forecast a massive war that begins when the Japanese fleet launches an attack in the Pacific against the US Navy—a sneak attack, no less, a plot point that was certain to catch Yamamoto's interest. In Bywater's imagined conflict, the Japanese surprise attack at sea is coupled with invasions of Guam and the Philippines, the goal being to create a Japanese empire in the western Pacific that is invincible. Bywater's plot then takes a turn: after Japan's initial hard-hitting blow, the United States recovers, counterattacks, and eventually crushes its enemy.

Upon its publication in July 1925, the novel became the talk of the town, especially in military circles. The *New York Times* showcased it on the cover of its Sunday book review section, where its reviewer, Nicholas Roosevelt, a cousin of future US president Franklin Delano Roosevelt, offered high praise: "If there is going to be a war between Japan and the United States, this volume may prove to have foretold accurately many of the most important incidents in it." Given the outcome of the fictionalized war, the Japanese reaction was outrage. The editor of *Kokumin*, a newspaper specializing in military news, denounced the novel's plot, calling it "mistaken" and "illogical" to think that Japan would lose. Some criticized the novel's premise, saying that Japan's leaders would never be so insane as to take on the United States, given the huge imbalance in military strength and in the raw materials the United States possessed to fortify its army and navy. Bywater replied by citing Admiral Togo's long-shot victory against mighty Russia in the Russo-Japanese War. Yamamoto followed the heated commentary from his perch at the Japanese embassy, and, not surprisingly, he and his staff read the novel to see for themselves what all the commotion was about.

Bywater's story certainly held a fascination, although its focus on battleships and destroyers as the core of a navy's strength was old school, and Yamamoto, throughout his second stay in the United States, stayed on task: aeronautics. "The aircraft carrier, the combination of sea and air power, was an obsession," Zacharias noticed. Back home the strategy was making headway; by 1927, four aircraft carriers had been added to the Japanese fleet. And in May of that same year, when Charles Lindbergh completed the first solo flight across the Atlantic Ocean to France, Yamamoto was on top of it; he assigned a staffer to study Lindbergh's success. The report that followed identified navigational instrumentation as playing a crucial role. Moreover, Americans were busy producing increasingly sophisticated instruments. "Naval aviation in Japan would find itself at a dead end unless it abandoned its reliance on the pilot's judgment and turned to instrumental flight," the report said. That would be especially true for planes taking off from carriers. Yamamoto sent the report home, urging Japanese naval leaders to adopt its findings. He was thinking big picture—always looking for advances in technology and military strategy generally, while mulling over the challenges of executing a long-range naval mission. His American adversaries took notice. "The naval attaché was interested in greater things; he was interested in war," Zacharias said.

YAMAMOTO RETURNED HOME IN MARCH 1928 AND ALMOST IMME-diately had occasion to address the topic that had continued to be hotly debated internationally, triggered by Bywater's celebrated novel: war with the United States. He was asked to address cadets at the Imperial Navy Torpedo School, and the talk he delivered seemed partly to channel Bywater's war thinking. Fresh from the United States, he was mindful, too, that the US Navy had begun overseeing a massive dredging of Pearl Harbor in Hawaii, removing nine million cubic yards of coral, so that within a few years a huge naval station at Pearl Harbor could serve as the home port for the United States' largest warships. Given those developments, Yamamoto told the cadets of a plan in which attacking Pearl Harbor, nearly four thousand miles from Japan, would provide the best shot at victory. It would be a surprise attack, a preemptive strike that would signal a bold break from the established military strategy of protecting Japan's Pacific holdings first and foremost. "Japan will lose if she adopts the traditional defensive strategy," one naval officer at the lecture recalled Yamamoto saying. But Yamamoto departed from Bywater's fictional war narrative in which Japan's battleships and destroyers led the way. He substituted his view that it was the aircraft carrier that would make possible a massive attack of Pearl Harbor, so far away from the homeland. The plan, which Yamamoto presented as a work in progress, would continue to evolve.

Once back in Japan, Yamamoto briefly commanded the light cruiser *Isuzu* and then took over as captain of the new and much larger *Agaki*, the very kind of warship he advocated. The 34,000-ton flattop aircraft carrier provided him a bully pulpit to push his views on the future needs of naval aviation, whether it was the development of torpedo planes or of faster, more maneuverable, long-range fighter planes. His prodding contributed to one aircraft company, Mitsubishi, eventually creating the plane that became the Japanese Imperial Navy's most feared carrier-based fighter, the Zero. Even though the 5:5:3 arms limitation agreement had constrained Japan— and increasingly rankled Yamamoto and his military peers—those years saw the Japanese defense industry racing to build aircraft carriers and destroyers to the limits of the treaty. Japanese destroyers were armed with new five-inch guns that would not show up on US destroyers for another decade.

In 1930, Yamamoto had a hand in loosening the arms treaty's restrictions. He was sent to England to assist the Japanese delegates at the London Naval Conference. The conference included the United States, Great Britain, France, and Italy and was convened to extend the arms treaty signed after World War I. Voicing objections to the original 5:5:3 formula, the Japanese won a concession giving Japan equal standing with the United States and Great Britain regarding production of light cruisers and submarines. It was half a loaf but better than returning home without any change to the treaty's overall inequity.

Yamamoto was promoted to rear admiral at the conference and, back home, assumed a position ideal to enact his agenda: he was named head of the Technical Division of the Aeronautics Department. For the next three years, he continued to expand the role of carrier aviation in the navy, even as he faced skepticism and hostility from battleship admirals—naval leaders who did not value fighter planes the way Yamamoto did. Then, too, he was caught up in the age-old rivalry between the navy and army, always at odds over military spending, with certain army leaders increasingly eye-ing Yamamoto's ideas for a bigger navy with suspicion. For his part, Yamamoto had a worldliness that was atypical. Many of his peers in the upper echelons of Japan's military, especially in the army, had never been to the United States. Yamamoto had been there a num-ber of times, including the two postings at the embassy, and he had traveled extensively to other countries. He'd had opportunities to evaluate on his own the world's powers, probe its leaders' thinking, and refine his own. And in Tokyo, while overseeing naval aviation development, he continued to move in diplomatic circles and make indelible impressions on Americans as they passed through. One of those American officers was Navy Lieutenant Edwin T. Layton. Layton was stationed at the US Embassy in Tokyo, where he was learning Japanese. Over his several years in Japan, he got to know Yamamoto in a variety of venues: at official embassy functions, at a theater party hosted by Yamamoto, during a daylong duck hunt, and even at a geisha party. Layton found Yamamoto to be "very human, a very real, and a very sincere man." Most Japanese officials, Layton said, were reserved and even aloof, "wearing false faces or masks to suit their role." Yamamoto was different. "I got to feel that on social occasions he did not wear a false face." Layton regarded Yamamoto as the real deal, "an official friend of mine."

Others were not so impressed. Lieutenant Joseph Rochefort, whose specialty was cryptanalysis, or code breaking, was another naval officer in Tokyo in the early 1930s. He was participating in an ambitious naval program in which those showing a knack in code breaking were sent to Japan for several years of exposure to Japanese culture and to gain fluency in the language. The US Navy during the 1930s was creating the largest cryptologic capability of any of the military branches—a necessity, its leaders had decided, given the growing threat of Japan's expanding Imperial Navy. In a few short years, Rochefort would take command of a radio intelligence unit in Hawaii—a station known as Hypo—with Layton as his boss.

In Yamamoto, Rochefort detected fire where Layton saw warmth. Based on a couple of encounters, Rochefort concluded that Yama-moto was a hothead "who might take violent action even without the consent of his superiors in Tokyo." From a US perspective, then,

Rochefort saw Yamamoto as a menace and potential warmonger—an assessment that, with the benefit of hindsight, was off. Yamamoto was strong-willed, to be sure, but he was no saber-rattling trouble-maker. He often worked behind the scenes to dampen the national fervor for war that was building steadily throughout the 1930s, a trend abetted by a media falling ever more under government control. In fact, Yamamoto's caution about exercising Japan's growing military might was at the heart of the widening rift between himself and the leaders of a chest-beating army. When the Imperial Japanese Army invaded Manchuria in 1931, the rear admiral went on record as op-posing the takeover. What the code breaker Joe Rochefort did not know—and which only emerged later—was that although Yama-moto sought a strong Japan militarily, especially cutting-edge naval armament, that did not a warmonger make. Yamamoto's resolve was aimed not at agitating for war but standing against it—at his own peril.

WHILE WORKING IN TOKYO RUNNING THE NAVY'S AERONAUTICAL technical division, Yamamoto was reunited with his family. A sec-ond daughter was born in May 1929, and a second son followed in November 1932. Reiko oversaw their home and the children—all under the age of ten when the fourth child was born—and although Yamamoto fulfilled his responsibilities on the home front he did so robotically. "Father seemed indifferent and undemonstrative," one of his sons wrote later, "but beneath this was a proper concern for our welfare." Both parents seemed determined to maintain a calm veneer, but when tension crackled Yamamoto was known to escape by going to his bedroom and pulling the covers over his ears. He rarely took Reiko to any social events and would occasionally ex-press envy when spotting the ardor of other couples. "You're lucky to be in love with your wife," he'd whisper to such an officer. "I threw in the sponge long ago."

His heart was simply not in a marriage that in 1933 was fifteen years old. But his time was coming, although it was hardly love at first sight. That summer Yamamoto attended a party—without Reiko—at the fancy Kinsui restaurant in Tokyo's Tsukiji district. The restaurant was popular with military brass, government officials, and the media's elite.

Yamamoto, dressed in a black-and-white-striped suit, his hair cut in the customary cropped style, was ushered to one of the guest seats. In attendance, too, was a young woman named Chiyoko Kawai. Chiyoko had had it tough, born into poverty in 1903, when nineteen-year-old Isoroku was completing the naval academy. By her early twenties, looking for a way forward in a patriarchal culture, she had followed the path of many young women in her position: she had become a concubine. Her new line of work had taken her to Tokyo. When she was twenty-five, both of her parents had died within months of each other, an unexpected loss that had driven Chiyoko, now more alone in the world than ever, to attempt suicide. She had recovered and by the summer of 1933 was settled into a life as a geisha girl.

Yamamoto did not take particular notice of the attractive Chiyoko, busy as he was bantering in the company of high-powered men. But Chiyoko noticed him, in part because he eschewed the sake that everyone else consumed in abundance, but mainly because of his presence and social ease. Instead of military stiffness, he exuded warmth. She also noticed that two fingers were missing from his left hand, and when she saw him struggle to remove the lid on a soup bowl, she approached and kindly offered a helping hand. Instead of the friendliness he showed his fellow partygoers, all Chiyoko heard was "I don't want help for my own business." The cold rebuke was off-putting, and it would be another year before their paths crossed again. But when it did, their interactions led to an altogether different outcome.

During the intervening year, Yamamoto was mostly away at sea as commander of the First Carrier Division. Then, appointed to the Naval General Staff in June 1934, he returned to Tokyo, where he found the workings of the Japanese government turning ever more tumultuous. The repression of naval ship construction required by the arms treaty was increasingly under bitter attack from naval and army leaders, government officials, and the public at large. Militarists, gaining further control of the government, were insisting that the treaty's imbalance had gone on long enough, and the second London Naval Conference was announced for the next year. Yamamoto, given his extensive work abroad as well as at the 1930 disarmament meeting, was named the chief Japanese delegate to preliminary talks held in the late fall to set an agenda for the new conference.

The posting was an honor, and that summer Yamamoto began preparing. It wasn't all work, though, and he continued to enjoy after-hours attendance at geisha houses. One night out, as fate would have it, he showed up at a party Chiyoko Kawai also attended. Chiyoko tried again, reminding Yamamoto of their meeting the summer before, even bringing up the lid and the soup bowl. But once again Yamamoto was gruff: "I scarcely remember such an event or you either." Several days later, however, things were different. Chiyoko ran into Yamamoto at another party, where Yamamoto was seated next to his old friend and classmate Zengo Yoshida, a naval officer who one day would become navy minister. The geisha and Yoshida already knew each other. With Yamamoto looking on, the two chatted easily, and Yoshida, surveying the spread of food, asked Chiyoko if she liked cheese. Chiyoko replied that she did, "very much." Suddenly Yamamoto blurted out, "Well, I'll treat you."

For Yamamoto, something had awakened. He invited Chiyoko to meet him at noon the next day at the Imperial Hotel. Yoshida feigned shock. "I have hardly ever heard this man say such a thing." He turned to Chiyoko. "You must go." And she did, and waiting for her in the hotel dining room was none other than Rear Admiral Yamamoto. He made good on his promise, too, ordering her a plate full of cheese as well as a cocktail. That date led to others. Soon they were intimate, and theirs became a summer of love. Yamamoto was fifty years old, Chiyoko was thirty-one, and the geisha had stirred desires Yamamoto had not previously known. By the time late September came around and Yamamoto left Japan for the long trip to London, he was riding high about his dual duties—to Japan and now to Chiyoko. "I was fired with high spirits and determination to hold myself responsible for the fate of my country," he wrote to her. But that wasn't all. "My passion burned with the excitement of seeing the rapid development of my intimacy with you."

CHAPTER 3

THE FLYBOY FROM ENID

—

John Mitchell, fighter pilot
Courtesy of the Mitchell family

SECOND LIEUTENANT JOHN MITCHELL, STRUGGLING FOR DAYS with a head cold, was more than happy when he spotted the runway at Hamilton Field down below. For two hours he'd been flying maneuvers in the Curtiss P-36 Hawk, a single-winged fighter that could reach speeds of 300 miles per hour but cruised most comfortably at 250 miles per hour. He'd taken it up to 20,000 feet and higher—the monoplane's ceiling was 30,000 feet—and it was in the thin air that his head had bothered him the most. He probably shouldn't have even flown that day, and now all he wanted was to get onto the ground as fast as possible. But as he descended, a bolt of pain shot through the right side of his head and bored into his temple. He had dropped down too quickly, and he nearly screamed as he pulled up to 10,000 feet as quickly as he could. He stayed there and slowed his breathing until the throbbing stopped. Then he tried again, this time taking the plane down ever so slowly. Not a good day in the sky, he later told Annie Lee. "My cold bothered me so much I really didn't enjoy it."

Which was unusual. Ever since he'd enlisted, John Mitchell had enjoyed flying to the utmost and had a hard time sitting around when he could be in the sky. Mitch, as everyone called him, had moved from Moffett to Hamilton Field in early September 1940 and for the next six months had trained at that Army Air Corps airstrip twenty miles north of San Francisco. He had settled into the Bachelor Officers' Quarters, or BOQ, a snow white stucco structure built on a hill that overlooked the base. From the small porch of his second-floor apartment he looked out onto the three tennis courts reserved for officers only, the base's outdoor swimming pool, and, in the distance, San Francisco Bay. The Officers' Club, a short walk away, was "something to write home about," which he did, describing the club in detail for Annie Lee—the bar, the game room, the Ping-Pong tables, and the expansive dining room. "It's just too wonderful for words." The living situation beat Randolph Field in Texas by a mile and reminded him of Hawaii. "Flowers are growing everywhere."

His raison d'être, though, was flying. He belonged to the 55th Fighter Squadron, one of several squadrons making up the 20th Pursuit Group. New pilots assigned to other flying groups continued pouring into Hamilton, eventually overcrowding the base, as the Army Air Corps moved to rapidly build up its overall pilot numbers. Likewise, the development of new fighter planes had taken on a sense of urgency. Mitch briefly flew the P-36 but by mid-October 1940 was switched to the Curtiss P-40 Warhawk, considered more agile in combat at low to middle altitudes, which the government had put into large-scale production. With war seemingly on the horizon, he was eager to pile up training hours. With each passing month, the fighting overseas had proved disastrous for the Allies. In early May 1940, the Netherlands had surrendered to Germany. In late May, British forces had retreated en masse to France's port city

of Dunkirk, and more than 850 civilian-owned boats had crossed the
English Channel to help with an all-out evacuation, the largest military evacuation in history. Belgium had fallen to the Nazis before the
month had ended. The US government, meanwhile, had earmarked
millions of dollars for a new massive military buildup. Then Nor-
way had fallen in June. In August, the German Luftwaffe had begun
bombing England, and in September, it had targeted London. By the
end of the month, just as Mitch was moving to Hamilton, the nations of Germany, Italy, and Japan signed a Tripartite Pact, which
stipulated that an enemy to one nation was an enemy to all three.
Around Hamilton Field there was a steady hum about war. "Yes, our
war situation grows more serious each day," Mitch wrote Annie Lee
days after the Tripartite Pact was signed. The next week he added,
"Really looks as though we are heading for a war of some kind." It
was a betting man's odds that US involvement was inevitable; the
question was when.

Mitch was part of the rush for new fighting men. He aimed to be
ready, but his eyes didn't lie; during exercises at Hamilton Field he observed the occasional failure of men and machines. Flying could be a
risky and dangerous business, especially at a time when hundreds of
untested pilots were hurrying to train in relatively untested planes.
One day it was a buddy of his, a pilot named Jack Jenkins, whose engine cut out right after he'd taken off. He was forced to land the P-40
in the bay. Luckily Jenkins wasn't hurt, Mitch told Annie Lee later,
"but this is the second motor failure we've had recently." Lucky, too,
was a pilot from another squadron, who, after landing, started on a
ground loop and gunned the engine, thinking it would straighten
out the plane. Instead, the plane stood up on its nose, probably because the pilot hit the brakes too quickly. The pilot came out unscathed, and the plane suffered only minor damage. Then there was
the pilot who came in too low. He scraped the embankment at one
end of the airstrip and sheared off his landing gear. He was forced to
keep going and fly to a bigger airfield in Sacramento, where he could
land without wheels. Mitch had his own scary moments as well. He
was flying a P-40 one day at 20,000 feet and had been at that high altitude for about an hour. It was thirty below zero outside, and he was
freezing. His engine suddenly cut out. Ice had apparently formed

in the carburetor. He feared he was going to have to make a forced landing. But once he took the plane to a lower altitude, the engine fortunately kicked in again.

Not so lucky was a pilot from another squadron at Hamilton. His P-40 stalled at 12,000 feet and went into a fast-descending spin. The pilot bailed out in a parachute, but as he exited the cockpit, he struck the plane's tail. The multiple injuries he sustained landed him in the hospital for six weeks. Some of the mishaps were worse. In late October, Mitch joined an aerial search party for two fliers who'd disappeared after taking off from an airfield in Oakland en route to Monterey, only a hundred miles away. He and the other pilots buzzed over the ground at about twenty feet high but found nothing. The two pilots were never seen again. Several months later, two rookie pilots learning to fly the P-40s collided in midair and were killed instantly. Neither pilot had seen the other. "You may have seen something about it in the paper already," he wrote Annie Lee.

Mitch was sympathetic to those calamities, but only to a point. He firmly believed that airborne troubles were mostly man-made. "Nine-tenths of all these accidents are due to actions of the pilot and not to some failure of the airplane," he said. "If a fellow is alert and keeps his head up, I see no more danger in flying than anything else. It's the fellow that goes to sleep at the switch that gums up the works." He was especially wary of the veteran flyboys at Hamilton who didn't hesitate to fly while hung over. "I sure would hate to have to ride with some of these older officers after they spent a night at the club," he said. "They are really not fit to fly and yet they do it." Mitch was hardly a prude, but flying was deadly serious. He'd stick to the "old rule of no drinking 24 hours before flying."

But beyond the periodic misfortune, Mitch embraced the training days, knowing full well he faced a steep learning curve. When he was on duty, the usual routine was to fly in the morning and attend ground school in the afternoon. One day not long after his arrival he saw he was on the list of pilots scheduled for "acrobatics," in which the pilots practiced a variety of stunts such as loops, slow rolls, snap rolls, and spins. Fighter pilots relied on aerobatic maneuvers to survive—and hopefully win—dogfights. Seeing his name, Mitch at first felt a pit in his stomach; he had not performed any acrobatics

in either a P-36 or the new P-40s. He awoke the next morning feeling anxious. But after only a week of drills, the early apprehension
seemed a distant memory. Suddenly acrobatics were lots of fun, and
he wrote to Annie Lee excitedly about one exercise on the morning
of Thursday, October 1, when he had flown in formation with two
other pilots. "We did loops, slow rolls, Immelmanns and chandeliers together, and it looked very good." The flight leader afterward
praised Mitch, telling him it was the best he'd done so far. Mitch
began looking forward to the sessions when he and other pilots were
performing the coordinated stunts, one of which was called the "rat
race," in which he and five other pilots lined up in the sky, one plane
right behind the other. It was a game of follow the leader, he told
Annie Lee: "The leading plane does all sorts of things and we have to
do them in order to stay with him." Mitch seemed like a kid at play.

Other mornings, Mitch took the plane up to 20,000 feet and
higher—nearly four miles into the blue—where he would fire his
machine guns to test their functionality, the worry being that below-
zero temperatures might hamper them. He found that most of them
worked fine. Or he'd see if he suffered from any of the sinus and ear
troubles high-flying pilots sometimes experienced when descending
sharply. He'd dive from 20,000 feet to 10,000 feet in a single dive, as
fast as he could. The instructors at the ground school had suggested
tips to avoid difficulties, but Mitch found he didn't need them. The
fast, deep dives, he said, "really had no effect on me." He liked the
high altitudes, the big sky with so much to see. During a longer flight
over interior California, he passed over snowcapped mountains that
rose 10,000 feet into the air, and later that same day he cruised over
miles of barren desert. "I fully enjoyed every bit of it," he told Annie
Lee. Flying off the California coastline, he'd spot a navy destroyer
or aircraft carrier heading for Hawaii, where construction at Pearl
Harbor was nearing completion. Hickam Field had been built specifically as a new base for the Army Air Corps, while the Navy controlled Ford Island. Following fleet exercises in the waters off Hawaii
in 1940, the Navy had decided to keep its entire Pacific Fleet perma-
nently based at Pearl Harbor.

With acrobatics, simulated dogfights soon followed, which Mitch
relished for their competition, even if he was a novice and, after a

mock dogfight against his squadron commander, had to admit he wasn't very good. The veteran commander got onto Mitch's tail within minutes of their start. Mitch maneuvered to shake him off, but the commander got right back onto him, bearing down. No matter what Mitch tried, he was outfoxed. "Naturally he has had a lot more experience," he conceded. In late October, when he got the chance to pilot a P-40, he wasn't afforded the luxury of taking the plane up for an hour or so to get used to it. He was ordered to get into a dogfight against one of the veteran fliers on the base. Once again his inexperience showed. "He got me the first time and then I got him. We decided to try the third time and I had him again, but relaxed for a couple of minutes and he got on my tail and I was unable to shake him off."

Initially Mitch wasn't impressed with the P-40—"Uncle Sam's prized possession," as he called the much-anticipated fighter plane. It handled okay and was faster, but "they have tried out the P-36 against the P-40s and every time the P-36 comes out the winner," he informed Annie Lee. Even so, he flew mainly a P-40 for the remainder of his time at Hamilton Field, and with repetition he became more competent and, just as important, more confident. He experienced a breakthrough of sorts during a ten-day training trip to Muroc Dry Lake, about four hundred miles away. From a training perspective Muroc was ideal—a dry lake bed about fifteen miles long that could be bombed and strafed relentlessly. From a lifestyle perspective Muroc was grueling. "Real dusty, rattlesnakes and varmints—worse than Texas," Mitch joked to Annie Lee. The daytime temperatures soared into the nineties, while the nights turned cold, Mitch and four other pilots cramming into a tent shaped like an upturned ice cream cone to fit a small stove to keep warm, with a stovepipe sticking out of the tent's top. Mitch was huddled there on election eve, November 5, 1940, struggling to hear the voting results on a radio with lousy reception that belonged to another pilot. President Franklin Delano Roosevelt easily outpaced his Republican challenger, Wendell Willkie, for what would be an unprecedented third term in the White House. Mitch was curious but didn't much care who won. "It doesn't particularly matter to me, though I think Roosevelt will lead us in to war," he told Annie Lee in a letter he wrote using his knees as a desk.

For ten days Mitch didn't shave or have to wear a tie, as he was required to every day at Hamilton, and he was able to go around just about any way he wanted to. He trained hard, waking up in the dark to get into the sky by 5:00 a.m. to practice maneuvers and shooting before the air got rough due to the heat. "I have never had so much fun flying since I have been in the air corps," Mitch said. "We buzzed the camp every day and then there were trains running by about ten miles from camp that we buzzed all the time, to say nothing of coyotes and jack rabbits." Formal training for strafing runs was virtually nonexistent; everyone was basically learning on the job. Mitch relied on instinct and, after firing the plane's .50-caliber machine gun at the desert targets day after day, found he had a knack for it. He ended up with the second highest gunnery score in his squadron. "I'm bragging," he told Annie Lee, "but I really can't help it as some of these fellows have been flying for years and firing every year and sometimes several times a year." Mitch came away from the exercises upbeat. "I am beginning to believe that I actually know a little bit about flying," he said. "We have finished our training and I guess we might now be considered full-fledged pilots."

His superior officers were beginning to take notice. One was a fellow southerner, a Virginian named Henry "Vic" Viccellio, whose grandfather, like Mitch's, had fought in the Confederate Army. The two had met not long after Mitch's arrival at Hamilton. Captain Viccellio, three years older, had been around awhile, enlisting as a private in 1934, the same year Mitch had concocted his unsuccessful scheme to get into West Point by joining the Coast Artillery Corps in Hawaii. Viccellio had served in the 20th Pursuit Group at Hamilton but more recently had been promoted to commander of a fighter squadron in the newly formed 35th Pursuit Group. Watching Mitch train from a distance, Viccellio sensed that the flyboy from Enid had the right stuff to lead. Indeed, early in 1941, as rookie pilots continued to swell the ranks at Hamilton, Mitch was one of the pilots tapped to help jump-start the new arrivals, even though he was only four months into his training. He complained to Annie Lee, "These new boys are as bad if not worse than we were with the planes." But they were crocodile tears. Mitch stepped up, knowing he must be doing right to be chosen for the added responsibility.

It was also in January 1941 that John Mitchell flew his most complicated training exercise to date, one he would never forget. He was part of a twelve-plane formation whose mission was to track down an enemy bomber that had penetrated their airspace. To make matters worse, the bomber was flying with an escort of fighter planes. Mitch's formation found the bomber easily enough, and the twelve planes went into a circle with one plane right behind the other so that the enemy planes could not get on their tail and shoot them down. By that time, Mitch had found flying in formation at once challenging and rewarding: each man had to do his part, and if he didn't the whole formation got messed up. In the exercise, the enemy fighter planes took runs at the formation, but they could not break it and would have to pull off again. The planes in Mitch's formation would try to get the attacking planes without pulling out of the circle too much. Both sides jockeyed that way for a few minutes. Then Mitch saw his flight leader suddenly make a move, pulling out to the side to catch the bomber. Mitch didn't hesitate: "I went with him." He and his flight leader zeroed on the bomber and got it, firing their guns' fake bullets. But two enemy fighter planes were on Mitch and his flight leader in an instant, and the four planes went around and around trying to hit one another. That went for a few more minutes, when everyone else called it quits. But not Mitch. He kept on fighting, locking into a dogfight, his P-40 and a single enemy fighter chasing each other. The radio crackled with instructions, but Mitch kept going, performing aerobatics to get an edge against his enemy. He finally stole a look around and saw that everyone else was gone except for him and the enemy plane. He was stunned and headed home.

The intensity he'd brought to the moment surprised him afterward. He realized that the radio call had been to tell him the exercise was over, return to base. But he'd been in a zone. "I didn't hear it and kept on dogfighting." In the end, he'd shown resolve and a focus proving that he was emerging as one of the squadron's best and brightest. And though he didn't know it at the time, he'd participated in an exercise foreshadowing a real-life mission that would someday come his way.

BEING STUCK ON THE GROUND WHEN HE WAS SUPPOSED TO BE flying was the worst. It might be due to unexpected bad weather, a bout of the flu, or a propeller malfunction. If he was unable to go—no matter what the reason—he was miserable. He'd kill time shuffling around the hangar, learning about supply, armaments, and the paperwork necessary for flying missions, but without flying it was all a big bore. When he got home at 4:00 p.m. he was in no mood to talk to anyone.

But if he was off duty, when flying wasn't even an option, that was different. He did his best to make the most of his free time. He took up badminton, tweaking the swing he'd used to smash a tennis ball across the clay courts in Enid to hitting a birdie. His main sport, though, was basketball, and in a matter of weeks at Hamilton he was on the court practically every night. He started on two teams, coached the 55th's squad, and refereed games in a town league in nearby San Rafael. Like a sports reporter, he kept Annie Lee updated on wins, losses, and his teams' run for league championships. He proudly reported leading the officers of the 55th Squadron to an easy victory over the officers from the 77th. "Boy did we lay it on them. Beat them 20 to 4. That certainly should be enough exercise for one day, but I have to call two basketball games tonight in the San Rafael league."

When the California weather was sunny and clear, Mitch would often joyride in the secondhand convertible he'd bought with a bank loan and had driven cross-country to Hamilton Field. He'd put the top down and whiz along the coastal roads. Sundays usually featured a feast at the Officers' Club—a dinner buffet of roast veal, fried chicken, fried scallops, potatoes, peas, giblet gravy, celery, tomatoes, and shrimp. Thinking of Annie Lee's mother's specialty at her Uncle Joe's family cookouts, Mitch said, "Needless to say I availed myself of fried chicken." Drinking was usually on the off-duty agenda. Three miles from the base, on the way to San Francisco, was a roadhouse—an eating spot with a bar—that became a favorite. On a Saturday night, Mitch might start with a few drinks at the base and then head out with friends. "Before the evening was over we were all drunk and wandering around everywhere," he reported to Annie Lee. Nights like that often ended in a fog of booze, with women sometimes coming into and out of the picture, then a wicked hangover and a promise. "I think that will be my last drinking for a while," he'd write.

Hard as Mitch tried to have fun—and he surely did—at some point during his partying his thoughts usually turned to Annie Lee. Often the trigger was the moon—little Johnnie Bill's moon, which, after he'd fallen for Annie Lee Miller of El Campo, Texas, had become their moon. It could be a half-moon. "The old moon is half full tonight," he wrote Annie Lee one night, "and it's funny how romantic a fellow gets when he sees a moon." Or it could be a disappearing moon. "The rain has shut out the moon completely," he wrote another time, "but not before I got a good look and could think of you." Or it could be a new moon: "The moon is beginning to be full again and whenever I look at it I think of you." Or, of course, it could be a full moon. "The moon shines so brightly out here," he wrote, "and I see that old moon and think how much I would like to see you tonight."

Annie Lee was never far from Mitch's thoughts.

AS SOON AS HE ARRIVED IN CALIFORNIA, HE WAS AFTER ANNIE LEE to send him a photo. "Hurry with that picture," he wrote at the end of one letter. Two days later he said, "I'm patiently waiting for that picture," and the next week, showing hardly an ounce of patience, he said, "I can hardly wait until it gets here." Annie Lee told him that she was having a photograph taken and would send one as soon as possible. When a big envelope arrived, Mitch melted. "Here I sit looking at the sweetest thing the Lord ever created! She's got the cutest dimple in her chin, to say nothing of a most bewitching pair of eyes."

Mitch and Annie Lee wrote a lot, keeping count of their correspondence like a couple of love-struck teenagers. If several days passed without a letter, Mitch complained, wondering if Annie Lee had fallen ill or, worse, that her feelings had changed. But then a letter would show up. "Just when I was beginning to think you didn't love me any more I got your sweet letters." In his missives Mitch described flying, sports, partying, and his love for her. They told each other what books they were reading and what movies they'd seen.

Mitch recommended the romantic comedy *Third Finger, Left Hand*, a new release starring Myrna Loy as a successful fashion journalist. "I thought it was really funny," he said. "They cut a couple of capers that had me almost rolling in the aisle." He decided that *Wyoming*, a western starring Wallace Beery, was lousy, and even worse was the movie *Dr. Kildare Goes Home*, starring Lew Ayres, Lionel Barrymore, and Laraine Day.

Mitch mostly went to the movies in San Rafael with his roommate, a pilot named Ben Bowen. The two had met during training in Texas. The outings turned into a threesome, however, once Bowen met and began dating the daughter of a marine captain. Mitch was now the odd man out, and he would end up sitting off by himself in the darkened movie theater, longing for Annie Lee. The upside—if there was one—was that he was taken with Bowen's girlfriend, and seeing her and Bowen together brought into focus his feelings for Annie Lee. He liked that Bowen's girl was down to earth, "not at all sophisticated and is really sweet," he wrote Annie Lee. "The reason I love you so much, you are what you are, sweet and friendly and try to get along with everyone, without any high-handed stuff."

Navigating long-distance love wasn't easy, though. Mitch and Annie Lee had known each other only a few months. Each might have fallen hard for the other, but by midsummer they were separated. Their relationship became a roller-coaster ride as they worked to make sense of their feelings while hundreds of miles apart. Intensifying matters for Mitch was that two of his best friends at Hamilton Field were on track to get married. One of them, Ben Brown, had been dating his girlfriend for quite a while, so when they announced a wedding date for January 1941, Mitch was not at all surprised. Plus he was asked to serve as the groom's best man. But Mitch was startled by the speed with which his roommate, Ben Bowen, and the captain's daughter went from dating to embarking on a road trip to Reno one weekend in early December to get married. And just as they had watched movies as a trio, Mitch accompanied them to Nevada, where he stood as his roommate's best man—further confirmation that he was someone his friends could count on. "I hope everything turns out o.k.," Mitch wrote afterward. "Seems like it is happening too fast for me."

Even so, talk of marriage began appearing in Mitch's letters, a topic that for him and Annie Lee was thrilling—or not—depending on the moment. "I become more firmly convinced every day that married life is the life for me" was the way he leaned into the matter in one of his first letters from California. In October 1940, he drove into town to buy Annie Lee a present for her November 2 birthday, and while at the jewelry store he also sprang for an engagement ring. He had no idea when he was going to give it to her, and certainly he had no plan for marrying, but he wanted to be ready when the time came. His plan for her birthday present, meanwhile, couldn't have turned out better. In her letters, Annie Lee had said nothing to him about her upcoming birthday. He didn't know why, but it turned out that she was testing him, to see if he'd remember on his own. Mitch did not disappoint. From the base, he made arrangements for a Waltham seventeen-jewel gold wristwatch to be delivered to her first thing the Saturday morning of her twenty-fourth birthday—as a total surprise. "It's very simple but I'm crazy about it," Annie Lee gushed to her Aunt Ludma when the package showed up at her apartment at 8:00 a.m. Mitch had passed the test with flying colors. The gift, said Annie Lee, made for "the happiest birthday I have ever had."

But thoughts of an engagement were tricky. They both saw other people and were fairly open about their "dates." Army and navy fliers stationed at bases near San Antonio called on Annie Lee and her cousin Josie at the apartment they shared. The cousins double-dated, or Annie Lee sometimes dated by herself. Mitch chided her, although he couldn't complain. He went out plenty. Off duty, he and his friends would pick up single women, hang out, drink, and dance. He insisted that it was all in good fun and wanted to blunt any thoughts Annie Lee might have that something more was going on. "I'm not stepping out on you that way," he said. At one point Mitch even floated a rationale for both of them going out: it was a way to cope with the loneliness of loving someone far away. "The gal can still be faithful even though she has a date every now and then." Yet while condoning dating, he continued to allude to their possible engagement. "It's a serious step to take, isn't it honey? But I'm sure I have never been in love before and I *am* now."

Most agonizing were Mitch's efforts to reunite with Annie Lee, only to see his plans come undone. In late November, he'd hoped to log what was called "cross-country flying time" with a trip to San Antonio. But the idea never came to fruition. The only flying he did around Thanksgiving was a quick round-trip to Muroc Lake to deliver mail. While Annie Lee visited with her many relatives at her parents' home in El Campo on Thanksgiving, Mitch had an uneventful day, driving for two hours in the afternoon with a friend who had a shack off in the woods. They took along beer and sandwiches, hiked some trails, and fooled around with a .22-caliber pistol, shooting at targets they set up on trees.

Christmas was even a bigger deal. In late October, Mitch announced that he expected to get a short leave and was aiming to catch a plane ride to San Antonio with some other pilots bound for Texas for the holidays. He excitedly mentioned Christmas in nearly every letter, but in early December his tone began shifting from confidence to worry. Then came the back breaker: the week before Christmas, he found out he'd been assigned duty as Officer of the Day on Christmas Day. Worse still, the commanding officers had ordered that no one was allowed to travel more than two hundred miles from Hamilton Field over the holidays, which meant that even if Mitch could find a replacement, he could not go to Texas. He got into an argument with his squadron commanders but made no headway. He was so mad he considered sneaking away, but then he thought better of it. "Though I am not too much in love with this man's army I still have no desire to get kicked out," he explained to her. Instead of being able to give Annie Lee her presents—a white bathrobe and matching slippers—in person, he rushed to mail them. Annie Lee, in turn, mailed him presents: a shirt, tie, and leather wallet. They managed to talk on the telephone, but that hardly healed the wound of not being together on Christmas Day. Annie Lee expressed her disappointment, especially after all the buildup and promises. Mitch became defensive. "You talked as though I could help it," he wrote to her afterward. "I am just as downcast as you." He was so frustrated by the busted Christmas that he blurted out something to Annie Lee he'd never said so directly: "I have a very potent suggestion: Let's get married!"

Annie Lee sidestepped the proposal. It hung in limbo through-out a winter of ups and downs, as Mitch pressed for a commitment and Annie Lee equivocated—until one night in early March 1941. Annie Lee had agreed to go out with an old boyfriend who was in the service and happened to be back in town. She saw much to ad-mire in her date and did enjoy his company at dinner. But while they were out, as she later told her aunt, she realized, crystal clear, "I miss Johnnie so much that I know for sure that it's Johnnie I want." From then on it seemed as though they were back on track. The Christmas debacle was behind them, and Mitch took to giving pep talks. "Just remember there is always tomorrow and let's play Scarlett O'Hara," he said, playing off the line from *Gone With the Wind*: "After all, to-morrow is another day!" He began talking up the summer, when he was certain he'd finally get time off. In fact, he had a trick up his sleeve to surprise Annie Lee on Easter: without telling her, he'd gotten permission for a ten-day leave and arranged to hitch a ride on a transport plane leaving from Sacramento for San Antonio. He couldn't wait.

Except that once again, fate intervened, because on the first Fri-day in April 1941, it turned out that his commanders had a surprise for him. He was called in and informed that he'd been chosen to go overseas on a special mission—he and three other pilots from the hundreds of fliers at Hamilton Field. They weren't told where they were going—although either England or China was a safe bet. And they were told not to leave the base, not even to fly, because their orders could come in at any moment.

"It was a great honor to be selected," he later wrote.

Mitch did not protest that time, even though his Easter trip was dead. He broke the news in an urgent letter to Annie Lee, including the fact that the mission had "turned upside down" the surprise visit he'd planned. But this was what he'd trained for, he said, "an actual taste of warfare." He hoped she'd understand. They would have to wait. "I told you that this was a here-today gone-tomorrow world we were living in, and today it really looks that way."

CHAPTER 4

NO ORDINARY STRATEGY

Vice Admiral Yamamoto in London
WW II Database/public domain

REAR ADMIRAL ISOROKU YAMAMOTO SET SAIL FOR THE UNITED
States aboard *Hie-maru*, departing Yokohama, the port city south
of Tokyo, on the afternoon of September 25, 1934. The ocean liner
followed a route to Seattle, where, for the second leg of the journey,
he and his small entourage boarded the Great Northern Railway's
transcontinental train to Chicago. In some ways, Yamamoto had
been there, done that—meaning he'd already traveled the United
States extensively and witnessed firsthand its breathtaking vast-
ness and natural resources. On those excursions in the prior decade
he had been merely a midlevel Japanese military man, unknown

beyond Washington, DC's military and diplomatic circles. Things were different now; he was Japan's chief delegate to the preliminary tri-power talks on arms control, to be held in England, that were shaping up as a showdown. Whereas Great Britain and the United States were looking to extend the London Naval Treaty of 1930, which had maintained a 5:5:3 ratio for naval shipbuilding, Japan was signaling that going forward, nothing less than naval equality among the three nations was acceptable.

The assignment was a breakout moment for Yamamoto, thrusting him for the first time into the international spotlight. Press interest in the fifty-year-old commander was constant, attention that grew during the train trip as Yamamoto rebuffed reporters and, even more mysteriously, rarely appeared in public. He was content staying inside his compartment indulging in the card games he so enjoyed: bridge, *shogi*, and poker. During a weekend layover in Chicago, he did venture out to watch big-time college football. On Saturday, October 6, he joined thousands of fans at Dyche Stadium in Evanston, a suburb just north of Chicago, where the Northwestern University Wildcats lost to the visiting Iowa Hawkeyes by a score of 20 to 7. But even at that highly public outing the admiral had no comment for reporters.

Yamamoto ended his press blackout once he reached New York City. He met with reporters the day after checking in at the upscale Astor Hotel. His main talking point confirmed all of the press speculation: Japan, he said, must be on the same naval footing as the United States and Great Britain in any new treaty. But he insisted that Japan was open-minded as to how equality could be achieved. He said, for example, that it could be reached not by addition but through subtraction; instead of redoing the ratio for new naval shipbuilding from 5:5:3 to 5:5:5, the nations could decide to scrap their existing naval armament. Whatever the way to parity, Yamamoto said, "My belief is that all nations are entitled equally to enjoy a sense of national security."

The admiral spent three hurried days in New York City prior to his departure for Southampton, England, aboard the luxury liner *Berengaria*, the flagship of the Cunard Line, which featured lavish suites, a grand ballroom, and a "Pompeyan" swimming pool. He huddled for hours during the day with fellow Japanese diplomats, among them the naval attaché from the Japanese Embassy, who'd ridden a train up from Washington, DC, just to see him. Pleasure followed business, as he dined one night at the town house of the Japanese consul general on the Upper East Side, and on his second night, he was honored at the private Nippon Club on West Ninety-third Street by throngs of fellow nationals living in the city. Reporters tracked his every move, writing glowingly that Yamamoto "is known as one of Japan's most brilliant naval officers. He has won remarkably rapid promotion."

On his second day in New York, Yamamoto felt the need to respond to news out of Washington, where retired brigadier general William "Billy" Mitchell was making headlines forecasting future troubles with Japan. Mitchell, an outspoken advocate of airpower, was no relation to young Johnny Mitchell from Enid, Mississippi, who earlier in 1934 had left Columbia University for Hawaii as a new army enlistee. The general, testifying before a congressional committee, had asserted that the United States was ignoring at its peril Japan's relentless military buildup, especially that of its Imperial Navy, and that the Pacific power should now be seen as the United States' greatest threat. It was not a dream, he said, to imagine a Japanese plane landing in Alaska, which "might be followed with disastrous results to New York, or some other important city." He said that the United States must commit to rapidly strengthening its own air forces.

Questioned by reporters the next day about the general's inflammatory testimony, Yamamoto calmly sought to reassure. "I do not look upon the relations between the United States from the same angle as General Mitchell," he said, "and I have never looked upon the United States as a potential enemy. The naval plans of Japan have never included the possibility of an American-Japanese war."

That was true at the moment, technically speaking. But it was the kind of comment that later would be cited as proof that Isoroku Yamamoto had been a conniving, treacherous foe all along, one who had smiled graciously while concealing a gestating war scheme.

ON OCTOBER 16, 1934, TWENTY-SEVEN DAYS AFTER LEAVING TO-kyo, Yamamoto checked into a sixth-floor suite at the Grosvenor

House Hotel in central London, his home away from home for the next three months. Japanese delegations tended to stay at the Grosvenor, and the hoteliers, ever gracious, hung out a rising sun flag upon the admiral's arrival. Formal talks were not expected to begin for another week, but Yamamoto immediately began making the rounds the next day, exchanging handshakes and niceties with representatives from the Western powers. The whole purpose of the Preliminary Naval Limitation Conference was preparatory—to lay the groundwork for the extension of the London Naval Treaty before its expiration the next year. That was the hope, but on his first day Yamamoto effectively drew a line in the sand when he reiterated to reporters that Japan would never submit to any continuation of the 5:5:3 ratio system. "I don't think there is any possibility of compromise on any plans based on a ratio," he said. "Japan objects to the ratio system and wishes to keep off the question of ratios altogether." For his hard-line comments, reporters began referring to Yamamoto as the "outspoken seadog."

The weeks that followed saw a round robin of diplomacy. Yamamoto spent his days conferring with the US delegates, then the British, then the Americans again, and so on. The international press covered each day's news—which, in effect, was that there was no news. "Deadlocked" and "inaction" emerged as the go-to terms in newspaper headlines and stories. Though each country's representatives projected a spirit of cooperation and appeared to be open to compromise, all were working to gain an edge without budging an inch. The Americans and the British insisted on the status quo—a naval building ratio of 5:5:3 for the three nations—to which Yamamoto found a myriad of ways to object. He quipped at one point, for example, "I am smaller than you, but you do not insist that I eat three-fifths of the food on my plate." He nonetheless continued to attend the daytime sessions in all earnestness, evidenced by the fact that he relied on an interpreter despite knowing English. "It takes twice as long when you have an interpreter," he explained to an associate, "and gives you time to watch the other man and consider your next move." And despite the difficulties, the negotiations remained cordial. "We sailors get on admirably together," Yamamoto told a reporter at one point. Evenings were often spent socializing at official

dinners or less formal cocktail parties. Yamamoto did not drink but
was hardly shy, always ready to initiate a poker or bridge game with
William H. Standley, the US chief of naval operations, and Ernle
Chatfield, Great Britain's first sea lord. He returned to his hotel one
night with twenty pounds he had won from Chatfield.

Yamamoto received a promotion to vice admiral on November
15, which enhanced his standing at the talks as the face of Japan,
particularly in terms of his strident opposition to naval quotas. But
behind the facade of Japanese unity things were more complicated.
Yamamoto was indeed firmly against the treaty's ratio system; he
bristled at its inequity, which put him into sync with the so-called
militarists back home who dominated the army and included a
number of younger naval officers. But their reasoning differed.
Though Yamamoto opposed the current treaty, he was genuinely
interested in finding new ways to extend its purposes: arms control
and world peace. In London he spoke firmly but hardly as a hawk.
In fact, he'd proposed that the three nations downsize their navies
to achieve parity, retiring their destroyers and cruisers. As an alternative to ratios, he favored what negotiators called an "upper tonnage limit," meaning that the overall tonnage of each of the three
countries' fleets would not exceed an agreed-upon maximum. For
him, a strong Imperial Navy—which he fiercely supported, especially its aerial capability—was for defensive purposes, a deterrent
to dissuade the United States or any other country from eyeing the
Pacific region for expansion. But the militarists saw the talks quite
differently; not only did they seek an end to a demeaning 5:5:3 ratio
that was insulting to every Japanese citizen, but they advocated an
abrogation of all treaties to unshackle Japan to set it free to build,
flex its military muscle, and challenge other world powers.

Privately, Yamamoto began to feel exploited, basically a pawn for
the militarists, misgivings he shared later with his lover Chiyoko. "I
couldn't help feeling very unpleasant," he wrote about the talks, "as
it seemed to me I was being used just as a tool." Even so, he continued to represent Japan's interests as best he could, wary of the rising
tide of jingoism at home but always loyal to his country. His position
in London created an impossible dilemma. For years he'd helped to
build a new and stronger Imperial Navy—and would continue to do

so—but whereas he saw the navy mainly as a peacekeeping deterrent, he increasingly understood that the militarists viewed it as an offensive asset against the United States. He understood, therefore, that it was his duty to consider the Imperial Navy as a strike force—and to plan for that future possibility. For several years now, he'd been tossing around options for a hypothetical war with the United States, first mentioning in a lecture to cadets in 1928 the idea that attacking the US fleet at Pearl Harbor might be Japan's best shot at victory.

When Yamamoto agreed to meet with a reporter one night in early December, those matters were certainly in play, if not explicitly, then as subtext. Hector C. Bywater, covering the London talks for the *Daily Telegraph*, was no run-of-the-mill reporter; he was the acclaimed author of *The Great Pacific War*, the best-selling 1925 novel dramatizing an imaginary surprise Japanese naval attack against an unsuspecting United States, with the aim, in Bywater's view, to build an empire.

Yamamoto welcomed the journalist into his hotel suite at Grosvenor House. They settled into easy chairs, and a butler served drinks—Bywater favored scotch, while Yamamoto went with seltzer water. Bywater's thinking about a hypothetical Pacific war had altered somewhat in the decade since the novel's publication. In fact, when he met with Yamamoto he was working on an article for *Pacific Affairs*, a scholarly journal, which had asked him to update his analysis. In the article, published a few weeks later, Bywater wrote that the broad contours of a Japanese-US war imagined in his novel had not changed—namely, Japan's surprise attack and the United States' eventual victory. But he now thought that Japan would radically alter its means of a first strike and rely on a growing fleet of aircraft carriers and bombers instead of the warships of old. It was a revision that certainly would have pleased Yamamoto, the leading voice in Japan's navy for airpower. Moreover, when Bywater said that the Philippines and Guam would "no doubt be gobbled up by Japan in the first weeks of war," he was speaking to the choir, for Yamamoto's own thinking was evolving along similar lines.

Whether the admiral and the journalist actually discussed *The*

Great Pacific War remains unclear. The only record of their December 3 meeting is the article Bywater wrote for the newspaper. The story's focus, not surprisingly, was the treaty talks, and it reflected cold pessimism. Yamamoto left little doubt about Japan's immovable rejection of the ratio system, which, to Bywater, guaranteed that the treaty talks would go nowhere. Worse, Bywater foresaw failure as triggering a tri-power arms race with an increasingly belligerent Japan in the pole position.

Pessimism indeed hardened as days passed. Fed up by the fruitlessness of the negotiations, the US delegation left several days after Christmas. Norman H. Davis, a US ambassador-at-large, decried the eventual end of the naval arms treaties that dated back to the Washington Naval Treaty of 1922 and had contributed to world stability. Pointing the finger at Japan, he said, "Abandonment now of the principles involved will lead to conditions of insecurity, of international suspicion, with no real advantage to any nation." Yamamoto managed to maintain his outward amicability and even met with Admiral Standley for a friendly lunch prior to the latter's departure aboard the steamship *Washington*. But it was clear that the talks were useless and, in hindsight, had actually always been so—dead on arrival in October. "The luncheon itself was a pleasant occasion," one reporter wrote, "as the two admirals have struck up a warm personal friendship regardless of their differences on naval questions. But after coffee they had a half hour's talk that again showed that Japan and the United States are far apart."

WHEN YAMAMOTO LEFT LONDON BY TRAIN IN LATE JANUARY 1935, German officials were hoping that as he passed through Europe he would meet with Adolf Hitler. The Nazi leader had seized power the previous summer by abolishing the offices of president and chancellor and declaring himself Germany's leader—der Führer. The thinking was that Yamamoto and Hitler could discuss the failed disarmament talks and their nations' mutual interests. It was a nonstarter, however. Yamamoto had displayed cordiality toward US and British military officials but felt nothing of the kind toward the Nazi dictator. Though he stopped overnight in Berlin, dined at the Japanese Embassy, and even met with the German foreign minister, he declined to see Adolf Hitler or talk of a possible alliance. First thing the next morning, he resumed his trip, passing through Poland and switching to the Trans-Siberian Railway once he reached Moscow. Then came the long last leg across Siberia, smack in the middle of the Siberian winter, during which Yamamoto occupied his time with marathon bridge or poker games with his aides, all the while heavily smoking the popular Cherry brand cigarettes he'd come to favor.

Yamamoto arrived in Tokyo on the afternoon of February 12. Gone five months, he returned home to newfound fame for his unyielding stance at the London talks. He now seemed to embody Japan's muscle-flexing might—for Japan first and Japan strong. Crowds greeted him at the station. Two days later, he presented his official report to the navy minister, in which he wrote that given the three nations' differences, it had been "impossible to reach any agreement." But although a growing majority in the military, the media, and the public cheered the breakdown in talks, Yamamoto did not, writing that he "deeply regretted that it was not possible to persuade Britain and America to accept the imperial government's views, and was convinced of the necessity for still further efforts in this direction." Yamamoto's concluding words were at odds with his popular image as strongman. Simply put, he favored arms control to belligerence. Most missed the point—except the militarists, who were openly cool to Yamamoto in the days after his return. They watched his approval ratings spike, but they knew he was not one of them.

Beyond public view, the military establishment had for years been splitting into two competing groups. There was the so-called treaty faction, to which Yamamoto and like-minded naval comrades belonged, and there was the hawkish "fleet faction," which controlled the army and parts of the navy. The balance of power throughout the 1930s tilted increasingly to the warlike fleet faction, so much so that within weeks of Yamamoto's return, its leaders managed to sideline him. He was assigned to the Naval Affairs Bureau, where he had a new job, and an office in Tokyo but was given little to do. They hoped he might resign and disappear.

The admiral faced a midcareer crisis. He was unaccustomed to being ignored and did consider calling it quits. He told a few close friends that maybe he should just "go to Monaco and be a gambler."

To help sort things out, he traveled in the spring of 1935 to his birthplace, the remote village of Nagaoka. His parents were dead by then, but an elder brother and sister were still living there. To them, the younger sibling was affectionately known by "Iso-sa." The visits revitalized him. The heavy snows had melted, and the flowering cherry trees along the Kaji River were in bloom. Yamamoto reunited with boyhood friends and spent a day with them in a boat on the river. He also enjoyed his favorite treats, broiled dumplings sold by vendors in the village. During the first visit in April, he stayed two weeks; he returned twice more before the summer was out. Meanwhile, in Tokyo, he had his mistress, Chiyoko, their budding romance not even a year old. He confided his frustrations to her. "To be honest, I feel utterly wretched working in Tokyo," he said. He also opened his heart further, breathlessly referring to her as "beguiling and beautiful." As he traveled between Tokyo and Nagaoka during the spring and summer of 1935, he spent most of his free time with Chiyoko and rarely with his wife, Reiko. "If it's really true that you miss me and have faith in me," he wrote to his lover, "then in practice I consider myself really fortunate." In another note he wrote about a dream. "I dreamed that we were driving along the coast of Nice, in the south of France. I thought how happy I'd be if only it were true."

Friends and supporters in the fleet faction lobbied Yamamoto to stay the course, insisting that the navy would be rudderless without him, and by late in the year, they managed to arrange his appointment as chief of the navy's Aeronautics Department. It was a midlevel position, but an important one that was right in his sweet spot. The department was responsible for all air programs in the navy: its carriers, seaplanes, and land-based airplanes. Japan was rapidly building its military, and he'd have a job where he could push his agenda. He took it.

THE LEADERS OF THE IMPERIAL NAVY MIGHT DEBATE THE POINT of its fleet—whether for offensive or defensive purposes—but naval expansion won support across the board. Working within the hierarchy, Yamamoto sought to build a fleet composed of ships carrying planes. In the early 1930s, he had had a hand in developing new attack planes, most notably the speedy, single-engine Zero fighter, and now he persuaded such leading aircraft manufacturers as Mitsubishi and Nakajima to devise newer aircraft of all kinds: fighters, torpedo planes, and long-range bombers. He also imposed tougher training for combat pilots. It was no longer enough for fighter pilots to learn to land on a carrier; Yamamoto insisted that training be ongoing, with the pilots constantly drilled in takeoffs, landings, and flight patterns, so that they would become the envy of the world. His efforts pitted him against "battleship admirals," who still measured sea strength in tonnage, big guns, and ever-larger ships. And those senior naval officials hardly sat still. Plans were well under way for the construction of two new giant battleships: *Yamato* and *Musashi*. *Yamato*, weighing 72,000 tons when loaded to capacity and requiring a crew of three thousand men, was set to become the pride of the Imperial Navy and to serve as the flagship of its Combined Fleet. Though Yamamoto argued for airpower as being central to the navy's future, his opponents saw planes as merely complementary to the new and supposedly unsinkable mighty warships.

In December 1936, after a year spent serving as chief of aeronautics, Yamamoto was offered the job of vice minister of the navy, a prestigious post that would strengthen his hand. But he was hardly thrilled about taking up what was mainly a political post. "What is there for a naval man to be pleased about in being suddenly shifted to political duties just when he's been trying his best to get naval aviation going?" he complained. To refuse the promotion would be a breach of military etiquette, however, and so he not only accepted the job but held it for the next two years and nine months, until the late summer of 1939.

He was now based mainly in Tokyo, and as he was largely a creature of habit, his days took on a predictable routine. He liked to start his workdays at dawn, arriving at the ministry before nearly everyone else. He'd work through the morning, often then conferring with his boss, Naval Minister Mitsumasa Yonai, a veteran admiral whose policies straddled a middle ground of sorts. Like Yamamoto, Yonai worried about the militarists' nationalism and warmongering, but, contrary to Yamamoto's fervent views on airpower, Yonai supported the building of mighty warships.

Given his early start, Yamamoto left work before many others—

departures that were shrouded in mystery, as he rarely revealed his evening plans to his staff. Sometimes he could be found at the Navy Club playing *shogi*, but it was Chiyoko and their affair that accounted for his discretion. The couple typically rendezvoused at her geisha house in a downtown shopping area known as the Ginza District. They'd have dinner and perhaps attend an art exhibition or some other cultural event. As his feelings continued to deepen, a potential conflict arose. Yamamoto learned that he was not the only important man in Chiyoko's life. A Tokyo real estate mogul was similarly enchanted with her; he spent money on her in ways an admiral could not. Chiyoko was skilled at juggling a love interest and a benefactor—or, as others at the geisha house observed, "keeping her heart and body separate." Yamamoto, married and the keeper of a longtime friendship with the geisha Masako Tsurushima, could hardly protest, and, sustained by Chiyoko's love, accepted the troika.

Becoming the navy's vice minister put Yamamoto squarely in the crosshairs of the conservative nationalists demanding control of Japan's military policy and practice. By 1936, the army junta had achieved dominance, and its agenda was focused on preparing to do battle to rule the Pacific region, which meant conquering first China and then all of Southeast Asia. The goal was access to other countries' rich natural resources—particularly oil—that Japan lacked but required to be economically independent and strong. "Without oil," noted one historian, "Japan's pretentions to empire were empty shadows." Over the next several years, the army more than doubled in size, from 20 to 50 divisions, while its air squadrons tripled from 50 to 150. Japanese newspapers and radio, increasingly falling under government sway, stoked a war fever. When Japan did instigate fighting with China in 1937, radio programs delivered cheerleading daily updates about the war front, while Nippon News newsreels that were jingoistic and less about the news than propaganda began playing nightly in movie theaters.

It was not Yamamoto's style to soften his positions. The navy was growing stronger but not necessarily in the best ways. He continued to argue forcefully and with moral courage against the admirals who belittled airpower and persisted in claiming that only a battleship could sink another battleship. "The fiercest serpent may be overcome by a swarm of ants," he retorted, using the ancient saying to stress the value of developing squadrons of planes flying from aircraft carriers. As if that comeback weren't enough, he mocked the admirals' obsession with building bigger warships as obtuse and backward looking. "They are like elaborate religious scrolls which old people hang in their homes—a matter of faith, not reality," he said. Further, he insisted, "In modern warfare battleships will be as useful to Japan as a samurai sword."

For those and other remarks, the number of Yamamoto's enemies in government grew. He was outspoken in opposing the war in China, for example, which quickly became a drain on the military at a time when so much emphasis was being placed on building, not depleting, it. The United States immediately condemned Japan's aggressions and demanded the complete withdrawal of its troops from China. In Europe, meanwhile, as Hitler ramped up for war, he and Italy's ruler, Benito Mussolini, invited Japan to join them in a tripartite pact. Yamamoto, along with Navy Minister Yonai, resisted, arguing that aligning with Nazi Germany and Italy was tantamount to a public proclamation that Japan now saw the United States as its enemy. To Yamamoto a pact would mark a big step toward a war with the United States, which his antagonists in the army might welcome but he saw as calamitous. "For Japan it would mean, after several years of war [with China] already, acquiring yet another powerful enemy—an extremely perilous matter for the nation," he wrote. "Japan should under no circumstance conclude an alliance with Germany." Given that the press portrayed Japan's leaders as unified in the war with China and for a pact with Germany, the nasty discord escalated with little public notice, a phenomenon Yamamoto complained about. "One can gain absolutely no information from the press or other such publications as to the true state of the nation's affairs," he wrote. "Personally, I feel that the crisis threatening the nation could not be graver."

By the spring of 1939, his personal safety was also a grave matter. His enemies in Japan were certainly capable of taking deadly action. Earlier in the decade, a bungled, ill-fated coup attempt by a cabal of young army officers had been crushed, resulting in the execution of thirteen rebels for treason. Now that Yamamoto was a leading voice for moderation, the worried navy minister, Yonai, wanted to assign him a bodyguard. Yamamoto, not one to flinch and fiercely protective of his privacy off duty, balked, but Yonai insisted. It seemed as though Yamamoto had a target on his back, and rumors continued to spread. He received a steady stream of hate mail and, worse, death threats. By midsummer, navy leaders believed that a plot to assassinate him—to blow up a bridge as he crossed it—was in the works. Yonai acted quickly, arranging for his deputy to be reassigned. Yamamoto was appointed commander in chief of the Combined Fleet. The thinking was to get him out of Tokyo and a possible assassin's line of fire and onto the safety of the open sea.

The promotion was big news. One headline announced, "Setting Sail Again on the Seven Seas After Six Years Ashore: Yamamoto, the Stern, Silent Admiral." During the press conference held at the ministry, a reporter described Yamamoto as "a sturdy frame clothed in spotless white uniform; manly features, solemn with emotion and resolve; confident stride." Yamamoto said that the new post was "the greatest honor possible" and pledged to do "my humble best in the service of His Imperial Majesty." In short order, he would be challenged to fulfill that very promise.

He left Tokyo by train to join the fleet on August 31, 1939. The next day Germany invaded Poland, a blitzkrieg signaling that World War II had begun.

YAMAMOTO'S DEPARTURE FROM TOKYO HAPPENED TO FALL ON HIS twenty-first wedding anniversary. His wife, Reiko, did not see him off, however; she was out of town. But Chiyoko was with him, although unknowing observers mistakenly thought she was his "maid." She stood among the well-wishers at the station—friends, military and government officials, and reporters—who watched as he boarded the train. Discreetly, she then climbed aboard, and together they traveled south to the port city of Osaka. From there Yamamoto continued southwest to Wakanoura Bay, where the entire Combined Fleet—the third most powerful fleet in the world with its seventy or more ships—was anchored, awaiting its new commander.

From that moment on, Yamamoto was essentially sea based. In the months to come, he'd visit Tokyo multiple times on business, but he'd never again call the city home. Instead, he lived mainly aboard the flagship, *Nagato*, and, once it was seaworthy, the mighty *Yamato*. Chiyoko went to see him when the fleet docked at Yokosuka, the port city thirty-seven miles south of Tokyo. Whenever Yamamoto went to Tokyo, he stayed at the Navy Club rather than with his family at their home, but his stays at the Navy Club soon became a cover for the fact that he actually stayed with Chiyoko. To accommodate their affair, she had moved out of the geisha house she'd been managing and rented a small home from a friend that was near the club. That gave the lovers the privacy they coveted for their trysts.

Yamamoto, as the fleet's top commander and with a promotion to admiral in the offing, was at the pinnacle of his powers. He was fifty-five years old (nearly his father's age when Yamamoto had been born) and, despite ongoing controversies, was the naval officer best prepared and better suited than any other to oversee the navy's development. He threw himself into sculpting the fleet to his vision, centering on aircraft carriers stocked with fighter planes, torpedo planes, and high-flying bombers. Tracking the war in Europe, for example, he paid special attention to Great Britain's use of new torpedo planes, and when torpedoes dropped from British planes sank three Italian battleships, he demanded detailed reports from the Japanese attachés stationed in Rome and London. For him, the feat validated his view that big warships were vulnerable, contrary to those who championed their invincibility.

Besides pushing for torpedo planes, Yamamoto imposed a new training program for bomber pilots. When he assumed command of the fleet, he was appalled by their poor performance in a bombing exercise; thirty-six planes trying to destroy an old decoy ship in the ocean a thousand feet below had hit the target only once. He spelled out his high expectations for fliers in a training pamphlet: "With tenacious and timeless spirit we are striving to reach a superhuman degree of skill and fighting efficiency." In the spring of 1940, he personally oversaw from the flagship, *Nagato*, war games he'd ordered to assess the outcomes of the more rigorous training. He witnessed an impressive turnaround, so much so that having for years contemplated a war plan against the United States in case war was ever necessary, he mused to his chief of staff, "An air attack on Hawaii may be possible now, especially as our air training has turned out so successfully." The tantalizing comment hung in the air, nothing more said. Except for a rare mention during a lecture to cadets years earlier or in conversations with the writer Hector Bywater, he had never presented the idea to peers and held out hope that he'd never have to.

But that hope was shattered September 27, 1940, when Japan signed the Tripartite Pact with Germany and Italy. Yamamoto angrily called the signing "irrational" and "impulsive," an alliance certain to pit Japan against the United States. Indeed, the United States immediately imposed sanctions and soon halted oil exports to Japan entirely. That all but ensured a direct confrontation, given that Japan was dependent on imported oil and its stockpiles could last only a year. Yamamoto called out other navy officials for assuring the emperor's advisers that their navy was superior and could win a war, even a protracted one. He kept up his unvarnished criticism of the army leaders he despised, most of whom had never been to the United States, telling them that they were foolishly underestimating the United States' will. "It is a mistake to regard Americans as luxury loving and weak," he would say in a talk in Tokyo a year later. "I can tell you that they are full of spirit, adventure, fight and justice. Their thinking is scientific and well advanced." Knowingly, he warned, "Remember that American industry is much more developed than ours, and—unlike us—they have all the oil they want. Japan cannot vanquish the United States. Therefore we should not fight the United States."

His remarks caused an uproar, with some accusing him of favoring the United States and Great Britain. Yamamoto did not back down. "I am Japanese," he replied forcefully. "I do only what is best for my country." He spoke with equal frankness when Japan's prime minister, Fumimaro Konoe, who'd signed the Tripartite Pact, asked for his take on a war with the United States. To be sure, the navy had made huge strides in modernizing and expanding, but Yamamoto had always believed that strength was to deter war, not instigate it. He said, "If we are told to fight, regardless of consequences, we can run wild for six months or a year, but after that I have utterly no confidence. I hope you will try to avoid war with America."

Yamamoto was torn; though he was against a war, his country was on a crash course toward one. But overriding everything was the fact that he was pro-Japan and supremely loyal, which meant it was his duty to devise a battle plan. "This moment is the critical time upon which the fate of the country depends," he wrote an old friend on November 4, 1940. He was ready to answer the call, and in order to give his beloved Japan a fighting chance, he began work on no ordinary strategy.

YAMAMOTO DISCLOSED HIS APPROACH LATE THAT FALL OF 1940 TO a few trusted senior staff, but it was early in January 1941 that he actually sat down at his desk aboard the flagship, *Nagato*, and composed a letter that, for the first time, outlined his long-considered idea. Titled "Opinions on War Preparations," the letter was addressed to Japan's new navy minister, Admiral Koshiro Oikawa, who, like Yamamoto, had served as a junior naval officer during the Russo-Japanese War of 1904. Oikawa, who was sympathetic to Yamamoto's antiwar position, continued in the coming months to push for maintaining peace through diplomacy—until the hawks dominating the government managed to oust him. Letter in hand, Oikawa hardly disputed Yamamoto's New Year's assessment, which was that conflict with the United States and Great Britain seemed "inevitable" and thus "the time has come for the Navy, especially the Combined Fleet, to devote itself seriously to war preparations."

Yamamoto got right to the point. He saw little hope for Japan following any traditional warship-based strategy. "The most important thing we have to do first of all in a war with the U.S., I firmly believe, is to fiercely attack and destroy the U.S. main fleet at the outset of the war, so that the morale of the U.S. Navy and her people goes down to such an extent that it cannot be recovered." He continued, "Only then shall we be able to secure an invincible stand in key positions in East Asia, thus being able to establish and keep the East Asia Co-Prosperity Sphere." The latter was a reference to Japan's one-sided initiative to exploit the natural resources of countries it controlled—by no means a "coprosperity" plan.

The unprecedented approach was now possible tactically speaking, given the advances of the previous few years: the greater distances the newer bomber planes could fly from aircraft carriers; the increased prowess of the fleet's pilots due to their relentless training; and advances in aerially delivered torpedoes, particularly one nicknamed "the long lance." The plan was basically a distillation of old lessons—including the surprise attack in the Russo-Japanese War—with the new: Yamamoto's reshaping of the Imperial Navy as carrier based. It echoed parts of Hector Bywater's tactics in the novel *The Great Pacific War* as well, albeit updated.

Given the opponent, Yamamoto wrote, "We should have a firm determination of deciding the fate of the war on its first day." Pearl Harbor should be hit hard, fast, and by surprise with a raid by a huge fleet of carrier-based planes. Nothing would come easily, of course, and there was no guarantee of success, "but I believe we could be favored by God's blessing when all officers and men who take part in this operation have a firm determination of devoting themselves to their task, even sacrificing themselves."

The mission became known as Operation Z, and Yamamoto entrusted the operational details to three top subordinates: Rear Admiral Shigeru Fukudome, Rear Admiral Takijiro Ohnishi, and Captain Minoru Genda. Genda served as chief planner for the massive task force that would employ six of the fleet's ten major aircraft carriers, two of its faster battleships, two cruisers, and a number of destroyers, tankers, and other supporting ships. The plan, revolving around the carrier fleet rather than the battle fleet as the central strike force, flew in the face of conventional wisdom. No navy in the world was thinking that way at the time, including Japan's. The mind-set for more than a decade had been "the great all-out battle," in which Japan would beat the bigger US Navy by first using its submarines to pick off US ships and then defeating the slimmed-down US fleet in an ultimate warship face-off in the Pacific. No surprise, some military leaders were initially dismissive of Yamamoto's bold plan, worried that it recklessly put too much of Japan's fleet at risk on one roll of the dice. Moreover, the skeptics believed that they already had a winning formula: the mano a mano approach of "the great all-out battle."

The reaction featured the hubris that had alarmed Yamamoto for years, what he viewed as the warmongers' ignorant miscalculation of the United States' resolve. A few weeks after submitting his Pearl Harbor plan to the navy minister, he addressed his concerns in a letter to an acquaintance who was one of the leading belligerents, an ultranationalist named Ryoichi Sasakawa. Yamamoto's main point was cautionary, a warning that too many decision makers did not seem to grasp the meaning of going to war with a nation as industrially powerful as the United States. It was folly to think that if Japan simply took Guam and the Philippines, or even Hawaii and San Francisco, it would convince a weak-willed United States to fold and leave Japan to its own devices, creating an empire in the Pacific. To the contrary, Yamamoto said, the United States would retaliate with all its might, which meant a bloodbath that could be won only by defeating the United States utterly and completely. "We would have to march into Washington and sign a treaty in the White House," he wrote. That was the cold reality the sword rattlers needed to confront.

Yamamoto left out one major point, though: he did not believe that crushing the United States was remotely possible. But to say that would come off sounding defeatist and perhaps even traitorous, so he kept his plan's true goal to himself. Others could think whatever they wanted about an illusory complete victory; his objective was to land a preemptive blow against US naval forces at Pearl Harbor so impactful that the United States' leaders, among them a potent bloc of isolationists, might prefer to negotiate an early peace before an all-out war got real traction. If the United States roared back from the opening punch, Yamamoto believed, Japan would not survive the enemy's superiority over the long haul. He struggled with that tension throughout the remainder of 1941, between his private beliefs and the duty to devise a plan putting Japan on its strongest war footing. "What a strange position I find myself in now," he confided to an old friend and classmate at one point during the year. "Having to make a decision diametrically opposed to my personal opinion, with no choice but to push full speed ahead in pursuance of that decision."

BY MAY 1941, YAMAMOTO, A DRAFT PEARL HARBOR PLAN IN HAND, ordered his fleet to commence training. The combined fleet used

Kagoshima Bay, whose entrance resembled Pearl Harbor's, on the island of Kyushu for its main exercises. Pilots piled up fifty and more practice runs taking off from and landing on carriers. High-altitude bombers worked on their accuracy, dropping dummy bombs onto a faux battleship drawn in a sandy inlet near the bay. Dive-bombers, meanwhile, worked on hitting targets in the bay with shallow-draft torpedoes. The extensive drills continued throughout the summer. In May, Germany ended months of bombing England to turn its sights on invading Russia, further escalating debate among Japan's leaders about whether to stay on the fence or advance its Pacific agenda. While one arm of a divided government continued developing war options, another continued negotiating with US diplomats to avoid war. Unknown to the Japanese, US intelligence agencies, having broken certain Japanese codes dating back to the 1920s, were able to monitor much of the internal turmoil. President Franklin Roosevelt and his aides were briefed on the intelligence, which revealed a cumulative if conflicted swing toward war, although US code breakers never managed to detect the navy's secret Pearl Harbor plan.

Yamamoto, bolstered by months of successful training, presented the plan's feasibility through a series of "tabletop maneuvers" at the Naval War College in Tokyo in mid-September. To justify targeting Pearl Harbor, Yamamoto even played to the militarists' bellicosity, characterizing the US fleet there as a "dagger pointed at Japan's heart." But the dress rehearsal met with stiff opposition, most notably from the navy general staff and its chief, Admiral Osami Nagano. Nagano and others preferred targets closer to home than Hawaii—in Burma, the Solomon Islands, and the Philippines, for example—given the ever-present worry of oil and fuel shortages. Ignoring the opposition, Yamamoto continued to train the fleet. By the next month, though, he'd had enough. He sent word to Tokyo that he and his entire staff would resign if he were not given the official green light. Whether a bluff or not, the move made by the consummate player of games of chance worked. The pressure for war was reaching a boiling point, and a reshuffling in government resulted in the militant war minister, Hideki Tojo—who was an unalloyed fan of the Nazi leader Adolf Hitler—being named as the new premier. The navy general staff swiftly came around and finally adopted Yamamoto's plan for a surprise carrier-based raid on Pearl Harbor. Nagano said that the naval leaders would put their trust in the commander of the Combined Fleet as the one who knew best how to tackle the US problem.

In private, Yamamoto continued to despair. Writing to a friend at the end of October, he expressed deep worry about his country and its "risky and illogical" fixation with confrontation. He hoped for a last-minute peace but was resigned that war, not peace, was in the cards. Less than two weeks later, on November 5, 1941, he distributed to his fleet commanders the plan that he and senior staff had been secretly working on aboard the *Nagato*. The 118-page document spelled out in every detail the surprise attack on Pearl Harbor and simultaneous attacks on Malaysia, the Philippines, Guam, Wake Island, and Hong Kong. It was titled "Combined Fleet Top Secret Operational Order Number 1." Final approval from Emperor Hirohito was still required, but for the fleet to be in position for the early-December target date, he ordered the strike force anchored in the Inland Sea to set sail.

Yamamoto stayed behind, aboard the fleet's flagship, *Nagato*, from which he would monitor and oversee the Pearl Harbor attack. In charge of the fleet assembling stealthily about eight hundred miles north of Japan in Hitokappu Bay in the remote Kuril Islands was Vice Admiral Chuichi Nagumo. Nagumo had not been Yamamoto's first choice; he had little experience in aviation and was an early doubter, a believer in warships not aircraft carriers. But Nagumo's seniority and the navy's strict adherence to protocol had given Yamamoto little option. Yamamoto did, however, surround Nagumo with his own trusted commanders, notably Captain Mitsuo Fuchida and Minoru Genda, the latter being one of the mission's key planners.

By the third week of November, the strike force was gathered in the isolated bay surrounded by snow-covered hills. The target date for the start of hostilities was December 7, and although Yamamoto's days were consumed with operational matters, he found time for a reprieve from the imminent war with the love of his life. Chiyoko traveled by train from Tokyo to join him on the scenic island of Itsukushima in Hiroshima Bay, where the couple spent the night of November 25 at an inn. The next day the Japanese Carrier Strike Force departed from its rallying spot and began the four-thousand-mile journey to the Hawaiian Islands. Events were unfolding at a hurried pace. Yamamoto left the *Nagato*, anchored in the Inland Sea, and traveled to Tokyo for a flurry of high-level conferences. Behind closed doors, the Imperial Council voted on December 1 to declare war on the United States and Great Britain.

Yamamoto, meanwhile, attended at once to his professional and personal affairs. On December 2, he issued the official order to his units traveling on the high seas: "Climb Mt. Niitaka," code for "Proceed with attack on Pearl Harbor." The next day he appeared in full uniform and medals at the Imperial Palace to hear Emperor Hirohito ratify the war and assert the nation's faith in Yamamoto's ability to secure "victory over the enemy." Yamamoto, the "humble servant," promised that the "officers and men of the Combined Fleet will swear to do their duty" and "with confidence we face operations." That was Yamamoto at his most dutiful and professional self—as one historian later described him, "the picture of hatchet-faced solemnity." His sentimental side showed afterward, when he surprised his wife, Reiko, and family by arriving unannounced at their home and, a rarity, staying the night. The next day, December 4, he hurried to the navy minister's official residence in the morning to toast the mission's success and then met up with Chiyoko in the afternoon. They strolled together in the Ginza District, the shopping area that included Chiyoko's former geisha house. Yamamoto bought her a beautiful bouquet of roses. He told her he must return to his ship immediately. He asked that she send him a photograph that he could keep close at hand, the way seamen and soldiers do when they don't expect to see their sweetheart for a while—just as Johnnie Mitchell had done with Annie Lee Miller. Bound to secrecy, Yamamoto could not tell Chiyoko why he was leaving, and so he spoke in a code of sorts, hinting at the shock that was about to hit the world. He said, "Watch for the time when the petals of these flowers fall."

CHAPTER 5

A TASTE OF WAR

London during the Blitz
National Archives and Records Administration

AFTER LEARNING ON APRIL 4, 1941, THAT HE WAS GOING ON A SPE-
cial mission, Second Lieutenant John Mitchell mostly sat around
Hamilton Field waiting. He and the other three pilots were not told

anything more—neither the destination nor the mission's purpose—just that they had to remain grounded, ready to leave at a moment's notice. The assignment had ruined his plan to surprise Annie Lee at Easter—although he wasn't complaining about being among the chosen—and he spent Easter Sunday, April 13, feeling "exceptionally lonesome."

Then, as expected, orders came on short notice. Two days after Easter, Mitch hustled aboard a train bound for Washington, DC. In the six months he had been at Hamilton Field, it had become clear that he'd moved to the front of the line, from being just another enlisted man who wanted to put in his time and go home to becoming a skilled pilot hungry for more. He'd scored second highest in gunnery exercises at Muroc Lake, coached the base's basketball team, trained rookie pilots as they came in waves, served as best man in back-to-back weddings, and, biggest of all, been chosen as one of four pilots to go overseas. Second Lieutenant John Mitchell was proving to be someone who could be counted on.

None of that maturity was particularly evident, though, during the train trip from San Francisco to Washington, DC, the second week of April. Bored out of their minds during the four nights and three days spent on the train, Mitch and his three companions livened up the ride across America by letting the booze flow. Mitch noted, "I can honestly say we didn't draw a sober breath from the night we left." With him was "Monty" Montgomery, a pal from his days as a cadet at Kelly Field, and two pilots he was just getting to know, one named Booth and the other named Ellery Gross. They had still not been told anything more about the mission, and it was only after they'd arrived in the nation's capital and had settled in for the weekend at the Officers' Club at Bolling Field that they were briefed. They were going to England to fly with Royal Air Force fighter pilots to learn as much as they could from the RAF. "Our job is to find out just exactly what methods they are using in combat," Mitch said in a letter he fired off to Annie Lee the night of the briefing. Unable to contain his eagerness, he continued, "This is a detail envied by everyone who is not able to go, and I wouldn't take a thousand, not even five thousand dollars for my part in it."

To avoid what British prime minister Winston Churchill had in

February begun calling the "Battle of the Atlantic," with German U-boats and warships blockading Great Britain, Mitch would travel a roundabout route, sailing from New York City to Bermuda, then to Portugal, and, for the last leg, flying to England. The US pilots would be in England between four and six weeks.

Though it was good to know, the news unsettled Annie Lee. England, ever since the fall, had been the target of deadly Nazi bombing raids, which the British press branded the "Blitz," the German word for "lightning." She wrote to Aunt Ludma with her concerns. "He says he's 'tickled to death' over the chance to go," she said. "And he doesn't want my worrying about him." But that was hard not to do. "It makes my heart stand still to think of them flying anywhere around England." Her fretting reached Mitch's father, Noah, who, overcoming his initial bias against Annie Lee's German and Czech heritage, had begun corresponding with her. "I want you to call me Dad—and mean it!" he'd written to her affectionately. He told her he'd tried looking for Annie Lee's hometown of El Campo but could not find it on a Texas map. "You will have equally as hard a time finding Enid." In that way, they shared the same small-town state of mind, and he encouraged Annie Lee to visit Enid. "We have a large, old-timey house—nobody here but Eunice and me—plenty of rooms. You can always dress as you please—go as you please—eat and sleep and read and walk or ride or hunt or fish as you please. No style—no ceremony—just whatever you care to do." And finally, as for their shared concern for Johnnie, Noah insisted that she not worry. "There are just dozens of things I could think of, but just to sum it all up in very few words—*Just Be Happy!*"

John William Mitchell will come home, he assured her.

TWO DAYS AFTER ARRIVING AT BOLLING FIELD ON THE POTOMAC River, Mitch and the others again boarded a night train—not for New York City, their eventual port of departure, but for Wright Field in Ohio for several more days of briefings. The overnight trip featured the first of a handful of occurrences that were unanticipated and forever reminders to Mitch that he was a long way from Enid. Just before dawn, around 5:30 a.m., a porter awakened the four pilots. The train's crew had detected a problem: their sleeping car had a "hot box," a term for when the bearings in a wheel axle overheated due to an oil leak. It had created a fire hazard. The car was going to have to be removed from service and the pilots relocated. To wait out the delay Mitch, Monty, Booth, and Gross went to the club car. They were only just seated when the door opened and a sleepy-eyed man in pajamas and a bathrobe—a commanding figure at more than six feet, two inches—wandered in. He was immediately recognizable despite his mussed hair and sartorial informality: Wendell Willkie.

Mitch had been training at Muroc Lake, listening to the radio on election night 1940 as the results were reported showing FDR trouncing Willkie, his Republican challenger. Now he and his cohorts greeted in person the popular lawyer and business executive, and Willkie plopped down in a club car chair to chat. Mitch and the others were eager listeners. Willkie had campaigned as an interventionist for increased US involvement in the Second World War. They were heading to England, and Willkie, just a few months before, had traveled there at FDR's request to demonstrate the country's building bipartisan US support for Great Britain. Willkie had seen firsthand bombed-out sections of London and other industrial cities, such as Manchester and Liverpool. He'd gone into the bomb shelters, stepped around the rubble, and reassured the Brits that the United States was with them. Once home, he had testified before the Senate Foreign Relations Committee in favor of FDR's Lend-Lease Act to supply military aid to the Allies, which had pitted him against the country's most famous isolationist, Charles Lindbergh. In all, Willkie was the pilots' kind of guy, and Mitch thoroughly enjoyed listening intently as Willkie made what he considered brilliant commentary on world affairs. The bull session ended when Willkie's wife, clearly annoyed, arrived to retrieve her husband and usher him back to their sleeping compartment. Before he left, however, Mitch pulled out his business card. Printed on the front was "John William Mitchell, Lieutenant Air Corps Reserve, United States Army." He asked for an autograph. Willkie turned the card over and signed his name in a swirling script.

Though delayed, the pilots arrived at Wright Field outside Dayton the next day and for the next twenty-four hours were put through their paces. The men were vaccinated for smallpox and given their tickets, passports, and whatever other paperwork was needed for the trip. They were taken on a tour of the army air base, where many new planes and aerial technology were being developed and tested. One particular fighter plane in the pipeline, which Mitch would continue hearing about in the months to come, was the P-38 Lightning, made by Lockheed Aircraft Company in Burbank, California. The twin-engine plane was said to combine speed, versatility, and shooting power greater than those of any other plane to date. It was supposed to be able to reach speeds of nearly 400 miles per hour, or 50 miles per hour faster than Japan's premier fighter plane, the Mitsubishi Zero; it was armed with four .50-caliber machine guns and a 20 mm cannon; and it could undertake longer-range missions of more than five hundred miles. The expectations for the P-38 were running high, except for the fact that test flights had revealed technical problems yet to be resolved: tail flutter occurring at high speeds and carburetors freezing up during dives. The pilots were briefed on the P-38 and other new planes and equipment. "They wanted us to find out all we could about our most up-to-date stuff before going over," Mitch said.

Flying from Dayton to New York, they boarded a ship on Saturday night, April 26, and sailed for Bermuda. Mitch made sure to dash off a quick letter to Annie Lee, again emphasizing the importance of the mission: "This trip is going to mean a lot to me in many ways and I am going to get all I can out of it. It will mean I have knowledge that *very* few people in the U.S. possess." He wasn't sure how much detail he'd be able to share—the mission being classified—but asked her to save his letters nonetheless. "Who knows, I might write a book when I return." He then lamented that he did not have a picture of her, having left the glass-framed photograph at Hamilton Field. "I was afraid of breaking it," he said, and would have to make do with her image in his mind's eye.

Traveling circuitously, it took Mitch, Monty, Booth, and Ellery Gross twenty-three days to reach England. The journey proceeded in fits and starts, which meant that along the way, while awaiting their next move, the men combined sightseeing and socializing, although at one point Mitch conceded that drinking "was getting a little tiresome." The first leg to Bermuda took forty-three hours by boat to cover the 650 miles. After a three-day layover, they left Bermuda on the USS *Siboney*, a ship chartered for war service for round-trip voyages between Lisbon and the United States to fetch Americans fleeing Europe. Whereas on its return the *Siboney* would be filled with hundreds of passengers, heading toward war the ship was nearly empty. Besides the four pilots, there were only twenty-three other passengers. Though the civilians on the ship suffered bouts of seasickness, none of the pilots did. Mitch even had one of his full-moon moments. "Last night after leaving the Captain's quarters at 2 a.m.," he wrote Annie Lee, "I stood on the deck and watched the full moon reflecting in the rippling water. Guess who was on my mind."

The men grew antsy as the voyage continued for more than a week, averaging about 375 miles a day across the Atlantic Ocean. They knew they were in for some more waiting once they reached Lisbon. The plane to England would carry only six passengers. Though they would have priority, there was a line of foreigners heading to Great Britain for any number of reasons. To make the best of their idleness, the pilots planned to stay in Estoril, the beach resort to the west of Lisbon that was a favored destination of Europe's elite and, during the war, spies. In fact, after the *Siboney* docked, they headed directly to the resort. Given Portugal's neutrality, Estoril was mobbed with people wanting to get away from it all. The pilots checked into the luxurious oceanfront Palácio hotel, with its grand white facade and extensive gardens. They sunbathed, drank, and gambled in the casino—where Mitch played roulette for the first time and lost. The men kept checking on their flight as three days turned into four and four days into five. Mitch had gotten a taste of the good life—unlike anything he'd ever experienced—but it was a taste of war he was after. Then, after six days, they finally got word they were leaving. "We were only too glad."

Not that the pilots would have recognized him had they stayed longer, but they left the Palácio a day before a British spy of particular note checked in. He was Lieutenant Commander Ian Fleming, the future creator of the fictional spy James Bond, or 007. Fleming was on his way to Washington, DC, to meet with his US counterparts.

THE FLIGHT TO ENGLAND LASTED NINE HOURS, THE LONGEST nonstop flight Mitch had ever flown. They arrived May 19, shortly after a climactic Nazi bombing raid. The previous September, Hitler had initiated the Blitz to soften up England for an invasion. But the Royal Air Force and other British forces had successfully thwarted the Nazis' aims. Even so, the air raids had continued at a relentless pace into early May 1941, when Hitler pivoted to another war front: Russia. That meant the Blitz was effectively over, but not before a final massive air raid on central London. On the night of May 10, more than five hundred German bombers had descended upon the city. RAF fighters and British antiaircraft units had fought back, but the raid, lasting until dawn the next day, had proved to be the deadliest yet—more than 1,300 Londoners had been killed and more than 1,600 others seriously wounded. When Mitch, Monty, Booth, and Gross arrived on May 19, the devastation still pockmarked the city. Bridges and railway lines were badly damaged, along with factories on the south side of the Thames River. Most notably, the House of Commons, the lower house of Parliament, had been hit hard, resulting in a fire that had caused the roof to collapse. The city seemed coated in plaster dust, and Tube stations were filled with Londoners who'd relocated their families underground for shelter. To all appearances London was a far different city from six years earlier, when the world's powers had convened for naval arms talks and where Japan, represented by Vice Admiral Yamamoto, had rejected the terms and left, signaling that the world order was unraveling.

Their first morning in London, the pilots reported to the US Embassy. Each was given a gas mask and a steel helmet, and they were told that in a few days they'd begin staying at a base near London to prepare for flying. In the meantime they met various US and British officials to discuss what Mitch described to Annie Lee as "military affairs, which would not interest you." Their first week ended with a surprise no one had expected: an audience with the queen of England. "That's right," Mitch wrote. "Friday afternoon we were taken to Windsor Castle." The men toured the castle, which Mitch called "beautiful, unique and ancient." One large room—an armor room—displayed thousands of ancient spears, swords, guns, knives, suits of armor, shields, lances. "I looked until my eyes were sore." They were taken to the royal family's living quarters, where Mitch, Monty, Booth, and Gross climbed a grand stairway and entered a large room. Waiting for them were the queen of England and her daughter Princess Elizabeth. The four second lieutenants from the US Army Air Corps stepped forward and were formally presented. Everyone shook hands, and the pilots joined her royal highness for tea. The queen was curious about the route the pilots had taken to reach London. Mitch described the trip, letting the copious drinking and gambling go unmentioned. "Both of them were most charming," Mitch said later. "Elizabeth is fifteen now and has certainly changed in the last couple of years from a gangling kid to a quite gracious and nice looking girl." He and the teenager hit it off. When the royal visit wrapped up, the princess handed him a card. On one side was a color photograph of the royal family—the queen, the king, and their two daughters—and on the back, printed in ink, was written "With Best Wishes from the King and Queen."

MITCH HAD THOUGHT THEY'D BE CLIMBING INTO A COCKPIT SOON after their arrival, but he was wrong. The US pilots were told they first had to attend ground school for several weeks to learn about the British aircraft. That bored him. He felt he already knew the material, and he didn't like having to listen to his British counterparts regaling combat "tall stories" and boasting about how British goods were superior to American. Brits rubbed him the wrong way. "I won't get started on that, though," he told Annie Lee. He much preferred hands-on learning, believing that the best way to get to know the British planes was to fly them. "I don't think we are going to learn a helluva lot until we get into an operational squadron," he said.

It meant that he had more downtime than he had expected, and so he wrote plenty of letters. He wrote to Annie Lee, of course, but also to his father in Enid. Noah replied quickly, saying that he and his stepmother, Eunice, were "mighty proud to hear from you." Noah shared the news of exchanging letters with Annie Lee Miller, how impressed he was with her handwriting: "Like copy plate writing (every letter distinct and clear) showing characters in every line—beautifully worded, plain and sincere." Noah was clearly smitten. "I'll lay a 1000 to 1 that she's just the gal you need." Attending ground school also meant that Mitch and the others had time to get out and tour London, especially with the stark decline in air raids. The pilots quickly found London a tough place for ordinary people; food, clothing, and other goods were scarce and rationed. The prices of everything were sky high. "I'll be glad when I can sit down at some table again, order me a two-pound steak, a pound of cheese, a pound of butter, and eat it all. Wishful thinking." He had his eye on a Harris tweed suit, but given the wartime price tag, it took him weeks before he finally decided to indulge and buy the suit made of Scottish wool.

No surprise, either, that Mitch splurged when it came to his twenty-seventh birthday on June 14, especially with Booth's birthday falling the day before. The two pilots lined up dates with "a couple of nice English gals" and planned for a big night: first a show at the Victoria Palace Theatre, where the musical revue *Black Vanities* was playing, then dinner at the Dorchester, one of London's fanciest hotels, before ending at a nightclub. For Mitch the birthday was mixed. The musical was lousy as far as he was concerned, and the dinner ended sourly when the waiter mumbled a complaint about the tip. "Whereupon I picked up the tip I had left and gave him nothing." Mitch tried to pick up his mood but confessed later that "the old spirit was lacking—I'm sure because I was thinking of you and you weren't there." It had been nearly a year since he and Annie Lee had seen each other, and Mitch, counting the days, seemed affected. There was an edgy undercurrent to his mood, and the fact that he'd flown only a couple of times since he'd been in England didn't help.

Things just didn't seem to be going his way. A few nights after his birthday, he was out with Monty, Booth, and Gross at a crowded dance hall when a Brit bumped into him. Even though he thought the stranger had done so deliberately, Mitch said it was no big deal. The Brit glared. Mitch saw he was "half-tight" and repeated that it was no big deal. But the Brit acted as if Mitch was the one who'd initiated contact. He goaded Mitch, told him they should step outside to settle matters. Mitch refused and told the Brit "to toddle along." He turned away but could sense the Brit lurking. When he turned to look back, a fist met his face—"Bingo, he let me hold five," he told

Annie Lee. He had been sucker punched. Before he could return fire, a dozen or so others got between them, "and I didn't get a crack at him." Blood flowed from a cut above one eye, which later took several stitches to close. "I later laughed about the whole thing but of course was quite angry at the time," he said.

Though he was writing to Annie Lee regularly, he hadn't heard from her, which left him "ranting, raving and cussing," until one day in late June he hit the mother lode: seven letters in a single day. "It is a wonderful world and today is the day of days," he cheered about the batch. Her letters had apparently been routed through Washington, DC, and carried by boat across the ocean instead of being sent via airplane directly to the embassy in London. The first one was dated May 2, before he'd even left New York City for Bermuda. He read and reread the letters, which were full of news, love, and questions about his trip, and before the night was over, he composed a six-page reply. He predicted that he'd soon be assigned to an RAF squadron but said he'd also been informed that he would not be flying any combat missions, "so you don't have to worry about anything along those lines." Finally out from under a cloud, he wrote mainly about the two of them, and the words came out strong: "I want you to have your mind made up—if you haven't already—and for us to lose no more time." Their time had come, he said, continuing in no uncertain terms, "War or no war I say let's get married as soon as I can possibly arrange it after I get back. What's the sense of waiting?"

Mitch was correct about finally getting the chance to fly regularly. To start, he, Monty, and Booth, along with three British pilots, were sent by train to Scotland to bring back six Spitfires. Ellery Gross did not accompany them; in mid-June he'd experienced severe abdominal pain and had undergone surgery for appendicitis. For his part, on the return flight Mitch discovered that he probably should have paid more attention in ground school. The Spitfire was much faster than any RAF training aircraft; the single-seat fighter plane was highly maneuverable, with finely tuned controls that responded to a light touch. Its long nose made seeing out the front difficult, so when landing, pilots had to learn to jockey the aircraft from side to side to gain a clear view. That was where Mitch ran into trouble. Due to bad weather, he and the five other fliers were forced down before they reached London. The airport they chose had a short landing strip, and as Mitch descended, he realized he was going to overshoot it. He gunned the engine in order to go around and try again. But the engine failed to respond. Mitch kept going down, beyond the strip and smack into a bunch of junk cars, a hedge, and finally a ditch. He wasn't injured seriously, just some bruising on his arms and left knee. But he'd wrecked the RAF Spitfire. "This is the first time I ever scratched the paint on a ship and I had to completely wipe it out!" he said. The thing that most surprised him was that he hadn't felt fear. "I'd always thought if I ever cracked up I would be scared to death."

His RAF hosts forgave him, and a few days later the four US pilots were split into pairs. Monty and Booth were dispatched to southern England to fly with a squadron there. Mitch and Ellery Gross, still hobbled but recovering, were sent north. It was about that time that the pilots got news of an official name change to their service branch: the Department of War had decided that, effective June 20, 1941, the United States Army Air Corps would become known as the United States Army Air Forces.

The afternoon Mitch and Gross arrived, they watched with envy as four squadrons of fighters took off to escort bombers on a mission over Germany. But combat, as Mitch had learned, was not part of their tutorial. Instead the commander basically gave him and Gross free rein to wander the base, talk to pilots, soak up what they could, and test fly the various aircraft. He met pilots from Canada and, given Annie Lee's lineage on her mother's side, made sure to connect with a number of Czech pilots. "They had great stories of how they managed to get here or into Africa or someplace when the Germans took France," he wrote to her. The base commander allowed Mitch the use of his private plane, a small biplane that had a cruising speed of only about 100 miles per hour but that Mitch discovered was perfect for sightseeing. He flew to nearby bases, taking Gross along for the ride, since Gross was still unable to fly alone. Then one day Mitch decided to push things a bit. Instead of the small biplane, he and a British pilot took to the sky in a couple of Spitfires. Maybe Mitch wanted to make up for wrecking one the previous week, to prove he could handle the pride of the RAF. Maybe he was just itching for adventure. Either way, he and his mate flew south toward the English Channel. He had been ordered not to fly the Channel, part of the directive to avoid combat. But that was where he headed anyway. They crossed the waters and flew just over the French coast. Mitch surveyed the war zone intently—until, that is, they were spotted by enemy ground troops. German antiaircraft guns began firing at them. Eluding the gun bursts, the two Spitfires turned and raced safely back to England. Mitch was electrified; he'd gotten a taste of the war he'd so far been denied.

"STOP THE PRESS! WE'UNS COMING HOME!" THE BREAKING NEWS was the lead in Mitch's July 4 letter—he, Monty, Booth, and Gross were returning to the United States sooner than expected. They wouldn't be shipped out until mid-July at the earliest, but the end was in sight. "If I arrive August 15 it will be a year almost to the day since I last saw you." In the glow of the news, he seemed more forgiving about his British tutors. He acknowledged that he'd learned plenty as an observer, had filled notebooks with information about aerial tactics and about firing aircraft weaponry while in flight. RAF pilots had even staged a rerun of the Battle of Britain for them on plotting boards, walking them through the strategic big picture. They'd learned small but key tips, such as not oiling machine guns before flying to high altitudes, where the oil might thicken and muck up the guns. They'd learned critical lessons about flying in formation: that a three-plane configuration was less effective than a more flexible two-plane formation, where the lead fighter could focus on attacking the enemy, knowing his wingman was watching for threats and protecting him. "We were doing a damn fool thing, flying in three-ship flights for combat," Mitch said. "We learned to keep two guys together, but not too close, but far enough apart to get wide vision. The Germans did this, too." Mitch was talking shop, but he also waxed poetic, reminiscing about the last time he'd been with Annie Lee in Texas. "Yes, last summer was the 'good ole days.' I was pretty happy then. I had you, a car, no money and plenty of fun. Plenty of chicken, barbecued ribs and juke music at Charlie's and Kline's." For Annie Lee, the unexpected news was semishocking. "Now that he says he's coming I can't actually believe it," she confessed to her aunt. "I'm anxious to see him and yet I'm a

little afraid. It's like meeting him for the first time all over again."
But she was excited, too. Noting that Mitch had bought a Harris
tweed suit in England, she was not going to be outdone. She went
and picked out a tweed coat with a fox collar. "I was extravagant,"
she told her aunt, but she wanted to impress Mitch.

Monty and Booth got lucky. They happened to be in London the
next week when a bomber with room for a couple of passengers
was returning to the States. They jumped aboard. But Mitch and
Gross were stuck taking the long way home, retracing their trip to
England—first flying to Lisbon and then sailing to Bermuda aboard
the passenger/cargo ship SS *Excambion*. They left England on July 23
and pulled into New York two weeks later, on August 5. Previously
Mitch had been scheduled to return to Hamilton Field, but with
war preparations ramping up, his assignment had changed. He was
now one of the more knowledgeable pilots, especially after his ex-
periences in England, and his commanders wanted him deployed to
Louisiana, Virginia, and the Carolinas for war games, training pilots
and getting them combat ready. He was named his squadron's operations officer, a kind of second in charge, overseeing pilot and plane
assignments. It meant being reunited with Army Air Forces captain
and squadron commander Henry "Vic" Viccellio. Mitch had always
liked Vic and was happy to be working directly with the popular
Virginian. He also learned that while they had been in England, he,
Monty, Booth, and Gross had been put in for promotions; the new
rank wouldn't become official until later in the fall, but First Lieutenant John W. Mitchell had earned his first promotion, with more
to come.

It was all good news that he got to share with Annie Lee when
he managed to fit in a quick stopover in San Antonio late in August,
between multiple flights from Washington, DC, to California and
elsewhere, all in preparation for his arrival after Labor Day at Es-
ler Field, Camp Beauregard, Louisiana. The hurried reunion did not
turn out as Mitch had expected, however. Things seemed awkward,
and Mitch figured that Annie Lee was not as certain as he was about
their future together. He left worrying that he'd misread her feelings. But he was wrong. She'd told him many times she loved him;
just look at the dozens of letters she'd written. She simply needed

more time than he did; an entire year had passed, after all, since they'd been face-to-face. Within a few weeks they were able to meet again, this time at an airport in Waco, Texas, to which Mitch had flown a plane that needed engine work. Annie Lee rode a bus up from San Antonio. They never left the airfield, just talked for a few hours. "Our visit this time seemed to me to be more like old times than ever before," Mitch said afterward. They reached a meeting of the minds or, more accurately, hearts, to marry. And they began making plans for when and where. "Seeing you this time," Mitch wrote, "I have felt gay, carefree and that I have nothing to worry about."

MITCH THREW HIMSELF INTO THE MANEUVERS ALL THROUGH the fall, first in Louisiana, then relocating for more combat training in Virginia, in Wilmington, North Carolina, and lastly in Charlotte, North Carolina. The "wars" were simulated battles involving hundreds of troops, fighter planes, and bombers that lasted anywhere from a few days to more than a week. In the backcountry of Louisiana, Mitch led a squadron of planes searching for a faux Red Army armored division, and when they finally found it after two days, they dive-bombed and machine-gunned it and then reported on their attack. When word circulated by radio that the war exercise was over, his squadron was given kudos for having located the enemy forces. Mitch's father and brother Phil had driven from Enid to watch the action, and afterward they told him they had gotten a kick out of seeing him in the operations. In Virginia, the "war" there lasted ten days, with Mitch stationed about five miles outside of Norfolk on a grassy field that had no runways but was hard and flat. The worst part was the mosquitoes. In the middle of one letter he was writing to Annie Lee, he quipped, "Pardon me—I was just interrupted by 9,836,431 mosquitoes—I counted them." For the second half of October, he moved on to Wilmington, where the living conditions were no better. "I'm fighting skeeters with one hand and slinging ink with the other—and let me tell you the skeeters are about to win the battle." Mitch wrapped a towel around his head to finish that letter. He was camped in a pup tent right beside a main road, with cars coming and going. "We even have to put our pants on in this little tent which is only two feet high anyhow. You have to be a contortionist as well as a pilot in this man's army." By contrast, Charlotte, North Carolina, was a five-star hotel—an actual army base with indoor cots, mattresses, hot showers, and food served in an officers' mess hall. "Seems like heaven compared to what we have had." Mitch was getting a ton of flying time, up at five and into the air by six, flying missions for a good part of the day. Early in the month, he trained junior pilots in dogfighting and acrobatics. "It doesn't take much of that stuff to tire you out." The exercises in early November 1941 were warm-ups for a simulated "war" that was to last two weeks. Mitch was eyeing December as a time to take a break—and to marry.

During the exercises he manned a P-40 Warhawk, the fighter plane he'd flown for the better part of a year and operated with complete comfort. But Mitch, and pilots everywhere, always gossiped in anticipation about the new interceptor pursuit planes in development—both the Bell P-39 Airacobra, the single-engine fighter with a new weapons system that was already in limited use, and especially the P-38 Lightning, the speedy, twin-engine fighter that was much ballyhooed but still undergoing extensive trials to work out its kinks. In fact, Mitch knew one of the test pilots, Ellery Gross, quite well. While Mitch had been deployed to the South after his return from England, his buddy Gross had been assigned to March Field in southern California to pilot the P-38s during trials. And on Thursday morning, November 13, as Mitch rested up at the base in Charlotte for the upcoming two-week war game, Gross was climbing into the narrow cockpit of a P-38 at March Field. He and a companion flier flew east through San Gorgonio Pass, circled Palm Springs, and then headed toward the nearby desert, flying at an altitude of about 10,000 feet. The other flier made a power dive, leveled off at 3,000 feet, and then sped off in the direction of Indio, California, twenty miles further east of Palm Springs. Gross followed in a power dive from 10,000 feet. But when he was at 3,000 feet, the plane did not pull up. Instead, it continued to plummet.

The first to report the crash was a telephone lineman named Boyd Moore, working atop a pole outside Palm Springs. Moore initially heard the roar of the motor, growing louder and louder as the plane descended at full speed, between 400 and 500 miles per hour. He looked out onto the desert to see the P-38 fall from the sky, hit the ground, and explode. The blast shook nearby homes—a red flash followed by a plume of dark smoke, then an eerie quiet. The force of the crash hurled parts of the airplane across a quarter mile of desert. The propeller was found twisted like a piece of wire, and pieces of the fuselage and wings dotted the cactus-covered sand. Lieutenant Ellery Gross was killed instantly and, like the plane, blown to bits; his wallet and other papers were found scattered over the crash site. Gross, just twenty-five, of Greenville, Texas—350 miles north of Annie Lee's family home in El Campo—was survived by his father and by his young widow in Riverside, California.

News of the tragedy traveled quickly in army circles. When Mitch was told the next day, he was shown a newspaper article. Stunned, he read the detailed account of Ellery Gross plunging to his death. Had Gross blacked out? Had the controls failed as he tried to pull out of the cometlike dive? Had he even tried to bail out? In a P-38 the pilot sat in a tiny control cabin between the two powerful engines. The only way to abandon it in flight was to flip the plane onto its back and drop out of the cockpit. But Gross had had no time for that, given the speed of his dive. He never had a chance. It also turned out his death had occurred exactly one week after another P-38 test pilot had crashed. That pilot, Ralph Virden, a veteran flier, had been returning to Lockheed's air terminal in a P-38 after a routine test flight. He had been streaking along at 3,000 feet at 400 miles per hour when the tail assembly had mysteriously broken off and fallen away. The plane had flipped onto its back, plunged earthward, and crashed into the kitchen of a Glendale, California, home. The homeowner, awakened by the fiery explosion, had tried to pull the pilot from the plane, but the flames had driven him back. Ralph Virden had perished. The tail assembly had later been discovered in a yard several blocks away. The twin tragedies—one personal to Mitch—were certainly enough to give him, or any pilot for that matter, pause about the new P-38 Lightning, which was described breathlessly in press coverage as "just about the swiftest thing in the sky." In a letter that night, Mitch shared the sad news with Annie Lee. "You remember me writing you about Ellery Gross, the boy that was with me while I was over in England?" Mitch told her how he had been killed in a P-38 crash in California. "He was a pretty swell guy, too."

WHILE MITCH WAS PREPARING HIMSELF AND THE JUNIOR PILOTS around him for war, the nation as a whole was doing the same, as its official position of neutrality proved increasingly untenable. In mid-September, President Roosevelt instructed that any US Navy ship facing a Nazi threat could shoot on sight. The order followed the first German attack on a US warship, a torpedo assault on the USS *Greer* in the waters off Iceland. In Europe the war raged on, and German forces began their siege of Leningrad. In North Africa around that time, British commanders gave approval to an unusual ploy in the hope of reversing Nazi momentum on that continent. The mission was dubbed Operation Flipper, and the target of "leadership decapitation" was Lieutenant General Erwin Rommel. One of Hitler's favorites, Rommel had previously built a legend as a cunning tactician with a string of battlefield victories in western Europe. Dispatched to northern Africa in early 1941, Rommel's Afrika Corps had confounded and defeated British forces time and again, earning him the nickname "the Desert Fox." Bewildered and frustrated, the British, who had secretly formed highly trained commando units within their armed forces, decided that getting Rommel was the very purpose of inaugurating the specialized units. The rationale for the targeted kill, Prime Minister Winston Churchill said later, was to eliminate "the brain and nerve-centre of the enemy's army at a critical moment." Despite weeks of careful planning, however, the attack on November 17 failed, due mainly to faulty intelligence. Rommel was nowhere to be found when commandos raided a villa near Beda Littoria in Libya. He'd left weeks earlier. Even so, Operation Flipper showed that taking out an enemy leader was gaining viability as a wartime move.

Meanwhile, in another part of the war-torn world, Admiral Isoroku Yamamoto's massive carrier strike force, having practiced maneuvers throughout the fall, was in late November mobilizing in secret at Hitokappu Bay in the Kuril Islands. Simultaneously, diplomatic negotiations continued between Japan and the United States. The so-called Hull note, named after US secretary of state

Cordell Hull, was delivered to the Japanese government. Officially titled "Outline of Proposed Basis for Agreement Between the United States and Japan," it was the United States' final bid in the endless back-and-forth talks. No one was optimistic, however, at that late point, and in early December, Yamamoto hurried to Tokyo for high-level operational meetings.

Mitch kept going, training by day and planning for his marriage by night. He was slated to return to Hamilton Field in December to break in P-39 Airacobras during new maneuvers scheduled for January 1942. "We have 109 new planes, P-39s, waiting for us," he said. But Mitch was owed a pre-Christmas leave before then, and the couple saw it as their window of opportunity. "Do you realize Miss Miller that in less than two months you will be Mrs. John W. Mitchell?" he wrote hopefully during the fall. They went back and forth about where to marry. Annie Lee thought maybe at her Aunt Golda and Dr. Joe's ranch just north of San Antonio, if Mitch got enough time off. Mitch said maybe they could meet up in Reno, Nevada, after a brief check-in at Hamilton Field. "We would be married in less than half an hour." He plotted their honeymoon: a road trip from Reno to Enid, a visit with his family, then on to San Antonio to pack up her things, then a drive west for a tour of a bit of California before settling in for good at Hamilton Field by year's end. "There are some beautiful sights on the way, such as the Grand Canyon, Boulder Dam, some of the parks." He ultimately deferred to Annie Lee, though. "I've always thought a wedding was the woman's show so whatever you want concerning it I'm agreeable."

They finally settled on a wedding in El Campo the week before Christmas. Mitch's timing abruptly changed when he learned that he would not be returning directly to Hamilton Field with the rest of his squadron. Instead, he was ordered northeast to Westover Field, just outside Springfield, Massachusetts, to pick up a plane another squadron had left behind for repairs. He'd return to Charlotte with the plane, then lead a flight of five planes across the country to Hamilton Field. Luckily the assignment didn't change the wedding plan; he'd still make it to Texas in time. Plus going north meant that Mitch would get to see one of his sisters, who lived within an hour of the base. "This is Edith, the one who has a little girl," he explained to

Annie Lee on November 30, as he readied to go. "I haven't seen her since 1938."

It was on Tuesday, December 2, 1941, when Mitch reunited with his sister, her husband, and his niece, Betty Jo, in Fitchburg, Massachusetts. Edith even called in a newspaper reporter, who interviewed him for a feature story in the local paper. That evening they feasted on a steak dinner, with Mitch going on and on about Annie Lee. Edith surprised him with a wedding gift: a handmade linen tablecloth and twelve matching napkins. Mitch, grateful, planned to be on his way the next morning; his marriage to Annie Lee was just a couple weeks away, so close he could feel it.

But at the same time he was at his sister's enjoying the warm comforts of her home, thousands of miles away a telegram that would change the course of history was sent through protected channels. It was from Isoroku Yamamoto to his massive strike force in the Pacific north of Hawaii, and it said simply, "Climb Mt. Niitaka."

CHAPTER 6

WHEN THE ROSE PETALS FELL

—

Admiral Yamamoto
WWII Database/public domain

ADMIRAL ISOROKU YAMAMOTO, COMMANDER IN CHIEF OF JAPAN'S
powerful Combined Fleet, prowled the operations room of his flagship, *Nagato*. Poster-sized maps of the Pacific region, particularly of

Hawaii, filled the four walls. Maritime charts were unfolded on a long table in the center of the room, with a large globe sitting on top. Next to the globe rested a speaker connected to the nearby radio room by a cable snaking along the floor and across the hall. It enabled Yamamoto to receive directly and immediately any new messages. A smaller table held files of operational orders and a growing stack of communications.

The battleship was at Hashirajima anchorage, about nineteen miles south of the naval base at Kure, Hiroshima Prefecture. Yamamoto was waiting for word from his carrier strike force as it closed in on Hawaii, having left its secret staging area off the Kuril Islands days before. The date was December 6, 1941, Hawaii time, or five days since Yamamoto had told the strike force commander, Vice Admiral Chuichi Nagumo, to sail for the Hawaiian Islands and to "Climb Mt. Niitaka." Nagumo's force, slicing through rough seas, included the Imperial Navy's six largest aircraft carriers, two battleships, and a number of destroyers, cruisers, and fuel tankers—thirty-one ships in all.

In the days since the order, Yamamoto had appeared before Emperor Hirohito in full dress, toasted the mission's success at the navy minister's residence, and hastily said his good-byes to his family, his longtime geisha friend Masako Tsurushima, and especially his lover, Chiyoko Kawai. Back aboard *Nagato*, Yamamoto wrote Chiyoko a quick note apologizing for being so preoccupied with official affairs while in Tokyo and asking for her to forgive him that "I couldn't stay with you even a single night, to my regret." He continued using the bouquet of roses he had bought for her as a metaphor for the imminent surprise attack on Pearl Harbor, this time framing it as a question that reflected the mission's existential uncertainty: "By the time when the petals of those flowers fall, what will happen?"

Yamamoto and his command staff were waiting to find out—waiting and worrying. They worried about the weather, which at that time of year in the northern Pacific was more likely to be stormy than calm. Rough seas could not only hurt the strike fleet's progress in general but also impede the refueling operations that had been so carefully planned and were critical if the fleet was to reach its destination. They worried about the fleet being spotted by ships or patrol planes of the

United States—or any other nation, for that matter—which would set off alarms and blow its cover. That was why Yamamoto's planners had chosen such a northerly, less traveled route—to minimize the chances of being detected through a random encounter. Lastly, they worried about US fleet commanders at Pearl Harbor deviating from their practice of having most of the ships at anchor Saturday and Sunday to give their crews time to rest and relax. What if the steady drumbeat of war had altered the weekend customs? Yamamoto, on the flagship, and Nagumo, his fleet commander, closely monitored the incoming intelligence reports tracking ship movements around Hawaii.

Yamamoto had also grown increasingly obsessed with wartime protocol. Negotiations had continued between Japan and the United States into early December even though Japan's leaders had chosen to go to war. Such talks were outside the admiral's bailiwick; they were handled by the Foreign Ministry. Even so, Yamamoto had demanded an assurance that no matter how great the strain and the rapidity of events, proper diplomatic practice would be followed, meaning that the United States would be put on notice of Japan's war declaration *before* the attack, not after. It was one thing to launch a surprise naval attack during a declared war; it was another, and a wholly unacceptable action, to instigate a sneak attack. Honor was at stake. Yamamoto was told not to be concerned, that the Imperial General Headquarters was drafting a message, in fourteen parts, that would break off talks with the United States and declare war. The note was being transmitted to Japan's embassy in Washington, DC, and the plan was for Japan's ambassador to deliver it to the US secretary of state, Cordell Hull, thirty minutes prior to the start of the bombing. The attack on Pearl Harbor would not be seen as cowardly or underhanded but rather as the surprising opening salvo in an officially declared war.

Unknown to Yamamoto and Japan's leaders at the time was that Hull and other top US officials, including President Roosevelt, were aware of what was afoot, at least partially. The deciphering of Japanese codes that had begun in the early 1920s had provided inside knowledge of Japan's war preparations. On December 6, code breakers were even intercepting the fourteen parts of the final ultimatum as they were cabled from Tokyo to Japan's embassy. Moreover, they'd intercepted messages instructing embassy staff to burn materials and prepare to leave Washington. But none of the intelligence landing in American hands specifically identified Yamamoto's assault on Pearl Harbor in the central Pacific. The code breakers picked up clues that spies in Hawaii were passing along information to Tokyo about US ship movements. There was also information that Japan's army was on the move much farther away in the South Pacific. Though several US defense officials had voiced worry about Pearl Harbor, no specific operational intelligence had surfaced. By December 6, as a result of the intercepted cables, the White House was in effect on notice, albeit surreptitiously, that war was in the offing. But as Congress adjourned for the weekend and the White House began to brace for war, the US fleet at anchor at Pearl Harbor was under no special alert. Its crews were looking forward to enjoying a charity football game, watching a "Battle of the Bands" at the Bloch Recreation Center, catching a movie, or taking part in any number of other fun activities.

Yamamoto, hovering in the operations room aboard the *Nagato* on December 6, eventually received word that the strike force's last refueling was nearing completion. The oiler ships were breaking off to rendezvous at a predetermined location. The strike force, now fully fueled and separated from the slower-moving tankers, picked up speed to cover the remaining six hundred miles south to Pearl Harbor. Yamamoto took that moment to send his final war message to every crewman, flier, and sailor at sea. The pep talk included the imperial rescript the emperor had given him, with the words "the responsibility assigned to the Combined Fleet is so grave that the rise and fall of the Empire depends upon what it is going to accomplish." Soon after, a signal flag was hoisted aboard the strike force's flagship, *Akagi*—a flag intended as a reminder of the "Z" pennant the heroic Admiral Togo had flown more than three decades earlier in the Russo-Japanese War. Nagumo, the strike force commander, repeated the legendary Togo's very words to his fighters: "The future of the Empire depends on this battle. All men must give their utmost."

Yamamoto then received a final intelligence report from Tokyo. Everything seemed quiet at Pearl Harbor, with no sign of unusual activity. Importantly, most of the US Pacific Fleet was accounted for:

some nine battleships, three light cruisers, three seaplane tenders, and seventeen destroyers at anchor, with another two destroyers and four light cruisers in dock. That was the good news. There was cause for concern, though; there was no sign of the fleet's aircraft carriers. But if Yamamoto, the fierce champion of a carrier-based navy whose entire Pearl Harbor operation was carrier centric, had any second thoughts about US carriers not being in his line of fire, he did not show it. Instead, he displayed the utmost coolness as the day turned into night and he and his staff officers awaited word. Yamamoto, as he did most evenings, played *shogi* with his underlings. This night, though, he stopped early. He headed to his cabin, where he bathed and rested. Most of his staff did the same. Whether or not they managed to catch a few hours of sleep, Yamamoto and his officers reassembled in the operations room after midnight. Yamamoto was standing by. It was approaching dawn at Pearl Harbor: December 7, 1941—the same day, as Chiyoko later told Yamamoto, that the petals fell from her roses.

ON DECEMBER 1, WHEN THE MASSIVE JAPANESE STRIKE FORCE HAD finally gotten word from Yamamoto to "Climb Mt. Niitaka," the anxiety that had dogged the fleet was dispelled. For weeks everyone had wondered: Would they be given the go ahead? Or not? With their commander in chief's telegram came resolve and certainty: the surprise attack would happen as planned on December 7. "We felt the apprehension that had made us worry so long disappear suddenly," noted the strike force's chief of staff, a vice admiral named Ryunosuke Kusaka. "I felt then that my mind was as clear as the autumn moon in the sky." Kusaka, who served alongside strike force commander Chuichi Nagumo aboard the *Akagi*, later composed a detailed account of the mission. When Yamamoto's final war message was broadcast aboard the strike force's ships five days later during the countdown to the attack, Kusaka wrote, the thousands of men serving on the force erupted in cheers. Said Kusaka, "Sensing the grave responsibility assigned to them, and also anticipating the bold enterprise to be carried out early the next morning, all hands were fired up in their blood."

On December 6, the night sky was cloudy, the moon only occasionally breaking through as the fleet churned through the open sea on its southerly route. Nearly four hundred planes were lined up wing-tip to wingtip on the flight decks of the six carriers. When the ships pitched, the planes carrying a maximum load of bombs and torpedoes rocked, their rubber tires bulging and squeaking. One of the pilots used chalk to scribble a message on the side of a bomb: "First bomb in the war on America." For the initial wave, the attack flight would include 183 planes: forty-nine high-altitude bombers, later nicknamed "Kates" by American fliers, filled with powerful bombs and equipped with a rear-firing 7.7 mm machine gun; another forty "Kates" outfitted for the mission solely as torpedo bombers; fifty-one dive-bombers, nicknamed "Vals," carrying bombs and three 7.7 mm machine guns, two of which were forward firing and the third in the rear; and forty-three fighter planes, nicknamed "Zeros," notoriously fast and agile and armed with two 7.7 mm machine guns and two 20 mm rapid-firing cannon.

About 5:30 a.m., the engines were started and the planes warmed up. Kusaka recalled that daybreak on December 7 was "still so dark that black and white could barely be distinguished." The deafening sound of so many propellers and engines competed with the noise of the roiling ocean waters below. Maintenance crews zigzagged through rows of aircraft, conducting final inspections, while pilots in the flying crew waiting rooms grew more restless with each passing minute. Map boards dangled from their chests. Given last-second instructions by operations officers, the flight crews then ran and climbed into their planes. The six carriers turned into the thirty-knot wind to achieve maximum lift. With visibility improving in the dim light now breaking on the horizon, Kusaka looked out from the *Akagi*'s bridge to take in the captivating sight of the strike force's vessels "making way in a gorgeous formation."

Hoisted atop the *Akagi*'s mast for the other carriers to see was a flag signaling "Take off." It was 6:20 a.m. The first to launch were the Zero fighter planes, then the Kate horizontal bombers and the Kate torpedo bombers, and last the Val dive-bombers. Within fifteen minutes, the entire first attack wave of 183 aircraft was airborne. Setting up in formation, the torpedo bombers climbed to a cruising altitude of about 9,200 feet; the high-level bombers were at 9,800 feet; the dive-bombers went higher, to 11,100 feet; and the Zeros climbed to 14,100 feet, where they were free to roam as lookouts. Led by the flight commander, Mitsuo Fuchida, who was also the lead pilot of the high-level Kate bombers, the planes began flying the remaining two hundred miles to Pearl Harbor on the southern coast of the Hawaiian island of Oahu. Kusaka was overcome. "I, who watched them take off at the bridge, was filled with deep emotion and couldn't help feeling my blood boil up," he said. The crews on the six carriers immediately began hauling more aircraft onto the flight decks from hangars in preparation for the next wave. Shortly after 7:00 a.m., the carriers began launching a second attack, and by 7:25 a.m., another 167 Zeros, Kates, and Vals were airborne. In all, in less than ninety minutes, the Japanese Carrier Strike Force had put 350 weapons-loaded planes into the sky, heading toward Pearl Harbor.

For the final approach, Commander Fuchida, flying with the big bombers at about 9,800 feet, was protected by cloud cover. Through a break in the clouds just north of Oahu, he spotted the white surf and coastline below. It was now 7:50 a.m. Fuchida signaled the start of the general attack to all pilots. Within minutes a dive-bomber dropped the first bomb—weighing 550 pounds—onto Wheeler Field in the northern part of Oahu. Within seconds, twenty-four more Japanese dive-bombers began dropping their payload onto Wheeler, the US Army's largest aircraft base in the Pacific.

The day had come.

THAT SUNDAY MORNING, MANY OF THE HUNDREDS OF SERVICE-men assigned to Wheeler Field were enjoying a weekend leave. Second Lieutenant Besby Holmes was no different. He had spent Saturday night partying hard and at dawn was paying the price. "Nobody had ever warned me about sweet rum drinks. One is fine, two is great, three is murder and four is death," he said later. He'd had plenty of reason to celebrate. The US bases on Oahu were on high alert due to reports coming out of Washington that war was imminent, so for the prior week or so his squadron had been flying daily patrols. When the alert had been called off on the weekend, the men had headed into town. "We'd been restricted for 24 hours a day, but the teacher let the monkeys out." Moreover, the San Francisco–born fighter pilot had turned twenty-four on Friday, and a pilot buddy whose cousin worked at the Royal Hawaiian hotel on Waikiki Beach had gotten them a deal on a fancy suite. Now, in the predawn darkness after a Saturday-night bash, Holmes was tossing and turning, unable to sleep. His head was splitting. He had one clear thought: to get himself to early Sunday Mass and, with his obligation to God out of the way, hit the beach to let the sun bake the poison out of his body.

Like Johnnie Mitchell of Mississippi, Besby Frank Holmes had been a boy when he had become enamored with flying. In Mitchell's case, he had gotten a taste of the sky when the barnstormers had come through his rural hometown of Enid, landed their two-seater biplanes in the cow pastures, and offered rides for $1.50. Besby Holmes, meanwhile, was a city boy, and his epiphany had come one bright summer morning when he had been fishing at the foot of Van Ness Avenue. A flight of shiny new fighter planes in formation had flown over so low that he could see the pilots—their helmets and goggles and their white scarves streaming in the wind. The planes had screamed by, and he had said to himself, "I just gotta fly one of those." He had gone to high school and junior college in San Francisco, where he had distinguished himself as a chess player, boxer, and swimmer, and he had been twenty-three when he had gone down to the recruiting station in March 1941 to enlist as an air cadet in the Army Air Corps. For the next eight months he had trained in southern California, then at Luke Field outside Phoenix, Arizona, and in November he had been assigned as a newly minted second lieutenant to Wheeler Field on Oahu. He had so far flown different planes: the single-engine AT-6 training aircraft, the Curtiss P-40 Warhawk fighter, and in Hawaii, for the first time, the sleek Curtiss P-36 Hawk, the Warhawk's predecessor.

Holmes was still essentially a rookie pilot as he pulled on his brown pin-striped suit with a green wool tie just after sunrise and staggered into a church across from the Royal Hawaiian, where he began praying to God that his headache would go away. The church was open all the way around for ventilation, and the first thing Holmes heard was a whirring, whooshing noise, followed by a commotion at the altar. The priest abruptly concluded the Mass while outside all hell was breaking loose. People were rushing around, military trucks were roaring down the street, and Holmes looked up and saw the sky dotted with planes dive-bombing the harbor and airfields beyond.

Holmes raced across the street to the hotel to find his friend freshly changed into his army uniform. The hotel manager was listening on a tiny portable radio to reports that the Japanese were attacking Pearl Harbor, but no one needed telling. Operating on instinct, Holmes and his pilot buddy commandeered the first car they saw outside, a red Studebaker Champion. They raced past the harbor, which was under attack, and up a hill to their base. Holmes was dumbstruck: Wheeler Field was in shambles. Dozens of P-40s parked in line were in flames. He drove toward his hangar, only to find it burning, and as he got close, the top melted and crushed in. The airfield was incapacitated and useless. Holmes kept going, driving as fast as he could to the nearby auxiliary Haleiwa Field with its single paved runway. He saw other pilots climbing into undamaged planes, and then a ground crew hurried over and said that an older P-36 Hawk was fit to fly. They handed him a parachute, a helmet, and a .45-caliber pistol. Holmes had flown a P-36 only once before, but he hustled toward the plane along with the line chief. He heard the rapid machine-gun fire first and next saw dust rising all around the plane. Over his shoulder he saw a dive-bombing Val strafing the airfield and the P-36. He aimed his .45 and fired, but the handgun was no match for the Japanese dive-bomber, just as the US fleet at Pearl Harbor, caught by surprise, was no match for the Japanese Carrier Strike Force.

The line chief boosted Holmes into the cockpit and then helped him start the engine and load ammunition into the .30-caliber machine gun. Holmes took off. Several other pilots had already gone up in P-40s, and Holmes was, comparatively speaking, late to the show; in the twenty or so minutes it had taken him to get from the church and into the air, the Japanese had mostly completed their devastation of Oahu. Two of the pilots who'd gone up earlier, Second Lieutenant George S. Welch and Second Lieutenant Kenneth M. Taylor, managed to engage the enemy and shoot down a total of six enemy fighters between them. Holmes searched all around but never encountered enemy fire. He did encounter friendly fire, however. He thought it absurd that everybody on the ground was firing at him while he never saw a Japanese plane in the sky. What he got instead of actual combat was an overview of the damage done. He flew to his base, Wheeler Field, where twenty-five unmolested Japanese dive-bombers had destroyed Wheeler's 140 unmanned fighter planes, and then to Schofield Barracks. Both were a chaotic mess. He passed over the Ford Island Naval Station in the middle of Pearl Harbor. The damage was massive: huge ships sunk and burning; fuel blazing; aircrafts burning. He swung quickly by Marine Corps Air Station Ewa, which he found was as bad off as Wheeler, and then flew to Hickam Airfield, likewise beaten up. He realized it was a good thing he hadn't run into a formation of enemy Zeros: "They would have creamed me."

Holmes's flight lasted about a half hour. He returned to Haleiwa Field, the little airstrip that had been largely overlooked by the Japanese. Pilots from other fields had been making their way to Haleiwa, wanting to go up in the handful of operable P-40s and P-36s. Holmes found the scene surreal, a mix of panic and chaos. Everyone was jumpy and, worse, paranoid. Rumors were rampant that a new air attack was minutes away, then not. There were rumors that a Japanese land invasion was under way, then not. That night Holmes and his pilot buddy were assigned to the beach as lookouts for a land invasion that was widely expected. They were instructed to fire three shots from their .45-caliber automatic pistols the moment they spotted the enemy. That would signal army pursuit planes to take off for battle. Holmes spent the night hiking up and down the length of the 1.5-mile-long beach, along the way passing his partner pacing from the other direction. Holmes was spooked several times, mistaking the moonlight jumping off a rolling wave as the bow of a landing Japanese infantry troop barge. Luckily he never fired his gun. And no further attacks came. Holmes could not remember a more welcome sunrise.

IT WAS AS IF EVERY SAILOR AND PILOT STATIONED UP AND DOWN the California coast and in the Pacific was nursing a hangover as word spread Sunday about the early-morning attack. Second Lieutenant Rex T. Barber, a thick-necked jock fresh out of cadet training school, was sleeping his off in a bunk aboard *President Garfield* when a fellow pilot shook him awake. News of the devastation was blaring on the ship's radio. The *Garfield* was a civilian cruise ship recently redeployed by the military to transport Barber, his squadron, and more than a dozen P-39 Airacobra fighters stored in the ship's hold from San Francisco to the Philippines in the South Pacific.

Barber was from Culver, a small town in central Oregon with fewer than a hundred residents. Rex, who had grown up on his family's 2,100-acre farm, was outgoing, athletic, and never lacking in confidence. He had starred in baseball and basketball at school and spent summers working on nearby cattle ranches, all the while being an avid hunter. He had a favorite uncle, a World War I pilot, whose frequent stories of war and of "hair-raising tales of flying and women," had been enthralling. Barber had known he wanted to fly someday.

Barber was twenty-three when he enlisted in the army in the fall of 1940. He became an expert marksman with the pistol, rifle, and machine gun—one of the surest shots training officers had ever seen. Training to be a pilot starting in early 1941, he quickly earned a reputation as fearless and, to some, reckless. Barber rejected the latter description, insisting that he was simply eager and always ready for action. He had only just completed cadet training school on October 10, 1941, when he was one of four graduates picked to become fighter pilots. He was rushed to San Francisco to join a squadron about to set sail for the Pacific. When he boarded the *Garfield* on December 5, he'd had only four training hours on a fighter plane, a P-40. He was a neophyte, basically a fighter pilot on paper who would learn on the job.

The past month had been a whirlwind. The fact that the *Garfield* had been a cruise ship meant it was outfitted for comfort, stocked with quality civilian food and waiters. Pilots sneaked cases of booze aboard, and for the first forty-eight hours at sea they turned the *Garfield* into a party boat—music blaring on record players, card games, dice games, and gambling, accompanied by plenty of drinking. Then came the wake-up call at dawn on December 7. Stunned, Barber and other hungover pilots gathered to listen to the radio broadcast. His brain marinated in whiskey, he struggled to comprehend the incomprehensible. Hours passed as the men waited in limbo, their party spirit crushed. Barber grew restless and worried that the *Garfield*, a big cruise ship wallowing at sea, would be a sitting duck for one of the notoriously accurate Japanese bombers. Finally his commanders got word to return to San Francisco, where they'd all learn their assignments for the war. It was now all hands on deck, a message not limited to the *Garfield* but for all of America.

SECOND LIEUTENANT THOMAS G. LANPHIER, JR., WAS FAST ASLEEP in a hotel room in San Francisco when his father called to tell him to get up and turn on the radio—right now. Tom Jr. obeyed the orders. He launched his wiry frame out of bed, his curly dark hair all tousled, and switched on the radio. His first reaction to the Japanese attack was shock; his next thought was what it would mean for his unit.

Like Rex Barber, Lanphier was a neophyte fighter pilot. He had just graduated from cadet training school and earned his wings the prior month and was newly assigned to Hamilton Field north of the city. And like Besby Holmes, he had recently celebrated a birthday, turning twenty-six on November 27. Unlike either of them, though, Lanphier was a military brat, his father a celebrated career army officer.

Lanphier, in fact, had been born in Panama City while his father was stationed there. He had then lived briefly in several locales until his teen years, when the family had moved to Detroit, Michigan. His father had been transferred to nearby Selfridge Field to take command of its prestigious 1st Pursuit Group, the country's oldest air combat group. Captain Eddie Rickenbacker had flown in one of the group's squadrons during World War I and become the top-scoring ace in aerial combat in France. Lanphier's parents had separated when Tom was fifteen, but Detroit had provided a certain stability in terms of living in one place for an extended stretch. He had finished high school in Detroit, excelling at academics and in baseball and earning acceptance to Stanford University in Palo Alto, California.

Unfortunately, his father's unwise investments had created financial havoc, and Tom Jr.'s college years were chopped up. He'd attend classes for a semester, then drop out to earn money, then return to school. He worked all kinds of jobs—one year as a grocery clerk in a small California town, another year as a ranch hand alongside his younger brother, Charlie. It was during one summer while visiting a classmate at Payette Lake in Idaho that he met his future wife, Phyllis Fraser. Probably his best job was as an apprentice reporter for the *San Francisco News*. Working nights and weekends, he juggled the reporting even after he returned to school, covering such breaking news as a longshoreman's strike and also getting a chance to review new books and theatrical plays. It took him eight years to earn his Stanford degree, and he was twenty-six when he did.

The first time he flew was as a teenager in Detroit. Without his father's knowledge, he pestered a couple of fliers into taking him on joyrides in the back seat of one of the training aircrafts, a Curtiss JN-4 known as a "Jenny." The rides turned into instruction, and one Saturday morning Lanphier, just thirteen, convinced the pilots to let him fly solo. He was frightened at first, but then exhilaration set in and he flew over Mt. Clemens and Lake St. Clair, looking down on his family's house. He assumed that at that early hour on a weekend—6:45 a.m.—his father would be asleep. He was wrong. Not only was his base commander father awake, he was already at the airfield to conduct a spot inspection. Lanphier, in effect, landed in his father's lap, and his father was not pleased. He began to flail at the boy right there on the tarmac, using a length of solid rubber shock absorber. The pain was worse than that of the coat hanger his father usually used. It seemed that his father was more frightened than angry, Lanphier later said, "but that didn't lessen the pain."

Despite the beating, his father was the one who would insist that he enlist in the Army Air Corps flying school to train as a pilot. The senior Lanphier traveled to the Stanford campus in the fall of 1940 with the sole purpose of persuading his son to see things his way. His argument was that war was imminent—not with the Germans in Europe, where Hitler was running amok and the United States was already openly allied with Great Britain through a program of "loaning" it US destroyers, but with the Japanese. In war, his father argued, pilots had it way better than soldiers on the ground. Lanphier had a standing job offer at the newspaper but followed his father's orders; he enlisted soon after his graduation in early 1941. Ten months later, Second Lieutenant Lanphier was at Hamilton Field outside San Francisco awaiting deployment to the Philippines.

During flight training he'd caught on faster than most, impressing his instructors as smart, skilled, and unafraid. Though he was friendly and fairly popular, some pilots nonetheless found him off-putting, a know-it-all quick to act as an expert on all things, even if he was well educated. He'd name-drop and seemed self-absorbed, with a salesman's personality selling not a product but himself. Lanphier certainly had a lot to live up to: a demanding father who had been a West Pointer, a classmate of Dwight D. Eisenhower, and a close friend of many of the flying heroes of the time, from Colonel Billy Mitchell to Charles Lindbergh. He had flown combat missions during World War I and was soon to fill the prestigious post of air intelligence officer for Army Chief of Staff General George C. Marshall in Washington, DC.

Now, in early December 1941, the senior Lanphier had come to San Francisco once more, this time to see his newly minted pilot son off and also to wish him well in his marriage. Like Johnnie Mitchell and Annie Lee Miller, Tom Lanphier and Phyllis Fraser had set their sights on a mid-December wedding. Then all hell had broken loose in Hawaii, and Lanphier Sr. was on the telephone, ordering Tom to get up and turn on the radio. The marriage, and most everything else in his life, would have to take a back seat to war. Racing out of his hotel on the morning of December 7, the only thing young Lanphier could think about was returning to Hamilton Field. He needed to find his squadron leader, Captain Henry "Vic" Viccellio, to find out, what next?

FOR HIS PART, ON SUNDAY, CAPTAIN VICCELLIO WAS HAVING AN early lunch with his wife in a motel restaurant in San Francisco. It was basically a good-bye meal, as Viccellio was shipping out to the Philippines and while he was gone, his wife and year-old son would be staying with her family in Longview, Texas. When he had first learned that his destination would be the Philippines, he had been glad. He and his men would be far away from the Germans and wartime combat, which was fine by him. The pilots under his command were presently scattered, although they would all reunite in the South Pacific. Some, such as Rex Barber, were already en route on the *President Garfield*; others, such as Tom Lanphier, were awaiting the next ship out. Viccellio had heard some of the complaints about Lanphier and his tendency to self-aggrandizement, but the squadron commander liked him. Lanphier might lack experience, but he possessed the traits Viccellio looked for in pilots: boldness and a nimble mind. Then there was his operations officer, Lieutenant John W. Mitchell. Mitchell was still across the country in Charlotte, North Carolina, preparing to lead a small flight of repaired fighter planes back to Hamilton Field.

Then everything changed. Viccellio and his wife had just started eating when the cook ran into the dining room screaming something about the Japanese, that the Japanese had attacked Pearl Harbor. Needing to hear for himself, Viccellio bolted from the table and followed the cook into the kitchen, where a radio was broadcasting news of the attack. Viccellio switched into automatic pilot, telling his wife he had to leave immediately and return to Hamilton Field as fast as possible. When he got to the gate, the guards refused to let him onto the field even though he was in uniform. It was as if the guards were expecting enemy spies to be pouring in from all directions. Viccellio was apoplectic that the paranoia was so blinding. He was Vic, the Virginia-born captain in full uniform, standing right before the guards' eyes—how could they possibly mistake him for a Japanese spy? It took arguing for nearly thirty minutes before he was allowed on the grounds. His mind was filled mainly with thoughts about his pilots, their whereabouts, and how nothing would ever be the same. More than ever, he needed his operations officer, who, at age twenty-seven, was one of his best pilots even if he hadn't yet seen combat.

The problem was that the lovestruck Lieutenant Mitchell was in North Carolina. He'd spent the last two days waiting impatiently to fly cross-country to Hamilton Field. Mitch was thinking about little else than his upcoming leave and his wedding to his fiancée, Annie Lee Miller, and the previous week had been nothing less than an exercise in frustration. Following his reunion on December 3 with his sister Edith and her family in Fitchburg, Massachusetts, he'd returned to the airfield in Westover, Massachusetts, expecting that the P-40 he'd gone north to retrieve would be good to go. The plane was ready, but bad weather had kept him from flying south to North Carolina. "It is a ground fog and is all over the Atlantic seaboard," he wrote Annie Lee. He was bored with nothing to do but sit around, read, and wait. "I've already read everything I can find, except the *Woman's Home Companion*." Two days later, on Friday afternoon, December 5, he caught a break in the weather and hurried down to Charlotte, but after he landed he was told that the lousy forecast for the next day, Saturday, December 6, would likely keep him grounded indefinitely.

Mitch was grinding his teeth, operating in a love bubble and seemingly unaware of anything else going on, obsessed with the weather and with getting himself to Texas in time for his pre-Christmas wedding. In a long letter he wrote to Annie Lee during that vexing week of idleness, he returned to an old theme: the moon, the symbol of their love, Johnnie Bill and the moon. During one subzero night, he'd caught sight of a full moon through the cloud cover and told Annie Lee, "This big ole moon shining outside my window isn't adding a thing to my mental stability. I keep looking at it, thinking of you and loving you always and forever."

To kill time that Sunday, he decided to catch a matinee movie to get his mind off his agitation and lose himself in other people's made-up melodramas. It wasn't until he was leaving the theater that he noticed real-life drama unfolding before him: the people of Charlotte were moving about in a nervous rush. A newsboy, his arms around a stack of the *Charlotte News*, came toward him, shouting something about an attack. Mitch heard the words "Japs" and "Pearl Harbor." He bought a newspaper and saw the huge headlines announcing the Japanese attack. He began reading, immediately recognizing many of the bombed-out locations as places he'd been to and trained at during the four years he had served in the Coast Artillery Corps before deciding to become a pilot. He'd loved his time in Hawaii, and now, according to the news accounts, Oahu sounded like Hell on Earth, with an eventual tally of more than 3,400 Americans dead or wounded and twenty-one vessels of the US Pacific Fleet, including seven battleships, sunk or damaged. Mitch tried all afternoon to reach his squadron commander, and when he finally got through that Sunday night Captain Viccellio told him that he had to get back, that everything was organized chaos and their next moves were uncertain. Except for one thing: they were going to war. Mitch knew that most pilots in the squadron were still works in progress, but the time for maneuvers and war games had run out. This was the real deal.

War was now at the forefront, and Mitch, as soon as weather permitted, was eager to reach Hamilton Field. But although Tom Lanphier and other engaged servicemen might call off their weddings, Johnnie Mitchell did not. Nothing, not even war, was going to interfere with his first getting hitched to Annie Lee Miller.

ADMIRAL YAMAMOTO, MEANWHILE, WAS A STUDY IN SOLEMNITY, seated in a folding chair at the big table in the *Nagato*'s operations room, his eyes shut as he calmly awaited news from the strike force four thousand miles away. It was after 3:00 a.m. Tokyo time on December 8 when the first messages began streaming in from his men attacking the US fleet at Pearl Harbor: "I hit enemy battleship"; "I bombed Hickam Field"; "Enemy warships torpedoed—outstanding results"; and, then, most reassuringly, "Surprise attack successful." In the next several hours, Yamamoto learned that only 29 planes in the strike force of 353 planes had been lost, far fewer than expected. More news came in, reporting the successful bombing and invasion of other targets from Malaysia to Hong Kong, Thailand, Wake Island, and the US bases in the Philippines.

In Tokyo, Prime Minister Hideki Tojo was on the radio by early afternoon announcing the surprise attack to a nation that, although it had been preparing for war, was stunned. He said, "The rise and fall of our Empire and the prosperity or ruin of East Asia literally depend upon the outcome of this war." Propagandists kicked into gear to calm a public shaken by the suddenness of the war against the United States. Central to their strategy was exploiting Yamamoto's renown. They had in hand the private letter Yamamoto had penned earlier in the year to the pro-war nationalist Ryoichi Sasakawa. The antiwar missive was Yamamoto's impassioned warning that it would never be enough for Japan to take over Guam or even Hawaii. He said that the United States could be defeated only by total annihilation, with Japanese forces marching into Washington. "We would have to march into Washington and sign the treaty in the White House," was what Yamamoto said—a scenario he expressly called impossible. That was why, in his view, the only hope had been creating the shock and awe of a Pearl Harbor attack to persuade the US leaders to settle for an early peace. But Sasakawa and others now saw a new use for the letter—as a vehicle to mold public opinion and to uplift the nation's combat fervor. They directed the Japanese news agency Domei to release a story that Admiral Yamamoto, in a personal letter to a friend the agency had obtained, had boasted that he was going to crush the United States. Converting the "we" Yamamoto had used in his letter to an "I" and making up new wording altogether, the news story had Yamamoto saying "I shall not be content merely to capture Guam and the Philippines and to occupy Hawaii and San Francisco." Rather, supremely confident, he was promising to go all the way: "I am looking forward to dictating peace to the United States at the White House in Washington." His previous antiwar statement had morphed into a bombastic rallying cry for the Japanese people. It was fake news.

Yamamoto knew nothing about the manipulation. He was busy sending his thanks to the strike force and its commander, Admiral Nagumo. But even as he handed out plaudits, he was indisputably the hero above all others, showered with praise and, in the coming days, receiving sacks of fan mail and postcards. With the help of a clerk he did his best to answer them, promising in his own brush hand to continue the good fight on behalf of the country. He took a break to write his lover, Chiyoko, telling her about the letters "pouring into my place." But, he added, "I am longing only for letters from you night and day." He also wrote to an old friend, Admiral Sankichi Takahashi, a past commander in chief of the Combined Fleet, discussing the factors leading to "our early victory in the war." He selflessly credited naval technicians for their breakthroughs that autumn that had made possible aerial torpedoing in shallow waters and for achieving vital improvements in the technique of horizontal bombing. Yamamoto continued, "We were blessed by the War God," and the Imperial Navy's "good luck, together with the negligence on the part of the arrogant enemy, enabled us to launch a successful surprise attack."

One bit of news stung deeply, however. Despite his insistence that the United States be given notice prior to the raid, the admiral learned that the delivery in Washington of Japan's official war statement had been delayed until after the bombing had begun. He was furious. He'd kept asking during the countdown about the warning and had been assured that it was being taken care of. But it wasn't—something about a slowdown in the document's transmission from Tokyo to the Japanese Embassy. No matter what the reason, as much as Yamamoto had wanted the raid to be a properly executed surprise attack, it would now always be cast as sneaky and underhanded.

But amid the crush of celebratory telegrams and congratulations the point seemed minor to most. The sailors aboard Yamamoto's flagship, *Nagato*, were giddy with the mission's apparently overwhelming success. One wrote home saying that his "dream was to go to San Francisco and there head up the accounting unit in the garrison unit after the occupation." It was as if total victory was at hand. "All of us in the navy," the sailor continued, have always "dreamed of going to America."

PART II

THE SOUTH PACIFIC

CHAPTER 7

WEDDING BELLS AND PACIFIC BLUES

———

Newlyweds John and Annie Lee Mitchell
Courtesy of the Mitchell family

LIEUTENANT JOHN MITCHELL AND THE OTHER PILOTS WATCHED as the noisy Greyhound buses emerged from the fog and slowed to a stop in a parking lot at Hamilton Field. The men were huddled against the January dampness and light drizzle that not only was typical for dawn in the San Francisco Bay area but seemed a fitting reflection of Mitch's mood. Downcast. For this was good-bye time. The buses had come to transport the thirty or so pilots from the base to the docks, where the pilots would go aboard the USS *President Monroe* and sail off for somewhere in the Pacific—none of them had yet been told where. The men were paired off with wives, girlfriends, and family, whispering departing words. Mitch was with Annie Lee.

It was January 12, 1942—five weeks after Pearl Harbor and four weeks since the December 13, 1941, marriage of John Mitchell and Annie Lee Miller. Theirs had been a pit-stop wedding. Within a few days of the Japanese attack, Mitch had finally gotten weather clearance to depart Charlotte, North Carolina, for Hamilton Field. He had flown the repaired P-40 Warhawk on a course that included a stopover in Texas. He and Annie Lee had headed directly to City Hall in downtown San Antonio to fill out a marriage license. By chance a street photographer had snapped a picture capturing the couple's hurried pace, an unplanned photograph that had become the next best thing to a formal wedding portrait. They're seen walking purposefully along the sidewalk, looking intently at each other, Mitch in full uniform, Annie Lee wearing the tweed overcoat with the fox collar she'd bought six months earlier while Mitch had been in England, the coat she'd told her aunt she'd loved and had splurged to buy. Annie Lee's parents had not made it in from El Campo in time for the vows at City Hall. Two days later, Mitch was back in the air on his way to California.

At Hamilton Field he had found everything in turmoil. His squadron commander, Captain "Vic" Viccellio, had told him that from their pursuit group, several dozen veteran pilots, mechanics, and armorers—the best men—had already shipped out to Australia aboard the USS *Langley*, the navy's first aircraft carrier, for later deployment to Java to assist in guarding the Dutch colony from Japanese incursion. It meant that the squadron Vic now commanded and in which Mitch served as operations officer, consisted of pilots who'd just earned their wings and had little meaningful flying experience. Pilots such as Tom Lanphier, Jr., the son of the demanding Army Air Forces colonel. And Rex Barber, back at Hamilton Field after the *President Garfield* had returned him and other rookie pilots to San Francisco. And Doug Canning, a skinny twenty-two-year-old Nebraskan who had a degree in education from a teacher's college but had dreamed of being a fighter pilot ever since his uncle, a World War I airman, had taken him up when he was only eight. Mitch learned that in a matter of weeks they would also be shipped out to the Pacific, and in the meantime, he got the novice pilots some training hours on any available planes. To him, Tom Lanphier and Rex Barber were quick studies as fliers with nerve, while Doug Canning, despite having a couple of near mishaps while landing, quickly displayed laserlike vision. Canning, it turned out, could spot targets either on the ground or in the distant sky long before most other pilots could.

With the clock ticking, Mitch wrote Annie Lee that "the thing for you to do is come on out as soon as possible." He wanted them to be together for as long as he was still in the United States. He hunted for an apartment for whatever time he would have left before he was gone. "We've got to make the most of it," he told her. Mitch knew that war and separation weren't going to be easy. "We're going to be called upon to make many sacrifices." And it wasn't as though they were the only ones; the rest of America was scrambling, too. His brother Phil and several in-laws were enlisting, and Mitch noted, "that makes us quite a fight family, now doesn't it? Let's see, one brother and three brothers-in-law all doing or dying for the good ole U.S.A." In a matter of days, he managed to find an apartment to rent by the week, and Annie Lee was able to get to the base by Christmas Eve—"The nicest Xmas I EVER HAD!" Mitch later wrote. They spent a few frenzied weeks together as Mitch was pulled in two directions. He and Annie Lee were newlyweds who hadn't been together for more than a year except for a handful of days. They needed time to get to know each other again. But it was wartime, and Mitch and his men were often on alert all day as rumors regularly cycled through the base that a Japanese invasion of San Francisco was imminent. One never came, but that didn't lessen the daily tension consuming the first weeks in the new year.

Then time ran out. Vic broke the news about the squadron's departure for the Pacific on the *Monroe*. Their "honeymoon" was over before ever getting started. It was like "Johnny, she hardly knew ye," and as the other pilots began boarding the buses in the fog of that January morning, Mitch leaned in and kissed his new bride a sad farewell. "I had been married only one month and we had lived together only about half of this time," he wrote later in his diary. "One of the most heart wrenching partings I have ever known."

THE USS *PRESIDENT MONROE*, A LUXURY OCEAN LINER BUILT FOR around-the-world civilian travel, had had its initial voyage out of San Francisco canceled once the Japanese had attacked Pearl Harbor. Like other cruise ships, she had immediately been taken over for emergency war service by the War Shipping Administration, which President Roosevelt had created through executive order. In early January 1942, as Mitch pushed his rookie pilots to get in some flying time at Hamilton Field, the *Monroe* was hastily outfitted as a troopship to join two other ocean liners, the SS *President Coolidge* and the SS *Mariposa*, in a convoy destined for the South Pacific. To escort the ocean liners and ward off possible Japanese interference, two destroyers and a light cruiser, the USS *Phoenix*, were added. In all, the ships constituted the first wartime convoy to the Pacific region, the beginning of a buildup that US military leaders needed desperately to accomplish as fast as possible. Mitch and his men hadn't been told anything yet about their itinerary, but eventually they would split off and sail to Fiji while the rest of the convoy continued on to Australia.

When the pursuit squadron of about thirty pilots boarded the *Monroe* on January 12, they joined a handful of radar specialists who made up an air warning unit, as well as nearly seven hundred ground troops. In the ship's hold, along with supplies and ammunition, were twenty-seven large wooden crates, each one containing a P-39 Airacobra, the new single-engine pursuit plane with its much-talked-about weapons system. The catch was that none of the pilots had ever flown a P-39, except for Rex Barber, and he'd done so only a few times. Mitch had been slated to start breaking in a P-39 following his return from maneuvers on the East Coast the previous autumn, but then war had broken out. They all faced teaching themselves how to fly the P-39, never mind assembling them by hand.

Even though the men had boarded at dawn, it wasn't until about 5:00 p.m. that the ship left the dock and passed through the submarine nets and underwater mines in San Francisco Bay protecting the city. Sliding under the Golden Gate Bridge, Mitch wondered, "How long it was to be before we again would see that bridge no one could tell, but I dare say that thought occupied everyone's mind at least once." Once free of the bay, the *Monroe* joined the other ships. The convoy was under way, full speed ahead at eighteen to twenty knots. Top heavy with other planes and artillery secured on deck under camouflage, the *Monroe* tended to roll in the open sea, causing bouts of seasickness. But the ship still had most of its luxury liner trappings—fancy food and waiters—and Captain Vic advised his men to live it up, enjoy themselves now, because he didn't figure much fun was awaiting them.

The men took his advice to heart. "We knew we had a long trip and settled ourselves in to enjoy it," Mitch wrote in a diary he began a week into the trip. Lanphier had brought along a phonograph, and Canning had a stash of Glenn Miller records. Others had smuggled liquor aboard. They played records, drank, dined on squab and baked Alaska, won and lost money at the seemingly nonstop poker and blackjack games. Lounging on deck chairs, they worked on their tans. "I browned but did not blister," Mitch reported. They sleepwalked their way through daily calisthenics and "boat drills," although on occasion Mitch actually had a sobering worry about the damage and fatalities that would result if ever they were torpedoed by the Japanese, given all of the mines and bombs on board. "It was not a pleasant thought." It was as if the *Monroe*'s maiden voyage, aborted at the outbreak of war, had been resumed, albeit with a boatload of army soldiers and pilots replacing the wealthy civilian passengers who'd paid top dollar for the life of leisure at sea.

The convoy crossed the equator eight days into the voyage. "My first trip down under," Mitch noted. He had been guessing they were bound for Australia, but Vic told him he was wrong, that although the rest of the convoy would end up there, they were about to break off and head to the Fiji islands. Mitch thought, "The Fiji Islands? I wasn't even sure where they were." That sent the young lieutenant to the ship's library, where he found maps and a book with a chapter about Fiji, which described the "cannibalistic tendencies of the natives." He was a bit alarmed; they were going back in time to a tropical island featuring tribal cultures and flesh-eating natives, "an interesting adventure," to put it mildly, Mitch predicted. He and the men spent the remaining week at sea mulling what to expect at Fiji, whether they would actually find cannibals, and, if so, what they would look like.

The *Monroe* arrived off the port city of Suva on January 28, sixteen days after leaving San Francisco. "We left a trail of hootch bottles all the way to the Fijis" was the way Doug Canning described the crossing. When Mitch first caught sight of Viti Levu, the largest of the Fiji islands, he was reminded of Oahu. There was so much lush greenery, like in a salad bowl, and Suva itself was hardly primitive, with its docks shaded by red roofing and a sprawl of stucco buildings and houses extending inland. There was no sign of a cannibal anywhere. Instead, the first day in Suva seemed an extension of the ease they'd experienced at sea, as several pilots sneaked ahead of the others, commandeered a bar, and started a party. Soon after Mitch stepped ashore, he was handed a scotch and soda. Taking a sip, he was left to wonder, This is war?

THE PARTY ENDED A FEW DAYS LATER WHEN THE SQUADRON SET up camp outside Nausori, about five miles north of Suva. The rains came and never let up. Shangri-La it was not. "One helluva mess," Mitch wrote. They might as well have renamed the inland village Mudville. "We are wading around ankle-deep everywhere we go." The only permanent structure was a weather station, and the squadron pitched two-person tents near the single unpaved airstrip. Mitch and Vic shared one, their new quarters set between a makeshift mess hall and the showers—"So called, but actually only one hole punched into a pipe, and you stand under that." The food was lousy. There were plenty of bananas, pineapples, and papayas but no fresh vegetables, only canned, and a daily serving of Corned Willie, the nickname for the army's canned corn beef. Tanks of water were hauled into camp on a tiny sugarcane train.

The twenty-seven wooden crates containing P-39 Airacobras followed, transported on trucks along a recently built dirt road that was hub deep in mud. The crates were deposited next to the grassy airstrip, where the squadron's immediate task was to unpack the planes and begin assembling them—in the rain. Vic and Mitch could then oversee the men's training to handle a P-39, get them ready for combat, and also make sure the ground crew of mechanics and armorers was up to speed in servicing the new planes. Mitch felt the need to hurry, as if they were working against the clock to get the planes set up before the enemy could strike. Week by week, Japanese forces continued to advance south largely unimpeded, expanding their control of a number of Pacific nations. Having captured Manila at the start of the year, they pursued retreating US and Filipino troops across the Philippines to the Bataan Peninsula. Elsewhere, Japanese troops had taken over Malaysia, Borneo, and the Dutch East Indies. Farther down toward Australia, Japan's navy and army had zeroed in on the Solomon Islands, starting with the invasion of Rabaul, on the northern tip of New Britain, on January 23. Rabaul was two thousand miles away from Fiji, but the possibility of a Japanese attack was an ongoing concern nonetheless. One of Mitch's initial duties in Nausori was to draw up plans for destroying documents and armaments in the event of an evacuation, which he hoped he'd never have to use. It wasn't constantly on their minds, but the squadron knew it had to get the planes going to do its part in the larger US effort—just getting started—to halt Japan's takeover of that part of the world.

Mitch, Vic, Barber, Lanphier, Canning, and the others all joined in to assemble the puzzle pieces that made up a P-39 Airacobra. They would have been left scratching their heads if not for the presence of a factory representative from the planes' manufacturer, Bell Aircraft. It took them more than a week, but by February 2 they had built two planes. Mitch, as operations officer, was going to be one of the first to test, or "check out," the plane, but he was suddenly sidelined. After waiting for the opportunity to fly, and with the planes finally ready, he carelessly flipped a jeep in the mud and was injured. He limped around on his bruised foot and was unable to fly for nearly a week. "Very restless," he complained in his diary. The others went up, and things were touch and go at first, in part due to the hard rain that fell each day. One pilot, cracking up a P-39 when he overshot the field, said afterward that the brakes wouldn't hold. Mitch derided the explanation, saying it had been man, not machine, that was in error.

"He came in too fast and only used half his flaps," he said. Another P-39 struck a hawk on takeoff, causing the pilot to jerk sharply into the front edge of another plane. "Looks like we are tearing 'em up faster than we are putting them together," Mitch noted sarcastically. It was no laughing matter, however, when a rookie pilot participating in a practice dogfight turned too sharply and lost control and his P-39 began to spiral downward. Mitch stood at the end of the runway watching. The plane crashed, and the pilot was killed instantly. "He was a helluva nice chap," wrote Mitch in a brief, matter-of-fact tribute in his diary. There were several other forced landings, but the men began to acquire a knack for handling the P-39 and, trading notes, discovered that despite its sleekness and enhanced weaponry, the P-39 seemed slower compared to other fighter planes they'd flown in their short time as pilots. They wondered about its effectiveness in a dogfight, especially against the speedy, faster-turning Japanese Zeros.

Between assembling planes and training, the men began finding outlets for their off-hours. They used the wood from the P-39 crates to build four walls as a handball court. They found a movie theater in Nausori—a tiny, hot venue with a single projector that required breaks in the action as the projectionist changed reels. Mitch saw *The Saint Strikes Back*, a crime thriller starring George Sanders as a private detective. He watched *Popeye*, the popular cartoon showcasing the spinach-eating sailor saving the day. In a few months new *Popeye* cartoons would begin featuring fight sequences with the United States' enemies. In the short "You're a Sap, Mr. Jap," the strongman sailor takes on a battleship crew all by himself, climaxing with the Japanese commander drinking gasoline and swallowing firecrackers to kill himself rather than face defeat at the hands of Popeye. The cartoon was part of a surge in Hollywood to produce war propaganda—filled with racist portrayals of the Japanese—as well as to document the war. The directors John Ford, William Wyler, John Huston, and George Stevens all jumped into wartime filmmaking service. Ford, a four-time Academy Award winner, had actually enlisted in the navy the previous fall and created the Naval Volunteer Photographic Unit in cooperation with the government's Department of War. The forty-six-year-old director trained film crews to work in combat and, following Pearl Harbor, headed to the Pacific carrying his 16 mm camera.

For Mitch, the thirteenth of February, a Friday, marked two months since his wedding. He hadn't heard a word from Annie Lee and hadn't bothered to write her a letter that would not go anywhere. There'd been no mail whatsoever since leaving San Francisco. Feeling lonely but wanting to acknowledge the moment in some way, he went across the river from Nausori to eat a real meal at the Suva Hotel. Vic and Rex Barber accompanied him. They dined on tomato soup, fresh corn, and grilled chicken. Mitch made a silent toast to his wife. The three finished two bottles of fine champagne, which Mitch decided was most fitting, as he'd been married two months and it was the first champagne he'd had since then.

Other times Mitch went into Suva to the Grand Pacific Hotel, located directly on the waterfront, a popular peacetime tourist spot that the Americans quickly adopted as a hangout. On each visit Mitch couldn't help but notice the additional warships anchored in Suva's harbor, and among the two destroyers, one aircraft carrier, and a heavy cruiser he recognized the *Achilles*, a Royal New Zealand Navy warship that had been involved in the first naval clash of the war against a German battleship. He also heard that twelve bombers, either B-17s or B-24s, were due shortly. "It's about time," he noted. They were receiving frequent reports that the Japanese were planning extensive operations in their direction. If their own buildup kept on, maybe they'd have a chance after all.

THE SQUADRON'S WORKING AND LIVING SITUATION IMPROVED dramatically when, at the end of February 1942, the men completed a move to Nadi on the western coast. Viti Levu was slightly smaller than Connecticut, and it was a quick flight in the P-39s to cover the seventy miles over the island's mountainous interior, a bit longer trip by truck convoy, more than three hours across a 120-mile coastal route on winding roads. In Nadi the climate was drier, the airstrip was macadam, or compacted crushed stone, with one end of the runway on a cliff above the ocean, and the housing was vastly improved. Mitch grumbled that upon first reaching Fiji the command staff should have sent everyone directly to Nadi. The disassembled

P-39 fighters could have been transported in their crates by barge, and they would not have had all the mud and rain to contend with. They set up temporary living quarters in two-man tents in a sugarcane field next to a big farmhouse and later moved closer to the airfield into four-man "bures," the Fijian word for "huts," with wood sides and thatched roofs.

Now that all of the P-39s were ready, the squadron began flying drills in earnest. Vic devised a regimen whereby his men practiced a wide range of skills, from instrument flying to acrobatics, gunnery, and dive and skip bombing. The plane's weaponry featured two .30-caliber machine guns on top of the nose, synchronized to fire through the propeller blades, two .50-caliber machine guns in each wing, and a 37 mm cannon located over the drive shaft between the pilot's legs. Lacking actual gunnery targets, Vic got creative and scrounged materials from navy boats offshore that he threw together to make targets. He positioned the improvised marks on reefs and had the pilots practice dive-bombing at them. Through the exercises, the fliers found that the P-39 was at its best in a ground attack.

The squadron was split into three "flights," or groups—A, B, and C—that rotated the use of the planes. Mitch, in charge of training all the men, directly oversaw B Flight, which included Rex Barber and Doug Canning. Most of the men gave up shaving, and with Mitch growing a mustache the flight was nicknamed the "Bearded Bastards of B Flight." In drills, Mitch championed boldness in combat, a go-for-it flying philosophy that stressed a readiness to strike over fearing failure and playing it too safe. He saw a critical difference between having a winning attitude and playing defensively, a distinction reminiscent of his basketball days as a player-coach. He always believed that the best outcomes resulted from a joy of winning, not a fear of losing. He drilled the men hard, teaching combat formations and acrobatics. "He was demanding," recalled Doug Canning, "and really put us through the hoops."

During training, Mitch got good news: a promotion to captain was in the works. Likewise, Vic heard he was soon to be a major. More weeks passed before the promotions became official, and when they did the men of B Flight made sure to celebrate. That meant heading to nearby Lautoka, which had fast become their go-to place to drink and socialize. The town, fifteen miles north of Nadi, was second to Suva as Fiji's largest port and was known as Sugar City given its location at the center of the island's sugarcane region. The men made do with New Zealand beer since no bar or hotel lounge could get any beer from the United States. They befriended New Zealanders who managed the Sugar Mill, the largest sugarcane operation on the island. The locals hosted Mitch and others for dinners and opened their tennis court to them. Mitch was one who jumped at the chance to play. His game was rusty, but tennis gave him something athletic to do when he wasn't flying.

During his downtime, Mitch ventured into a nearby village as well. He no longer expected to face spear-carrying flesh eaters. The grotesque practice belonged to a distant past, "completely eradicated," he theorized, by "missionaries who came here years ago to stop the cannibalism then going on." He was told that it was now safe to visit any of the villages. Joining Mitch were a couple of other pilots, including a Texas native Mitch had gotten to know. Lieutenant Wallace L. Dinn was from Corpus Christi on the Gulf of Mexico, equidistant from Annie Lee's hometown, El Campo, and San Antonio. It seemed that Mitch had a thing for Texans—not just his wife and now Dinn but also Ellery Gross, his buddy on the stint in England who had crashed the previous fall. Mitch and Wallace Dinn were both twenty-eight, born a month apart. He found Dinn easy to like: lean, tall, and friendly, a Gary Cooper look-alike who claimed as a great-uncle the folk hero "Bigfoot Wallace," the nineteenth-century Texas Ranger who had battled Indians and bandits on the Texas frontier. Mitch wasn't sure whether that was true or not; either way, he saw in Dinn the nervy flying attitude he was trying to instill in all the pilots he was training.

They arrived by jeep over a dirt road, drove into the middle of the village, and parked next to one of the fancier huts, which proved to be the chief's home. The village was one of the prettiest Mitch had ever seen. It overlooked the ocean to the west, with open fields of high grasses to the north and south. Coconut and breadfruit trees were all around. The chief, dressed in a dark shirt and bright skirt, emerged from his hut—hard muscled, at least six feet, four inches tall, and weighing about 210 pounds. Stately and somewhat reserved, he greeted them graciously. Mitch, Dinn, and the others climbed out of the jeep to look around. The villagers yelled, giggled, and clapped their hands, which Mitch took to mean that they were happy that Americans had come to visit. The villagers reminded him of the blacks in Mississippi. "Black as the ace of spades," he wrote in his diary—an observation, however benign, imbued with cultural racism. His stab at field anthropology continued: "Very bushy kinky hair just like the southern negro. Happy go-lucky, singing a lot and working and worrying but very little." The Fijians spoke in a "native tongue" Mitch couldn't follow but also in the King's English as a result of foreign influences. After what Mitch described as "the customary Boolas," or greetings, the chief gave them a tour, showing off a one-room schoolhouse with an attendance board listing sixty-three names of children. In his hut—unlike the others in that it was longer and had large mats on the floor—the chief offered them English tea in china cups. The chief's daughter served and fanned flies away with a palmetto fan. Mitch thought it "most odd that we should be sitting in such primitive surroundings and drinking tea in this manner." The Fijians sang native songs, which Mitch found soothing, as they sounded like church hymns. He speculated that they'd learned the tunes from missionaries. The chief admired their jeep, so before departing, the Americans gave the village leader and some other high-ranking Fijians a ride. They were thrilled, and all ended well, with the chief extending an open invitation to come back. They did, and, having learned that the natives were crazy for cigarettes, Dinn next time took a carton of Camels along.

Mitch was able to share all of that with Annie Lee because, finally, after more than two months, the mail kicked in. He began writing to her on March 17, and ten days later, on the morning of March 28, a small mailbag arrived from the States. "When I got to it, after fighting off about half the squadron, I found three letters from my honey," Mitch told her. When a second bag arrived later the same day, it held nineteen more. Annie Lee had smartly numbered each so that he could track the letters sequentially and notice if one was delayed or didn't arrive. As it was his day off, the timing was ideal. Mitch spent the afternoon reading and rereading her letters, like a novel he could not put down. She had put a touch of her perfume on some of them, which sent Mitch's head spinning. "The perfume really brought back some beautiful memories," he said. She included the photograph taken on the street outside City Hall on their way to get a marriage license. "It's certainly a candid shot," Mitch said. "I didn't know people looked so serious when they were fixing to get spliced."

From then on they could correspond, with occasional interruptions. In one letter, Mitch apologized for the incompatibility of marriage and war. "You have been cheated out of a lot," he said, "as we didn't have an opportunity to take our honeymoon, only lived together less than three weeks, in which time you had to adjust yourself to entirely different surroundings, married life etc., and I usually came home tired, didn't have a car etc., and now that I think back upon it I'm rather surprised you didn't throw me out or something. It's you who are the brave soldier." In their exchanges, Mitch learned that Annie Lee was keeping busy. She was riding a bike for exercise, learning to drive, and taking up bowling. He told her about his promotion to captain and began sending home money for her to buy defense bonds and to save for their future family. She began sending him stacks of magazines to read. She told him she was looking for work, but he insisted that she take only a job she liked. In all, he found comfort in her reports. "You just don't know how good it makes me feel to know that you are so well situated in San Antonio," he said. "I know you have a nice apartment in a nice neighborhood. I know that you have lots of friends around you and that Aunt Golda and Uncle Joe are always on hand to assist you in any way."

MITCH WASN'T FEELING SO GOOD ABOUT THE SITUATION IN NADI, though. He'd grown restless as March flowed into April and April into May. The squadron was ready as far as he was concerned—well prepared, alert, and on the ball. The problem? "No action, no bombs and no enemy." The Pacific blues were setting in, "one day just the same as the next, and they just roll along." He and his men kept at it, training and exercising, trying to keep from getting into a rut, but with Japan seeming to be having its way in the war, Mitch was edgy. "All of us are anxious to get onto something." One rumor had it that the Japanese Imperial Navy's southwestern drive in the Pacific would next target Port Moresby on New Guinea. In fact, unknown to Mitch and the Americans, the Imperial Navy's commander in chief, Admiral Isoroku Yamamoto, was indeed crafting a plan to capture the island to use as a base to launch an attack on Fiji and Samoa. There was another rumor that Japan was mobilizing to attack New Caledonia, just 840 miles west of Fiji. That island was prized for minerals Japan lacked (mainly nickel and chrome) and also because it was within striking distance of Australia and New Zealand—900 and 1,000 miles, respectively.

Then came a moment when Mitch figured he was going to kill his "first Jap." It was a Sunday night. He was relaxing playing poker, and a call came over the radio that a flight of planes had been spotted 130 miles away. Mitch threw his cards down and raced from his quarters. "I tore down the line, grabbed my equipment, untied my plane and was warming her up." But Vic reached him before he took off: the call had been a false alarm. Later in the month, there seemed to be another chance. Near dawn one morning, the siren went off, signaling enemy planes. Everybody ran to the field. Mitch climbed into the cockpit of a P-39. He led four planes up and scoured the sky for Japanese planes but found only a B-24 Liberator bomber that had strayed off course while returning to the States. "It's a waiting game," he complained to Annie Lee. "I can't help but wish for action—we all do."

Mitch certainly heard about the fighting elsewhere. "We receive news flashes here and hear news reports on the radio so we keep up with things pretty well," he said. Word had spread quickly about the surprise bombing raid that Army Air Forces Lieutenant Colonel Jimmy Doolittle had commanded on April 18. Sixteen B-25 bombers— "Doolittle's Tokyo Raiders," as they were later nicknamed—flying off of an aircraft carrier off the coast of Japan, had hit targets in Tokyo, Yokohama, Osaka, and Nagoya and then headed for China as they ran out of fuel. The raid's shock value was immense, as the Japanese people awoke to a new reality—that the mainland was not invincible, despite their military leaders' insistence that it was. In the United States, the message delivered by the 500-pound bombs was uplifting: You attack Pearl Harbor, we bomb Tokyo. Mitch and his army fliers were in awe of the mission's boldness. Said Tom Lanphier, "We were thrilled and inspired by the imagination, guts and skill of the

Air Forces and Navy crews who created and executed that morale-boosting adventure."

It made Mitch eager, hoping he'd be sent closer to the fighting. Stories about the enemy's barbarism added to his fervor. The Pearl Harbor attack was itself Exhibit A; it had been a despicable sneak attack, the product of "Oriental treachery." Then, as Japan swept through the Pacific, reports came of the horrific treatment of US prisoners. The surrender of Allied forces in the Philippines in early April after months of fighting was followed by the sickening brutality of the Bataan Death March. With more than 20,000 Americans among the 76,000 prisoners, the forced sixty-five-mile trek to a detention camp was a crime against humanity; the heat, starvation, and beatings along the way and the fact that POWs who couldn't keep up were bayoneted, shot, and even beheaded by Japanese soldiers. The Doolittle Raid had been uplifting to the nation, but the treatment of the eight pilots captured was nothing short of a wartime atrocity. Their torture included finger crushing, knee hyperextension, and the "water cure," in which a POW was forced to lie on his back, his hands pinned to the ground and his mouth forced open with a round stick. Then a bucket of water was poured into his mouth, followed by another at his captors' pleasure, as the POW gagged, gasped, and was utterly helpless, feeling as though he was drowning. Just as sickening, the Japanese army, slashing through the China countryside in search of Doolittle's pilots, had killed thousands upon thousands of Chinese who were even remotely suspected of aiding the Americans.

Mitch began repeatedly telling Annie Lee that he wanted a go at the Japanese. "We are ready for anything," he said. Unlike Japan's top naval leader, Isoroku Yamamoto, who had traveled to the United States and mingled for years with Americans, Mitch had never been to what was now the homeland of the enemy and had never known a Japanese person. And when President Roosevelt on February 19 approved the internment of 120,000 Japanese Americans living on the West Coast, he got no argument from John Mitchell. Like many Americans, Mitch saw the enemy as monstrous, so that when Annie Lee voiced alarm in June at Mitch's craving for combat, he practically snapped at her. "You say you disagree with me about my screaming for action of some kind," he said in a harangue. But, he continued, sitting around on alert for five months, just waiting, was driving him nuts. "Occasionally I catch myself jumping down someone's throat for practically nothing." Japan's Pacific advance, he said, had to be halted, and he was fed up hearing about the fighting elsewhere while he sat on the sidelines. "It makes me powerful anxious to get my share of those yellow devils."

CHAPTER 8

UNFINISHED BUSINESS

Yamamoto on a US propaganda poster
WWII Database/public domain

EARLY ONE EVENING IN LATE FEBRUARY 1942, AN OIL FIELD WORKER by the name of G. O. Brown spotted a Japanese submarine surfacing off the coast of Santa Barbara, California. Though it was a mile offshore, Brown was taken aback by the size of the dark mass as it floated idly atop the ocean. "It was so big I thought it might be a destroyer or cruiser," he later told a *Los Angeles Times* reporter. "I have seen many submarines and this was larger than any of those in the United States Navy." It was as if Brown was describing the Moby Dick of subs—a frighteningly huge and dangerous ship the likes of which no one had ever seen before. Brown watched helplessly for the next thirty minutes as the submarine's deck guns fired a dozen or so five-inch shells at the Ellwood oil refinery near the seaside community north of Los Angeles. None of the oil wells was hit, and many of the shells fell into a nearby pasture where horses were grazing. "They went mad," Brown said. "Shells were exploding in their pasture and the horses screamed and raced about."

No one was hurt and the damage was minor, but the attack further fueled a skittish nation's fears that the Japanese were mighty fighters possessing stunningly powerful weaponry—giant submarines, no less. (In reality, Japanese submarines, although differing in certain features, were comparable to US subs.) Ever since Pearl Harbor, any movement on the high seas off the West Coast—ranging from fishing boats to floating logs to sharks and whales—was frequently mistaken for attacking Japanese ships. And Secretary of State Henry Stimson warned the anxious citizenry that it must buckle up and brace for "occasional blows" from the enemy.

Then, just thirty-six hours after a Japanese sub had actually attacked Santa Barbara and put all of southern California on edge, the skies over Los Angeles were lit up. Trigger-happy coastal artillerymen pumped more than 1,400 rounds of antiaircraft fire at what they thought were incoming Japanese bombers. Sirens blared as the region went into a blackout. For the next hour, searchlights sliced up the night sky while US gunfire burst in midair. One artilleryman later said he had spotted six planes, while another said he had seen two dozen. Civilians reported seeing Japanese paratroopers descending from the sky, and there was even a story of a damaged Japanese fighter plane crashing on a Hollywood street.

The city was under siege—except that it wasn't. There were no enemy bombers, no paratroopers. In the light of day, it was determined that there had been no attack. Secretary of the Navy Frank Knox concluded that jittery nerves had resulted in a false alarm. It was also revealed that weather balloons had been released earlier and, seen on radar, had likely been mistaken for planes. The mirage attack was not without consequences, however. Five people died of heart attacks during the blackout, while shrapnel and unexploded shells that had fallen to earth broke windows, damaged homes, and caused multiple injuries, including slicing a six-inch cut into one man's head. Further, local police had detained more than fifteen Japanese Americans believed to be signaling enemy aircraft using flares or other lights.

The enemy was everywhere and nowhere. Rage and paranoia had erupted in the aftermath of Pearl Harbor. On the Hawaiian island of Kauai, a manhunt for a Japanese pilot who'd put down there after the December 7 aerial attack resulted in the burning of residents' homes and the apprehension and even the shooting of Hawaiians suspected of shielding the pilot. On December 8, US attorney general Francis Biddle, citing national security, had directed the FBI to round up 737 Japanese Americans in California who had been determined to be "enemy aliens"—the first roundup of many more to come. Japanese all over the country, many of whom were citizens and had been born in the United States, were targeted as possible spies feeding information to the enemy or, worse, as saboteurs and terrorists. "The Japs live like rats, breed like rats and act like rats," Chase Clark, Idaho's governor, said. Racist stereotyping abounded. In a stab at grisly satire, quasi-official-looking "Japanese hunting licenses" began circulating, while *Time* magazine tried a more serious means of being helpful: Its editors published a handy guide titled "How to Tell Your Friends from the Japs," containing ten tips for patriots to differentiate Japanese from Chinese immigrants. Japanese men were said to walk "stiffly erect," for example, while Chinese men were more relaxed. Facial expressions could be especially revealing: "Chinese expression is likely to be more placid, kindly, open; Japanese more positive, dogmatic, arrogant."

Compounding the fear, the submarine attack of February 23 revved into overdrive the removal of Japanese living on the West Coast that President Roosevelt had authorized the prior week—an executive order the US Supreme Court would uphold as constitutional. Families up and down the coast were forcibly removed from their homes—120,000 people in all, taking with them only what they could carry—and taken to internment camps that were being hastily constructed. Their new "homes" would consist of barracks-style structures surrounded by barbed-wire fences and towers manned with armed guards. "Don't kid yourselves and don't let someone tell you there are good Japs," California congressman A. J. Elliott told his House colleagues in a speech following the attack. "We must remove the Japanese in this country into a concentration camp somewhere, some place, and do it damn quickly."

For all the widespread fear of the Japanese everyman, first and foremost the face of Japanese evil became that of Admiral Isoroku Yamamoto. Until Pearl Harbor, few outside of US military and government circles had known of him, even though he'd spent considerable time in the United States, most recently traveling cross-country by train en route to the London disarmament talks in 1934. He had lost that anonymity virtually overnight. Once the Japanese press released for domestic propaganda purposes the falsified Yamamoto letter that had him boldly predicting total victory, the American press pounced on his alleged inflammatory statements. "'I'll Capture White House,' Jap Admiral Bragged Year Ago," read a *Washington Post* headline on December 17, 1941. For most Americans the news coverage was a first look at the Japanese admiral who'd devised the Pearl Harbor attack—a preening warrior making jingoist pronouncements. "I am looking forward to dictating peace in the United States in the White House in Washington," the *Post* story quoted him as saying in its second paragraph. From that point on, the quotation—or some form of it—was attached to Yamamoto like a middle name whenever he was mentioned in the war's press coverage.

Time magazine made Yamamoto its top story in the pre-Christmas issue of 1941. The demonic sketch of Yamamoto on the cover was more fitting for Halloween, however: yellow skinned; dark slits for eyes; misshapen head; lips pursed and turned downward, as if readying to spit; cannon looming over each shoulder, aimed directly at readers. Every man, woman, and child could study the repulsive-looking face of the foreign fiend who'd promised to take the White House and imagine a bull's-eye in the center of the caricature. Yamamoto was, as the caption read, "Japan's Aggressor," the evil mastermind of Pearl Harbor, "a hard-bitten professional with a sixth-sense—hatred."

Then came the major profile of Yamamoto in prestigious *Harper's Magazine* by Willard Price, the writer who had dusted off his notes from his meeting with the young Yamamoto in 1915 to produce a seething, bigoted portrait that built upon the *Time* magazine account. Yamamoto, wrote Price, was a "hard chunk of man, hair cropped as short as the bristles on a beaver-tail cactus, lips thick, jowl heavy." He had been a hater since boyhood with a "conquest complex"; according to Price, his principal motivation in life was to hate and destroy the United States.

Yamamoto was the United States' archvillain. He'd exploded into the national consciousness in the weeks and months following the Pearl Harbor attack. As one historian noted, "For no other enemy, not even Hitler, did Americans hold such a bitter hatred. Yamamoto was the man who had planned the treacherous blow at Pearl Harbor. And as if this were not enough, he had added insult to injury by boasting that he planned to dictate peace in the White House. To all Americans he was a peculiarly personal foe."

MEANWHILE, IN JAPAN DURING THE REMAINDER OF DECEMBER 1941 and continuing into the new year, Yamamoto was applauded, honored, and idolized. He'd become a supreme national hero. The smashing success of the Pearl Harbor raid, followed by the string of quick victories elsewhere in the Pacific, had elevated him to the status of the nation's foremost naval icon Heihachiro Togo, a forerunner as commander in chief of the Combined Fleet, who had decisively defeated the larger and more powerful Russia in the Russo-Japanese War at the start of the century. The raw numbers told the story: the US Navy had lost 21 ships at Pearl Harbor, 7 of which had been battleships either sunk or seriously damaged. Ninety-two of the Navy's 169 planes had been wrecked; another 39 needed significant repair. The Army Air Forces' three major installations had been ravaged, with another hundred planes allocated to the Hawaiian Air Force also ruined. The total count of American dead and wounded had come to more than 3,400. In contrast, the Japanese losses had turned out to be minimal—and much lower than Yamamoto's planners had predicted. Just 29 of the strike force's 353 planes had been downed and one submarine and five midget submarines lost. Going in, Yamamoto had been prepared to lose half of the mission fleet's six aircraft carriers, but all had remained unharmed. In terms of fatalities, the Japanese had lost fifty-five airmen and nine submarine crewmen. In a way, the admiral had achieved by brute force an end to the despised 5:5:3 ratio for naval armament that he'd failed to obtain as chief delegate to the tri-power talks in London in 1934. The new calculus showed not just parity but actual superiority for Japan's Imperial Navy. To the Japanese public, Yamamoto was invincible.

He was the hate-filled warmonger in the United States and the indomitable naval patriot in Japan, but neither caricature truly captured Yamamoto. He was, above all else, a reluctant warrior. But in Japan the news coverage was gushing day after day, reporting the fall of Manila in the Philippines on January 4, 1942; the paratrooper assault on Celebes, an Indonesian island in the western Pacific, on January 11; the fall of Singapore on February 15. The last was heralded as a huge victory and led to yet another rescript, or official accolade, for Yamamoto, this one including a banner the empress had embroidered picturing an airplane against the rising sun. In late February, he received word of the Imperial Navy's success in the Battle of the Java Sea, a prelude to the invasion of the Dutch colony. Over several days of fighting, enemy Allied forces had lost three destroyers and two light cruisers, including the USS *Langley*, a carrier transporting a group of veteran pilots trying to reach Java to man P-40 Warhawks to defend the island. But they had never made it. They had been among the roughly 2,300 men killed by the Japanese during the fighting. Yamamoto would not have any particular interest in thirty-odd pilots, but a squadron now stationed on Fiji did. The dead were the pilots who'd left San Francisco in December ahead of the less experienced fliers in the squadron that was commanded by Major Henry "Vic" Viccellio and Captain John Mitchell. The flyboys on Fiji were shaken. Vic and Mitch knew it could have been them on the *Langley* and were left to think "There but for the grace of God go I."

Japanese radio stations began playing music to accompany the nationalistic news updates—rousing military pieces such as "Battleship March," the official march song of the Imperial Navy. The patriotic song was the equivalent of the "Navy Hymn" in the United States. In January, when the text of the Imperial Declaration of War was broadcast again from the Yoyogi parade grounds in Tokyo, as the emperor reviewed troops, the nation listened with undivided attention. "The town was filled with rising-sun flags," one woman wrote in her diary later that day, "and the sky was clear. We passed a day with deep gratitude." Yamamoto continued to receive a steady stream of adoring fan mail aboard his flagship, so much so that his aides, knowing their commander's preferences, began to prioritize the stack so he'd find on top anything arriving from his favorite and most cherished correspondent—his mistress, Chiyoko. "When your letter reached me to make me so happy I wrote *kwu kwu* [bell of fawn]," he said affectionately in one of their frequent exchanges.

With few exceptions, the war continued going great guns for Japan during the winter of 1942. Its forces chased General Douglas MacArthur from the Philippines in March and then by May had overrun the remaining US Army troops there. Military leaders began talking up new and ever-more-ambitious actions: Go for India? Australia? Hawaii? Maybe Alaska? Why not? In one strike after another, Japan had defeated every European colony in Southeast Asia. By springtime the Imperial Empire in the Pacific had bulged so that it now extended east from China to Wake Island and south from the Aleutian Islands near Alaska to Indonesia atop Australia. With the Imperial Navy flexing its muscle, many of Yamamoto's officers urged more, as the national government had succumbed to what later was labeled a "victory disease."

Yamamoto had a different perspective. "It is easy to open hostilities, but difficult to conclude them," he told a friend. Others may have begun to see the world as their oyster, but Yamamoto's position had not changed; the war's aim, he felt, should not be to take over the world but to secure a Pacific cordon protecting Japan's regional interests. The surprise Pearl Harbor attack had not been intended to start a slugfest continuing until the very last man was standing but to stun a United States caught off guard into settling for an early peace. He was therefore deeply dismayed that the political leaders in Tokyo seemed more interested in making new conquests than in pushing for peace talks through diplomacy. Well before the December 7 attack, he'd predicted that "for a while we'll have everything our own way, stretching out in every direction like an octopus spreading its tentacles." But he'd also predicted that Japan's good fortune would last a year at best, and "we've just got to get a peace agreement by then."

Key tactical shortcomings had come into focus in the raid's aftermath. By chance, the three US aircraft carriers based at Pearl Harbor—the *Enterprise*, the *Lexington*, and the *Saratoga*—had not been in their berths at the time of the attack. That meant that even though Yamamoto's strike force had ravaged several hundred thousand tons' worth of US naval shipping, the critical pieces of the US fleet—the aircraft carriers—remained at large. In addition, the strike force, which had focused on ships and planes, had not gone after the base's repair facilities, power stations, and above-ground fuel storage tanks. Had that been done, ship repair at Pearl Harbor would not have been possible. Tugboats would have had to tow the damaged US fleet to California, a procedure that would have consumed months and further delayed a US rebound. Even if some ships could have been fixed at the Hawaiian base, there would have been no fuel to power them had the oil tanks been destroyed. In a sense, the attack force had gone after the tools—the ships—rather than the basis—the support systems—of the navy's sea power. It meant that the US Pacific Fleet's recovery could happen faster, especially since six of its eight battleships had not been entirely wrecked. Finally, even the timing of the attack could have been such to make it more damaging. Though on Sunday a plethora of battleships and destroyers was generally on hand, a majority of the men were not; most were on weekend leave. If the strike force had hit on another day, the casualties would likely have soared well beyond several thousand dead and injured.

The way Yamamoto saw things, there was unfinished business that too many people were ignoring while caught up in the hoopla of war. In a letter to his older sister in hometown Nagaoka he confided, "The war has begun at last, but in spite of all the clamor that is going on we could lose it. I can only do my best." Not helping matters was the way the war was being hyped by Japan's politicians and militarists, as well as by the press, which was under their control. "All they need do really is quietly let people know the truth," he said. "There's no need to bang the big drum." The wartime propaganda, which included the distortion of his letter so that it contained his now-infamous promise to march into the White House to dictate peace, disgusted him. "All this talk of guiding public opinion and maintaining the national morale is so much empty puff," he complained to his staff. "Official reports should stick to the absolute truth—once you start lying, the war's as good as lost."

YAMAMOTO SENSED, TOO, FROM A MIX OF PRESS ACCOUNTS AND military intelligence, that the United States was waking up—that following the deadly Pearl Harbor debacle, instead of curling up into a fetal position, she was climbing to her feet, raring to fight and seek vengeance. Yamamoto had no way of knowing its full extent, but the winter of 1942 saw the United States hastily and effectively establish its wartime footing. "No matter how long it may take us to overcome this premeditated invasion," President Roosevelt had said the day after the raid, "the American people in their righteous might will win through to absolute victory." He'd already initiated the first peacetime draft in the fall of 1940, and by the time of Pearl Harbor the active military personnel—airmen, seamen, soldiers, and marines—totaled nearly 2.2 million. Then that winter came the push to expand the military on every possible front while overhauling the US economy and changing the nation's way of life to accommodate a global war. The president said that fifty thousand new aircraft were needed annually—at first a shocking number to many in Congress—yet within a few years more than double that number had been produced. Likewise with ships, tanks, submarines, and guns—all the myriad weapons of war were needed in great quantities and as fast as possible. Americans went to work, and they also began rationing and recycling—salvaging scrap metal from old pots and pipes, bed frames, and beat-up cars to help meet the war machine's insatiable need for steel and aluminum. "The average American cannot buy a new car, or a new tire, or a reconditioned old tire, or be sure of repairs of those he has," observed the journalist Arthur Krock, the Pulitzer Prize–winning columnist at the *New York Times* whose "In the Nation" column was a must read. "He cannot repair his broken plumbing with copper. Steel, his own contribution to industrial science, is denied him. In the six months since Pearl Harbor the national spirit has shown itself ready and able to counter the hardest blows of destiny. The sacrifices made thus far, and the cheerful manner in which they have been made, are impressive."

To pay for it all, federal defense spending skyrocketed—from $1.5 billion in 1940 to $81.5 billion by 1945. And to subsidize the increase, the government began to sell "war bonds" to the public, even utilizing the demonic Yamamoto to promote sales. "The Knight of the Double Cross!" barked the headline of one such advertisement beneath a cartoonish portrayal of the treacherous Japanese admiral: thick lipped, with hanging jowls, looking rodentlike. The ad continued, "There's only one effective way to fight this type of Skunk. Take a good firm grip on your nose, haul out your checkbook, and lay your money on the line: BUY WAR BONDS." Most Americans loyally responded to the plea—indeed, the Pacific fighter pilot John Mitchell, like so many others away at war, began sending money home with instructions to buy as many as possible. "We should have a right nice nest egg when this is all over," he told Annie Lee that April about their wartime investment in their country.

For Yamamoto, there was a feeling of urgency, and amid the overall success of the war's initial phase, he spent the winter considering his next operational moves. "The 'first stage of operations' has been a kind of children's hour, and will soon be over," he wrote a friend. "Now comes the adults' hour, so perhaps I'd better stop dozing and bestir myself." Some navy leaders argued for advancing westward across the Indian Ocean; others thought driving south would be the best idea—to capture Fiji and then Noumea in New Caledonia and, by doing so, cut Australia off from the United States. But by late winter, Yamamoto had his mind set on an altogether different target—Midway Island—and assigned trusted staffers to devise a plan of attack. As the Japanese had taken over the US islands of Guam and Wake within a few days of the Pearl Harbor raid, the tiny atoll of Midway—two small islands, actually—was now the closest US military base to Japan, at 2,800 miles away. It was a decent-sized one, with an airfield and a naval port providing anchorage to submarines and aircraft carriers. Yamamoto favored Midway for several reasons. For one thing, by eliminating the nearest enemy outpost, he'd be expanding Japan's defensive perimeter in that region. More important, he saw a sea battle for Midway as a way to complete the leftover work from Pearl Harbor, decimating the remains of US naval power in the Pacific, the aircraft carriers and other ships that had gotten away. He knew that for the United States, the stakes were sky high. Midway was just a thousand miles west of Hawaii; the Americans would fear that if they lost Midway, Japan would use it as a stepping-stone to attack a battered Hawaii and, from there, use Hawaii as a base to attack the West Coast of the United States. But conquest of that sort was not Yamamoto's primary intent; he was still preoccupied with his original Pearl Harbor logic—of winning an early peace, a goal now more urgent than ever given the enemy's massive military buildup. The way he saw it, Midway could be the tipping point. Crushing the US Navy might induce the United States to settle for peace with Japan. He'd achieve the early finish to war that had been his aim for Pearl Harbor but that had proved elusive, even though his forces had destroyed more US ships in that raid than had been lost by all other nations up to that point in the war. Yamamoto even confided to a close underling, Mitsuo Fuchida, the navy captain who had led the first wave of bombing attacks on Pearl Harbor, that that was the purpose of the Midway operation. Once completed, he promised to "press the nation's political leaders to initiate overtures for peace."

The plan Yamamoto and his staff drew up was initially met with opposition, mainly from the army but also from a navy general staff troubled by its bold, offensive orientation, a departure from the traditional naval strategy of having the fleet stick close to home and await an enemy. Yamamoto spelled out his rationale forcefully, arguing that Japan's success in the Pacific would hinge on finishing what it had started at Pearl Harbor, "destroying the United States fleet, especially its carrier task forces." The Midway plan, not any of the other proposals under consideration, would accomplish that goal. "The proposed operation against Midway will draw out the enemy's carriers and destroy them in decisive battle." If it turned out that the enemy did not take the bait and engage its fleet in the so-called decisive battle, Japan would still come away having taken over Midway, a key Pacific asset.

The haggling continued throughout March but ended quickly

after the unexpected US bombing attack on Tokyo and other locations—the Doolittle Raid of sixteen B-25 bombers on April 18. The nation was shocked and shaken by the attack. Yamamoto wrote, "Even though there wasn't much damage, it's a disgrace that the skies over the imperial capital should have been defiled without a single enemy plane shot down." He was so unnerved by the news of the raid that he spent the day alone in his cabin taking stock. He knew that the US bombers had flown from an aircraft carrier—which would have been impossible had the US carriers been at Pearl Harbor during the Japanese raid instead of at sea. The navy general staff and army leaders, suddenly worrying about the security of the homeland's borders, gave the green light to Yamamoto's Midway plan.

Yamamoto and his staff began working toward an early June date to attack Midway, rehearsing the ways their ships would engage the residuum of the US fleet. In the meantime, the admiral kept up with several other actions in the Pacific that were focused to the northeast of Australia, including an occupation of one of the Solomon Islands—Guadalcanal—to construct an air base there. But the action Yamamoto was monitoring most closely was Operation MO, the invasion of Port Moresby on the south coast of New Guinea that was planned for early May. After taking Port Moresby, Japanese bombers and fighter planes would have only a short flight across the Coral Sea to Australia. The idea was for an occupying force of troops to take over the island, protected by ships from the Imperial Fleet in the surrounding Coral Sea. Yamamoto had committed two large aircraft carriers loaded with highly trained navy fighter pilots to be part of the covering force.

Operation MO did not unfold as expected, however. The Japanese were met by an Allied task force consisting mainly of US and Australian ships and planes. Japan's troop transports were bombed relentlessly by US planes from nearby aircraft carriers, and then for two days, May 7 and May 8, the navies squared off on the high seas. The confrontation became known as the Battle of the Coral Sea, and it marked the first air-sea battle ever, meaning that the warring ships never saw or fired on each other. The fighting involved aerial strikes from the respective carriers. The US resistance proved so unexpectedly fierce that the Japanese commander pulled back, and Yamamoto, realizing that his carriers could not provide the necessary cover for an occupation force, called off the invasion on May 18. They'd have to try again another time.

Yamamoto and his commanders were flummoxed. Tactically speaking, Japan had come out on top. Its planes had sunk the US carrier *Lexington*, heavily damaged the carrier *Yorktown*, sunk a destroyer, and damaged a fleet oiler. But Japan's invasion had been rebuffed, the country's first setback in the Pacific. Yamamoto had lost the small aircraft carrier *Shoho*, which had sunk during the fighting, and, more important, the larger carrier *Shokaku*, so badly damaged that it limped into dry dock in Japan. Lost, too, were forty-three aircraft and their crews, most of whom were seasoned—flying experience that was difficult to replace. They could no longer be part of the equation in planning for the upcoming Midway operation.

Yamamoto was not happy. The turn in events heightened the importance of Midway. He was going to direct the attack himself and be completely hands on, unprecedentedly for him. And as he turned his full attention to finalizing the preparations, one had to wonder. The enemy had seemed ready and waiting in the Coral Sea. Had they somehow known the Imperial Navy was coming?

CHAPTER 9

MIDWAY: YAMAMOTO'S LAMENT

Japanese heavy cruiser *Mikuma* burning at Midway
National Archives and Records Administration

THE COMMANDER IN CHIEF OF THE US PACIFIC FLEET, NAVY ADMI-
ral Chester W. Nimitz, did in fact know plenty about Admiral
Yamamoto's intended May 1942 move on Port Moresby, thanks to intelligence provided largely by US code breakers stationed in Hawaii.
For weeks and with increasing specificity, the team had decrypted

Japanese naval transmissions indicating that warships and transport ships filled with troops were assembling at the Imperial Navy's main base at Truk Lagoon in the central Pacific and that the force was preparing to head south to the Coral Sea. Many of the initial messages contained the sign "RZP," which code breakers had previously established was a geographical designation for Port Moresby. But cryptanalysis is never a straight line. On April 24, a new term appeared in the Japanese transmissions: "MO." The term was used in several contexts, as in "MO fleet," or "MO occupation force." The code breakers were stymied, uncertain if the term was connected to Port Moresby or referred to another operation altogether.

The unit's chief, Commander Joe Rochefort, put the mystery term into the hands of a pair of star linguist-cryptographers—Major Alva "Red" Lasswell and Lieutenant Commander Joseph Finnegan. Finnegan insisted that "MO" meant Port Moresby, not to be confused with its geographical code name (which was RZP) but as an operational code name for the Port Moresby invasion. When Lasswell separately reached the same conclusion, Rochefort felt they had unraveled Admiral Yamamoto's next move. Further confirmation came five days later, on April 29, when Yamamoto's Operation Order No. 1, transmitted to Vice Admiral Shigeyoshi Inoue, was intercepted. Rochefort's men decrypted the order, which read as a pep talk from Yamamoto to the vice admiral charged with leading the Fourth Fleet's imminent conquest of Port Moresby. "The Imperial Navy will operate to its utmost until this is accomplished," Yamamoto wrote. The message left no doubt about the decryption of "MO" as the code name for the invasion of "RZP." The code breakers rushed their intelligence about Port Moresby to Nimitz, who was able to plan accordingly. Yamamoto's Imperial Navy was anticipating a surprise attack in early May 1942—and yet another Pacific triumph—but was instead the one surprised. Nimitz's forces fought Yamamoto's to a standstill on May 7 and May 8 in the Battle of the Coral Sea.

SINCE THE PEARL HARBOR ATTACK THE PREVIOUS DECEMBER, RED Lasswell had emerged as one of the most valuable and productive talents in the decrypt unit known as Station Hypo. For all of his linguistic skill and affinity for numbers, the lanky, sandy-haired

Lasswell, who had grown up in rural Piggott, Arkansas, had never earned so much as a high school diploma. The lone marine in a code-breaking unit largely made up of naval officers, he'd had to take a correspondence course in 1925 to gain acceptance into the Marine Corps's Officer Candidate School. Now, at Station Hypo, he had established himself as a measured and precise analyst, a workaholic among workaholics who always wore a green eyeshade to block the glare of fluorescent lighting and smoked either Cuban cigars or a pipe as he huddled over his work at a tidy gray metal desk. Unit commander Joe Rochefort liked pairing Lasswell with fellow Hypo linguist Joe Finnegan, an intense chain smoker from Boston whose desk was usually piled high with intelligence notes, a mess only he could fathom. Finnegan was more freewheeling in his approach to cracking a Japanese message, relying on intuition and an infallible memory for retrieving data from prior transmissions. Rochefort saw Finnegan as complementary to Lasswell, the yin to his yang.

They worked in a windowless room in the basement of a nondescript navy administration building at Pearl Harbor nicknamed "The Dungeon." Rochefort's desk was at one end, the way a teacher's desk is positioned in front of a classroom. To his right were the traffic analysts, whose job it was to study the "externals" of radio transmissions—the volume, location, and senders—and based on those determine the Japanese fleet's whereabouts. To Rochefort's left sat the language officers Lasswell and Finnegan, while directly in front were desks for the information specialists with tables of maps, files, and books. Seated at the other end were the cryptanalysts, whose uncanny mathematical skills were key to deciphering coded groups of numbers into text that Lasswell, Finnegan, and their staff then translated into English. Though the cryptanalysts and translators worked hand in hand, Rochefort considered the latter to have the most essential function. "You can assign values and all that sort of thing," he said later about the important first step performed by cryptanalysts, "but unless you do a good job of translating the whole [effort] is lost."

Station Hypo was one branch in the navy's cryptologic program, the largest signals intelligence program of any of the US military services. Observing at a distance Japan's rapid naval expansion during the 1930s, prescient US naval leaders had seen a corresponding need to expand their service's intelligence-gathering capability and to exploit the advantage US code breakers had from years of deciphering various codes used by the Imperial Japanese Navy. Radio intercept stations were set up on tiny islands throughout the Pacific, including Hawaii, that, like a net, captured Japanese radio transmissions. Three code-breaking outposts, up and running by decade's end, then worked on deciphering the messages. The unit in Washington, DC, known as OP-20-G, served as headquarters. A second outpost, known as CAST, was in Cavite on the island of Luzon in the Philippines; it would have to relocate to Melbourne, Australia, when the Philippines fell to the Japanese. Station Hypo at Pearl Harbor was the third location. Over time, and especially after the Pearl Harbor attack, intense rivalry flared as the units competed to be the first to decipher messages quickly—and accurately. In addition, throughout the 1930s, the navy sent prospective code breakers to Japan for three-year tours to gain fluency in the language, mingle with Japanese naval officers socially, and develop a sense of the local customs and culture of a possible future enemy. Joe Rochefort, Station Hypo's eventual commander, had participated in the program, as had Joe Finnegan. In the shadow war for intelligence, the navy program would prove itself far superior to its Japanese counterpart.

Red Lasswell had also participated in the navy's language immersion course. His colonel, noting that the program was seeking a marine in the mix, had pushed him to volunteer despite his lack of education credentials. Lasswell had spent 1936 to 1938 in Tokyo, fast becoming an expert in the language and forging a bond with Finnegan during the two years they overlapped. Finnegan, he later said, "was my good right arm during World War II." In the late 1930s, Lasswell followed Finnegan to the outpost in Cavite, and it was there he was "first indoctrinated into the work that later occupied me during the war." By that Lasswell meant he began work in earnest as a code breaker—not just translating but also developing the ability as a cryptanalyst to break coded number groups into words. That was why he and Finnegan became so valued; they were hybrids, both linguists and analysts. In the spring of 1941, Lasswell was transferred to Pearl Harbor to work for Rochefort at Station Hypo, officially known as Fleet Radio Unit Pacific Fleet (FRUPAC).

That fall, US code breakers at Hypo and elsewhere had had a sense that war with Japan was imminent, but no one had detected the coming of Yamamoto's surprise attack on Pearl Harbor on December 7. Rochefort later unblinkingly bemoaned the collective failure, saying "an intelligence officer has one task, one job, one mission. That is to tell his commander, his superior, today, what the Japanese are going to do tomorrow." Lasswell was on overnight duty the morning of the attack. "I was at the office and watched the bombers come in," he said. "I was just as surprised at the moment in which it started as anybody else there."

DURING THE WINTER AND SPRING OF 1942, IT WAS AS IF THE CODE breakers at Station Hypo were determined to make up for Pearl Harbor. Japan's navy was old school in the sense that it favored codebooks over emerging machine technologies. The words and information contained in the books were converted to number groups—each of five digits, of which there were some 50,000 in total—that communication specialists strung together to create a coded message. The coded number groups were then embedded in other groups of numbers, a process called superencipherment, to make it more difficult for Allied code breakers to determine which of the number groups made up the actual message. But during that period, code breakers had penetrated the Imperial Navy's main code, which was called JN-25, short for "25th Japanese naval code." The code-breaking challenge was adjusting to changes the Japanese introduced periodically for security reasons, when they altered the numerical equivalents of words. The change to the latest edition, known as JN-25(b), had not thrown off the Allied code breakers, however. That winter, Rochefort's men at Station Hypo and code breakers at the other Allied outposts had been all over JN-25(b). It was the successful deciphering of JN-25(b) messages that had given them a heads-up in early spring about Yamamoto's bid to attack Port Moresby. And if Yamamoto was surprised by the outcome of the Coral Sea battle, he was in for an even bigger surprise regarding the critical naval operation he was developing and personally planning to direct: invading Midway.

By early May 1942, Japanese naval radio traffic was off the charts, an explosion that practically overwhelmed the US Navy's radio intercept nets throughout the Pacific. The radiomen at the Wahiawa station in central Hawaii, for example, found themselves delivering huge bundles of raw intercepts by courier to Station Hypo at Pearl Harbor, and the arrival of five hundred to a thousand intercepts a day radically altered life in the Dungeon. "From a monastery the place turned into a pressure cooker," one historian noted. In a bid to cope with the surge, Lasswell and Finnegan set up a two-watch system in which one of them would work a twenty-four-hour shift overseeing the less experienced translators and the other would then take over for the next twenty-four hours. That way one of Hypo's star linguist-cryptanalysts was always on duty. Even though they were working around the clock, they still didn't have enough time to process everything coming over their desks. Lasswell set up a cot behind his table. "I would use it to doze until I felt refreshed enough to go back to work." To maintain the brutal work pace, a number of the men began popping amphetamines.

One worrisome discovery was the Japanese navy's plan to implement a new version of its main code, replacing JN-25(b) with JN-25(c), effective May 1, 1942. For the code breakers, the intelligence amounted to good news, bad news. On the one hand, the tightening of communications security suggested that the Imperial Navy was gearing up for something big. On the other hand, until the code breakers managed to crack the new edition of the code, they'd be at a loss to figure out Yamamoto's plans. What Rochefort, Lasswell, Finnegan, and the others were not aware of, however, was the logistical hurdle the Japanese navy faced phasing in its new code—a consequence of its wildly successful expansion throughout the Pacific during the winter. With naval operators scattered all over the Pacific, the distribution of new codebooks was no easy task, so much so that May 1, 1942, came and went and the Japanese navy was still relying on JN-25(b)—and would continue to do so until the end of May, when the new version was finally activated.

Rochefort's men hardly complained about Japan's failure to switch to its new code come May 1; they and other Allied code breakers in Australia and Washington, DC, continued to pile up inside information on Yamamoto's developing operational plans, practically looking "over Yamamoto's shoulder as he moved his forces around the Pacific," as Rochefort's biographer later described the US advantage. In early May, the Hypo code breakers were picking up repeated references to an upcoming campaign. On May 5, they intercepted a message from Yamamoto's fleet to Japan's Navy Ministry to speed up delivery of fueling hoses, which told the Americans that Yamamoto's attack fleet would be needing to refuel at sea. Then, on the afternoon of May 13, the decryption of a message from Yamamoto's 4th Air Attack Squadron to the transport ship *Goshu Maru* provided a vital clue. The small ship, containing equipment to establish an air base and ground crews to service it, was ordered to Saipan, a Japanese stronghold. The ship was told it would join Japanese carriers and warships gathering at Saipan to sail toward "AF."

The analysts recognized AF as a geographic symbol, the same way RZP stood for Port Moresby. They had previously deduced that AF was the designator of Midway Island but had not seen AF in the enemy's radio traffic for at least two months. Other messages pouring in revealed that Yamamoto had assigned the same commander who had overseen the Pearl Harbor attack, Vice Admiral Chuichi Nagumo, to lead this strike force. (The code breakers never learned, however, that Yamamoto himself would be sailing behind Nagumo's carrier fleet aboard his flagship, *Yamato*.) That Japanese ships were meeting at the central Pacific island of Saipan also fit as the most likely location from which to stage an attack on Midway Island. The Hypo cryptanalysts ranked decryptions from A to WAG. The highest, A, ranking meant that they were positive about their analysis— "As having been determined by more than one source and having turned up in radio traffic numerous times, removing any possibility of doubt." The lowest, WAG, ranking stood for "wild assed guess." Rochefort and his men put their assessment of AF at the highest ranking of reliability. They were certain that they'd figured out Yamamoto's next target: Midway Island.

But their rivals at headquarters in Washington—at OP-20-G— didn't buy it. No one at any of the code-breaking outposts disputed that Yamamoto was plotting a major operation, but the code breakers at OP-20-G did not accept that AF stood for Midway Island. The disagreement grew bitter. It wasn't as if Washington could agree on an alternative. Some argued that Yamamoto's target was Hawaii;

others said San Francisco. Some floated the idea that even if AF did stand for Midway, it was part of a hoax Yamamoto had concocted to disguise his true intention: attacking Hawaii or the West Coast. Rochefort and his men, firm in their decryption, considered the opposition idiotic. Tension persisted at the highest levels, and for Admiral Nimitz the stakes on where to deploy his fleet were huge; a mistake could be catastrophic for the Pacific war. "Our fleet was then limping back from the battle of the Coral Sea," Lasswell said, "and Admiral Nimitz had to make fast decisions about repairs to ships, refueling, re-ammunitioning, adding personnel."

Lasswell started a shift on May 19 by going through a stack of raw intercepts freshly delivered to the Dungeon. He had flipped through fifty or so messages considered high priority when he detected what looked to be an important message. Two things jumped out at him. The first was that the message was from Yamamoto. The second was the list of people to whom the message had been sent: it was directed to just about every important Japanese naval command. Lasswell dug in. He worked all night to identify the unknown code groups and the next morning distributed a deciphered message that was nothing short of eye-popping: operational details of the Japanese navy's attack on Midway, including the location of Nagumo's carrier fleet northwest of the island. "I was able to give them the longitude and latitude from which these carriers were going to launch their planes." But once again Washington did not buy Hypo's conclusion that AF was Midway. Lasswell was furious. "I was sure of myself but others thought I was wrong."

What to do? Others at Hypo began working feverishly to further backstop Lasswell's decryption. Meanwhile, Rochefort, under mounting pressure from his superiors to put the controversy to rest, put into play a plan he and a small group at Hypo had devised the previous day. The idea had been Lieutenant Commander Wilfred J. "Jasper" Holmes's. Holmes was an information officer who was an engineer by training. He had never been to Midway Island but knew that drinking water was a constant concern at the navy base there. He suggested that the Midway base transmit an urgent message reporting that the desalinization plant had broken down and a freshwater barge should be sent from Pearl Harbor immediately. It was

MIDWAY: YAMAMOTO'S LAMENT

the kind of intelligence that Japanese radio signal units would pass up the chain of command right away. Rochefort made the necessary arrangements, and the dispatch went out. To make it easy to read, the faux water shortage at Midway was reported in plain text, not coded.

Within hours, on the morning of May 20, Hypo's Joe Finnegan was deciphering a message sent from the Japanese outpost on Wake Island to Tokyo, reporting a water emergency at a US Navy base in the Pacific. "The Japanese took the bait like hungry barracuda," Holmes wrote later. Most important, the Japanese message used AF as the geographical designation to identify the island where the US base was located.

The ruse had worked. Beyond any doubt, the designator AF had been verified to mean Midway Island. Five days later, on May 25, the Japanese Imperial Navy finally altered its code to JN-25(c). "There was a big change," Lasswell said, "so that we were back in the dark again." But it was too late. The code breakers had gotten what they needed. With the treasure trove of intelligence, Admiral Nimitz began amassing a force of ships, the idea being to ambush the ambushers.

ON THE FIRST DAY OF JUNE, ISOROKU YAMAMOTO STOOD ON THE combat bridge of the 72,000-ton *Yamato*, the new flagship of the Combined Fleet and the most powerful battleship ever built by the Japanese. Sailing eastward toward Midway Island, *Yamato* was at the forefront of the Main Body, an armada under Yamamoto's immediate command consisting of thirty-three other ships. Included were two more battleships, *Nagato* and *Mutsu*, a destroyer division, a cruiser division, and two light aircraft carriers. Yamamoto looked out over the open sea knowing that six hundred miles ahead of his armada was the attack force commanded by Vice Admiral Nagumo. Known as Kido Butai, meaning "Mobile Force," Nagumo's fleet included destroyers, battleships, and light cruisers that spread out in a circle surrounding its key assets: four of the Imperial Japanese Navy's six large aircraft carriers. Elsewhere in the Pacific, transport ships carrying three thousand army soldiers and two battalions of the navy's special land unit were on route to meet up with Kido Butai closer to the final destination. In all, Yamamoto and his staff had put together a force that was larger than the one used in the surprise Pearl Harbor attack six months earlier.

The Midway plan called for more than a hundred bombers from Nagumo's carriers to pulverize the US base at Midway starting on June 4, crippling its defenses and grounding its planes ahead of an invasion by five thousand troops two days later. When the US Navy, alarmed at the prospect of losing Midway, dispatched aircraft carriers to resist the Japanese invasion, Yamamoto and his Main Body would ambush and obliterate the carriers. It was Yamamoto's second chance to force an early peace in an unwanted war.

Coordinating the massive force had taken weeks. In early May, Japanese warships, including *Yamato*, had begun gathering in southern Hiroshima Bay. The deep waters of the bay's Hashirajima anchorage were a favored staging area; in fact, Yamamoto had monitored the Pearl Harbor attack from Hashirajima. From the anchorage, ships requiring last-minute repairs and fueling sailed into the navy base at Kure, nineteen miles away. Yamamoto's flagship did so on May 13, and the next six days in Kure were a kind of lull before the storm, a period when, as was customary, officers and staff summoned their wives and family to the seacoast town. Yamamoto was no different, although instead of his wife and family he sought the company of his longtime lover. Before arriving at Kure, he called Chiyoko in Tokyo to arrange for her to be with him. He had turned fifty-eight the previous month; Chiyoko was about to turn thirty-nine; they'd been together now for eight years, ever since their first summer of love in 1933. But the couple had to face a new hardship in addition to the already constant strain of separation that a war imposed on any relationship, familial or otherwise: Chiyoko had fallen ill, since mid-March suffering from pleurisy, an inflammation of the tissues surrounding the lungs, causing her severe chest pains and breathing difficulties. She had been receiving treatment, although at one point her doctors had worried that she might not recover. Of course, she longed to see Yamamoto, and she eagerly took his call "but hardly spoke by telephone due to much coughing," she wrote in her diary. "Lost patience with tears."

Even so, the couple made it happen. Though weakened, Chiyoko boarded an overnight train in Tokyo, accompanied by a doctor who administered injections of antibiotics. The next afternoon, May 14, Yamamoto waited in anticipation on the platform of the Kure station. He did not wear his uniform but was dressed in a disguise consisting of civilian clothes, glasses, and a mask—the kind of gauze mask worn over the nose and mouth to avoid spreading or inhaling germs. "My darling awaiting me," Chiyoko wrote later. "I was wild with joy." Yamamoto scooped up the depleted Chiyoko, put her onto his back, and carried her to a car waiting outside to take them to his hotel. "My thrilling body was carried in his arms," she said. "Having difficulties in breathing, I had repeatedly been given injections, and it was only with this pain that I could be there."

They spent four nights together at a hotel in Kure, with Yamamoto attending to his fragile lover. "Your spiritual strength in overcoming illness day by day is amazing indeed," he told her afterward. Taking her to the train on the morning of her departure, Yamamoto watched from the platform and then reached up to take her hand through the window. Chiyoko later wrote, "Though I was so weak that I could not hold your hand strongly, you were very, very strong in holding my hands." She felt a new pain—heartsickness—and did not want to bid him farewell. "I wished I could get off the train and remain beside you. When the train started to move, I hated to loose our firmly held hands."

Yamamoto was likewise heartsick. His flagship left Kure and returned to the Hashirajima anchorage, the countdown to the Midway operation having begun in earnest. He poured out his feelings to Chiyoko—about war and love—in a May 27 letter written from *Yamato*. While she fought to regain her health, he said, he would "exert myself to fulfilling my duty for the country." He even told her about Midway. "I will make an early sortie in the morning of the 29th to command the whole fleet at sea for about three weeks." He also confessed that a part of him wanted none of it. "I wish, if I could, to desert everything in the world to live alone with you." He ended his letter with a short verse pining for her, "with hot kisses on your lovely picture." Chiyoko eventually recovered, and in the months to come Yamamoto continued to write and, on occasion, even called, but the four days and nights in Kure would turn out to be the last time they would be in each other's arms.

JUST AS HE HAD SAID IN HIS LETTER, YAMAMOTO SET SAIL FROM the waters off Hiroshima on May 29, two days after Vice Admiral Nagumo's fleet had departed—and nine days after Station Hypo code breaker Red Lasswell had decrypted a message detailing the Midway operation. The weather was clear starting out, and the atmosphere aboard Yamamoto's flagship was upbeat, so much so that during the early going sailors belted out war songs. The further along they sailed, however, and as June 4 approached, a quiet tension set in. Yamamoto, manning the bridge, certainly appreciated the patriotic melodies, but his face was often grimacing in discomfort. Though nowhere near as ill as Chiyoko, he was struggling with his own health issues. He suffered shooting pains in his abdomen along with diarrhea, which was eventually diagnosed as worms.

It seemed to Yamamoto as he crossed the Pacific on the first of June that every part of the Midway plan was in place. One aspect had a smaller fleet that included another of the Imperial Navy's aircraft carriers; it would attack the Aleutians. The barren, frozen islands were part of the United States' Alaska territory; they were so remote that they had little strategic value except that whatever nation held them gained an edge in controlling northern Pacific transportation routes. Yamamoto and his planners were using the attack mainly as a diversion to pull the US Navy away from the main Midway operation—even if it defied the long-standing naval practice of concentrating sea power, not scattering it. Because Yamamoto had ended up dispersing his naval assets to such long distances, they would not be able to back each other. Then, in another bid to confuse the enemy as to the whereabouts of the Imperial Fleet, Yamamoto had instructed the radio communications staff to compose false messages making it seem as if the bulk of the Combined Fleet was training west of Japan in the Inland Sea. The trouble with all of that from Japan's perspective, however, was the huge disadvantage it had in the shadow war of intelligence. Close to a hundred Japanese warships might be roaring toward Midway, but no amount of Japanese trickery mattered. Unknown to Yamamoto, his counterpart, Admiral Chester Nimitz, knew what was coming and had already patched together an opposing sea force.

As planned, Yamamoto's attack force commander, Vice Admiral

Nagumo, commenced the raid early on the morning of June 4. Nagumo ordered the launch of 108 planes—horizontal bombers and dive-bombers with Zero fighters as escorts—to attack and soften up Midway for a land invasion. But nothing went right. Nagumo was aboard his aircraft carrier, *Akagi*, positioned northwest of Midway, when his fleet found itself facing fighter planes coming at them from the United States' Midway base. Nagumo's forces managed to repulse the first US counterattack, but reality set in: instead of the Japanese fleet pummeling the US base with little resistance, the enemy was ready and waiting to do some pummeling of its own. Nagumo realized that he could not try landing ground troops yet and was urgently ordering a second aerial attack against the US base when new information arrived that turned everything upside down. A Japanese reconnaissance plane had discovered Nimitz's warships on Nagumo's left flank at a spot northeast of Midway known as Point Luck. The fleet included destroyers, cruisers, and, most surprising, three aircraft carriers, including *Yorktown*, outfitted with wooden braces to shore up its bulkheads. The Japanese had thought that *Yorktown* had been destroyed in the Coral Sea, but in fact it had been patched together at Pearl Harbor in time to join the US force at Midway.

The news was shocking. Yamamoto's ultimate goal was to engage and crush the US carriers that had eluded the attack on Pearl Harbor in December, but the plan called for engagement to occur once the Midway attack was well under way, as the US carriers rushed to the scene. That three enemy carriers were already on location was off script—and Yamamoto's Main Body was hundreds of miles away, nowhere close enough to help Nagumo. Minutes later, dozens of dive-bombers from the three US carriers began descending on Nagumo's flagship, scoring direct hits that detonated the ship's own bombs. The explosions rocked the *Akagi*, fires broke out, and black smoke spewed into the sky. "The scene was horrible to behold," one officer later wrote. The devastation came swiftly and stunningly. That *Akagi* and the three other Japanese aircraft carriers were clustered so close together, never having expected any resistance, proved to be costly. They were easy targets.

Yamamoto's face turned ashen as reports arrived from Nagumo's staff. "Attacked by enemy land-based and carrier-based planes. *Kaga*,

Soryu and *Akagi* ablaze," one such report read. The messages came at a rapid-fire pace; in less than six minutes, Yamamoto learned that *Kaga* was hopelessly afire, *Soryu* was likewise engulfed in flames, and, stunningly, Nagumo was abandoning *Akagi*, escaping the flames by climbing through a window on the bridge. Nagumo transferred the attack force's command post from the crippled *Akagi* to a light cruiser. He and his staff put their heads together in confusion and apprehension in the tiny bridge operations room. Yamamoto and the others had looks of "indescribable emptiness" in their eyes, a clerk on the *Yamato* later recalled. By late afternoon, both *Kaga* and *Soryu* were sunk, and not long afterward the carrier *Hiryu* was reported out of action. The four aircraft carriers that Yamamoto had committed to the Midway operation—ships that had been so central to his Imperial Fleet's overall success to date—were gone. Yamamoto was notified that Nagumo's flagship carrier, *Akagi*, though abandoned and listing, somehow remained afloat. Worried that the enemy might tow it back to the US mainland to show it off as a trophy, he made the agonizing decision to sink it. His staff protested the notion of torpedoing their own ship, but Yamamoto was insistent. "I'll apologize to the Emperor myself."

Yamamoto had a destroyer carry out the sinking of *Akagi*. He considered forging on and ordering Nagumo to stay in the fight, perhaps even land troops on Midway, in order to give the Main Body time to catch up. But then, reluctantly, he decided to call off the Midway operation. Without carriers to provide aerial support, he knew there was no chance of success. Japan's naval forces began to withdraw and limp home. It was nearing midnight, less than twenty hours after fighting had begun—a nightmarish finish to a raid built on the high hopes for a decisive victory against the US fleet.

Japan's largely uninterrupted expansion in the Pacific since the beginning of the year had been halted in its tracks. Yamamoto sternly ordered his staff not to criticize the Nagumo force. The mission's failure, he said, was his responsibility. The Combined Fleet's commander in chief retreated to his cabin, where he stayed for several days. A funereal silence fell over the *Yamato*. In the weeks to come, the Imperial Navy's immediate challenge would be to rebuild its carrier arm. Beyond that, Yamamoto and the Japanese high command would turn their focus to the Solomon Islands, northeast of Australia, where they already had a stronghold at Rabaul on the island of New Britain. On June 7, Japanese troops made a landing on Guadalcanal, and in the wake of the massive setback at Midway, the decision was made to build an air base there. The thinking was that control of the Solomons would put the Japanese navy and army into a much stronger position to isolate Australia and disrupt any US plans to use it as a base.

IT WASN'T AS IF THE JAPANESE HADN'T INFLICTED DAMAGE AT Midway. The battle had been a bloodbath for both sides. For one thing, the Japanese had managed to finish off the carrier *Yorktown*, along with a US destroyer and 150 aircraft. But Nimitz's outnumbered and outclassed forces had pulled off a David-versus-Goliath victory. Yamamoto had lost four carriers, the heavy cruiser *Mikuma*, and more than 240 aircraft. More than three thousand Japanese sailors and airmen had died, including many of his most experienced pilots and, just as important, many aircraft maintenance workers on the carriers that had sunk. Yamamoto and Japan's military leaders were left to second-guess the plan, pick apart its execution, and find fault, but one game-changing factor lay beyond their awareness: the role of the US code breakers. Nimitz himself later wrote that "Midway was essentially a victory of intelligence."

Few in Japan were told about Yamamoto's loss at Midway, the grim truth suppressed in favor of propaganda aimed at maintaining the nation's war fever and its veneer of invincibility. The nationalist press, loyal to Imperial General Headquarters, falsely claimed that the US fleet had lost two aircraft carriers and 150 planes to one of Japan's cruisers and only 30 planes. "Navy Scores Another Epochal Victory" was one banner headline. To help keep lips sealed, most officers, sailors, and soldiers who'd served at Midway were immediately sent to different bases around the Pacific instead of being allowed to return home to be with their families, as was the usual custom.

The US media told quite a different story. "Jap Fleet Blasted in Midway Battle: Battleship and Carrier Damaged, Raid Repulsed," screamed the front page of the *Boston Globe*. Every major daily newspaper in the United States carried similar headlines. "Pearl Harbor

has now been partially avenged," Admiral Nimitz asserted in a statement issued from his headquarters in Hawaii. "Vengeance will not
be complete until Japanese sea power has been reduced to impotence." One newspaper's story gave US military leaders pause, however. The front page of the *Chicago Tribune* on June 7 had an article
headlined "Navy Had Word of Jap Plan to Strike at Sea." The article's
subhead went on to say that US officials had even known beforehand
that Yamamoto's move on the Aleutian Islands "was a feint." The
reporter who had written the scoop was war correspondent Stanley
Johnston. It turned out that he had been aboard a transport ship returning from the Battle of the Coral Sea when he had gotten a look at
the classified dispatch Nimitz had sent to commanders in the Pacific
outlining Yamamoto's plan to seize Midway. Johnston's bombshell
article was packed with specifics, saying that US naval officials had
known which ships Yamamoto had deployed as well as his overall
game plan: "It was known that the Japanese fleet—the most powerful yet used in this war—was broken into three sections: First, a
striking force, next a support force, and finally an occupation force."

Though the article did not say so explicitly, it certainly suggested
that US code breakers had cracked the Japanese naval code. How else
could Nimitz have known Yamamoto's plans in advance? In Washington, the chief of naval operations, Admiral Ernest J. King, along
with other top military officials, was livid, especially after other
newspapers picked up the story. They pushed for a Justice Department leak investigation and possible charges of treason against the
reporter and his editors under the Espionage Act of 1917.

The huge intelligence advantage provided by US code breakers
suddenly seemed at risk. Washington held its collective breath, waiting for any sign that the Imperial Japanese Navy now realized that its
code was compromised. One sure indication would be if Yamamoto's Combined Fleet junked its JN-25 code system for an entirely new
one, rather than following its usual practice of periodically modify-
ing the JN-25. Japanese naval leaders, licking their wounds after Midway, were indeed suspicious. Yamamoto, for one, told Chiyoko that
correspondence from the fleet had briefly been suspended, "as it is
said that secrecy of the operations and movements seems to have
been leaked out to the public and foreign countries." But whether

the Japanese somehow missed the newspaper article or whether out of hubris they could not imagine Americans being able to crack their code, the crisis passed. The Imperial Navy stuck with its basic JN-25 code system, and US officials detected no other evidence that the Japanese were onto the US code breaking. In the United States, a grand jury investigation was scuttled, in part to stop further public attention being given to the leak.

The glow of the Midway victory spread across the United States and especially to its forces all around the Pacific, including Fiji, where squadrons of Army Air Forces fighter pilots were training in waiting for their chance at combat. The pilots celebrated the news, with no one summarizing the triumphant moment better or more succinctly than did Captain John Mitchell. "We waxed the hell out of them," he wrote Annie Lee the week after Midway, "and that's just the beginning."

CHAPTER 10

MITCHELL ON THE MOVE

———

Lanphier, Holmes, and Barber
Courtesy of the Mitchell family

CAPTAIN JOHN MITCHELL AND THE REST OF THE FIGHTER PILOTS in Major Henry "Vic" Viccellio's squadron decided to stage a special celebration of America's Independence Day and, in the days leading up to July 4, 1942, got all caught up in preparty preparations. The way the pilots saw it, a big reason—or excuse—for a bash was to return the favor to local Brits and New Zealanders who had befriended them while they had trained at Nadi on Fiji's western coast. The locals had hosted them plenty of times to dinner, drinks, and other leisure events, and the army fliers decided it was their turn to host the hosts, and what better time than July 4? With plenty of Texans around, including Wallace Dinn of Corpus Christi, who'd become Mitch's closest friend, everyone liked the idea of throwing a Texas-style barbecue. Just as in flying, the proud pilots were eager to show their skill—at partying hard.

The Texans took the lead, along with Second Lieutenant Rex Barber. Though not from the Lone Star State, the country boy from Oregon with his big, bullish, derring-do personality might as well have been. The organizers lobbied the other men to chip in money to cover the feast, lined up a small orchestra, and tapped local contacts to gather a flock of girls to join the dance they had in mind for the day's end at a nearby farmhouse. The men persuaded the squadron's physician to hand over ten gallons of medicinal alcohol to create a "Fiji Punch." With whiskey in short supply—about as rare as a snowflake in the tropics—securing the 180-proof alcohol was critically important. The beer that was occasionally on hand was never cold, causing Mitch to make a promise: "When this war is over I'm going to buy a *big* refrigerator and keep nothing but beer in it." Last, the party planners negotiated with a nearby rancher for the purchase of two cows.

The other big reason for the bash was that when the pilots weren't in the sky training or on an occasional reconnaissance flight, they had plenty of time to spare. In their five months on Fiji they'd flown countless drills in the P-39 Airacobras, logging some four hundred hours. During June, they'd even had the benefit of being tutored by navy pilots who'd fought at Midway. The pilots' aircraft carrier had been damaged, and they were stuck on Fiji for several weeks. Like other Midway combatants passing through Nadi, the fliers had lots to say about the unexpected victory against Yamamoto's Imperial Navy. "They really scattered yellow meat all over the ocean" was the way Mitch enthusiastically summarized their combat tales in his diary. Even so, Navy Admiral Chester Nimitz and other top military leaders back home began to strike a slightly more cautious note, warning the public not to get carried away. One story that ran in many stateside newspapers was headlined "Jap Sea Power Blunted,

But Still Mighty." It said, "Japan has suffered a terrible defeat in the battle for mastery of the mid-Pacific, naval experts agreed today, and her striking power has been badly blunted, but she still has left great strength with which to attempt a comeback." Complicating matters—and Nimitz was not about to share this publicly—was a dramatic falloff in intelligence. Not only had the Japanese navy changed its version of JN-25 in late May to confound the code break-ers, it had made another switch in early summer, which, as one historian observed, "plunged Allied crypto teams into almost total eclipse." In July, the United States was left mostly to guess the next moves by Yamamoto's navy.

For several weeks, the visiting navy pilots watched Mitch, Bar-ber, and the others handle the P-39s, offering tactical tips on fighting against the Japanese Zeros. The two groups performed exercises together, the navy pilots in their Grumman F4F Wildcats, which in some ways resembled the speedy Zeros. "We'd been dogfight-ing with other P-39s and now had a chance to learn combat and weaknesses against other planes," Rex Barber said. The navy pilots advised how best to take on the Zeros in combat by flying in pairs. "They taught us how to scissor—back and forth with two planes— and keep the Zeros off our tails." Worrisome was the realization that the heavily armored P-39s were cumbersome and slower than the nimble Wildcats-cum-Zeros, prompting anxious talk about when the faster, high-performing P-38 Lightning, still in develop-ment back home, might be ready.

Mitch and Vic were now convinced more than ever that their pi-lots, nearly all of whom were only ten months removed from flight school, were combat ready. But although they continued to hear about Midway and other conflicts, no action had come their way. "Outside of the mail coming we have had absolutely no excitement here at all," Mitch complained to Annie Lee in mid-June. Weeks later, he sounded like a broken record. "Things are absolutely quiet here," he said. "Only mail breaks the monotony, so don't feel your time is misspent." Vic was pushing his superiors to get them moved, as Mitch put it, "to somewhere where the fighting was go-ing on," even to Europe if need be. The challenge was now less about training than about maintaining a high level of readiness.

Fiji had become a quiet and comfortable stopover for a variety of bigwigs. Politics even seemed to come into play, as Texas congressman Lyndon Baines Johnson, a lieutenant commander in the Naval Reserve, inspected the forces there as part of a two-month Pacific tour of US bases. Johnson had asked President Roosevelt for the assignment. For many observers, the trip seemed to be more about the ambitious thirty-four-year-old wanting to buff up his résumé for future campaigns than about FDR's need for eyes and ears on the ground. Indeed, Johnson earned what became a controversial army Silver Star as a result of his brief swing through the Pacific. He was aboard a B-26 as an observer during a bombing raid of New Guinea and, according to himself, came under heavy fire from Japanese Zeros. Others, including members of the bomber's crew, later said that had never happened. Either way, Johnson proudly wore the medal en route to the presidency.

The army fliers at Fiji certainly saw more combat-hardened brass than LBJ. One was an Army Air Forces general named George C. Kenney, a decorated World War I veteran who was on his way to Port Moresby to take command of the Fifth Air Force. Another was Army General Henry "Hap" Arnold, a senior officer and longtime proponent of airpower who'd actually learned to fly in a Wright Brothers biplane. Arnold, the commanding general of the Army Air Forces, was making the rounds in the Pacific to assess frontline needs. In Washington, the Joint Chiefs of Staff's main focus was on Europe and fighting Hitler's Germany, which meant that more efficient coordination of the limited military resources in the Pacific was essential to halting Japanese expansion and protecting Australia and New Zealand. In that vein, the next to come through Fiji was a group of high-ranking officers under the command of Army Major General Millard F. Harmon, Jr. They were on their way to New Caledonia, about 840 miles west of Nadi, where Harmon would start bringing order to the chaos that had followed Pearl Harbor, when Army Air Forces units had been deployed hastily throughout the region. As the commander of Army Air Forces in the South Pacific, with headquarters at Noumea on New Caledonia, he would pull together under a single command the various fighter squadrons scattered about the different islands.

Mitch, Vic, and the others—such pilots as Rex Barber, Tom Lanphier, Wallace Dinn, and Doug Canning, all of whom had been
training together for months—were left to wonder, what about us?
Mitch and Vic knew Harmon from Hamilton Field in San Francisco. Before that, Harmon had been the officer who had shaken
Mitch's hand on the podium at flying school graduation in Texas
in 1940. They hoped that their personal connections might help get
them off the sidelines. "It may be that he can do something for us,"
Mitch wrote in his diary. Vic also knew one of Harmon's top aides,
Lieutenant Colonel D. C. "Doc" Strother, and began working on
him. The pitch was that Mitch and Vic "honestly believed that this
squadron is better trained right now than anywhere in the U.S." For
his part, Mitch thought, "I feel as though I am being wasted. Not
that I'm so valuable, but at least I have two years' service, about 800
hours, have been to England and have observed and learned all I
can here. I have done the biggest part of the training of the younger
pilots and know them to be as well trained as they will ever be, until
we get some aerial gunnery and actual combat."

In the meantime, as spring turned to summer, the squadron continued to cope with the monotony. Some found novel outlets. Late
in the summer, Tom Lanphier talked his way aboard a B-17 that took
part in a bombing run at Truk Lagoon, the main forward base of Yamamoto's Combined Fleet. When he returned, he elatedly told others about the raid, claiming that he had even climbed into the waist
gunner's position and shot down a Zero. The reaction of the other
fliers was mixed—the kid from Stanford, a son of a decorated army
general now working in Washington, was seen as either ballsy or out
of his mind. Vic, the squadron leader, was angry, not just because
the ride had been unauthorized but because Lanphier, at this point
emerging as one of the best pilots, had risked his life. To Vic, the
life of a trained fighter pilot had far more value than that of an airman second class. Red Barber asked Lanphier afterward why he had
done it. Lanphier replied, "Rex, you are over here because you are
patriotic. Well I'm here because I'm patriotic, but I have another reason." He was looking ahead to a future in politics—the presidency,
even—and figured that a sterling war record would be a prerequisite
to fulfilling that goal. Lanphier years later denied that explanation,

saying that his goal had been "to do as well as I could as a military pilot in order to at least approach the standards my old man had set as
a flier." But it wasn't as if a political calculus never entered a service-
man's mind. Lyndon Johnson's brief time in the Pacific that spring
was seen in part as politically motivated, and when a National Guard
division showed up on Fiji Island, Mitch looked down at the reserves
as "one of the God damnest messes I have ever seen. Nothing but a
bunch of politicians." Their first night at Nadi, he said, they made a
beeline to the shortwave radio to touch base with the governor of
Ohio. "They are all from Ohio and all have political aspirations."

Beyond Lanphier's adventure, the pilots at Nadi occupied themselves doing what GIs everywhere did during downtime: played
sports of all kinds, played poker, drank whenever beverages were
available, and passed around magazines as they turned up: *Time, The
Saturday Evening Post, Newsweek, Flying Magazine,* and especially *Esquire: The Magazine for Men.* Mitch received letters numbers 105 and
106 from Annie Lee in a package that included tobacco for the pipe
he'd begun to smoke, as well as the latest issue of *Esquire.* The magazine was always in high demand for the full-page pinups inside. Mitch
had tacked up several from prior issues on the wall of his bunk, next
to a photograph of Annie Lee standing on her porch in San Antonio
wearing a Dutch cap and a flirtatious expression. Mitch had com-
plimented her pose, admiring her "very shapely pair of pens" and
insisting that she easily put the *Esquire* models to shame. The new
June issue he received was an instant hit at Nadi and, as it turned
out, with military men everywhere. The issue featured Gypsy Rose
Lee and a foldout of the Hollywood starlet and sex symbol Jane Russell lying on her back in a hayloft. Most compelling, though, was a
luscious portrait of a model named Jeanne Dean in a shimmering
gown. In her right hand she held a white orchid, a symbol of in-
nocence and love. "Jeanne," as the pinup was titled, was done by
Alberto Vargas, a well-known Peruvian painter. The portrait was
accompanied by a short verse, "Victory for a Soldier," including such
lines as "My heart's with a Boy who heard a call to March and Fight
for all that's real on Earth." The combination portrait-poem struck
a chord with soldiers stationed in the European and the Pacific theaters who were separated by war from lovers and wives. "The *Esquire*

meets with everyone's approval," Mitch told Annie Lee. "The poem with the Varga [sic] drawing was excellent—Thanks darling for sending them."

Mitch had last partied on his birthday, June 16, when he had turned twenty-eight. The weather was miserable—heavy rain all day—but he and Wallace Dinn finally managed to get through to a nearby club, where they spent the night at the craps table. Mitch started out losing but went home a winner—and sent the $75 home to Annie Lee. With the Independence Day bash next up on the social calendar, he watched the planners kick into high gear in the days beforehand. Slaughtering the cows turned into something of a gory horror show. Rex Barber slashed one cow's throat with a big knife, releasing a torrent of blood that made men turn away; one actually fainted. The second cow killing was worse. One of the Texans shot the cow in the head several times with a .45-caliber service pistol, but the cow just glared at the shooter and bellowed. They next tried an ax, but the cow did not even blink. The animal finally succumbed when someone retrieved a .22 rifle. Barber watched, later reporting, "The boy came up, shot her through the eye with a .22 and she dropped." The cows were carved up, basted, and slow cooked. "Basted the beef all night," Doug Canning said. Next up was making the Fiji Punch—emptying the ten gallons of medicinal alcohol into tubs and mixing in juice from fresh oranges, pineapples, and limes, along with juice from canned fruits—anything to help render the alcohol tasteless so it could be swigged easily.

The party went off spectacularly. In contrast to the rain on Mitch's birthday, July 4 was sunny and hot. The two main Texan hosts strutted around in cowboy boots with .45-caliber pistols slung low on their thighs like gunfighters, as partygoers played baseball and cricket, devoured the barbecued beef, and drank all through the afternoon. Twilight signaled a second act, as the men showered and changed into formal attire before heading over to the farmhouse, where an orchestra and dozens of women awaited. The men toasted Independence Day and the need to protect the nation's freedom. They danced and drank as the band played on. "We carried on all night, a hell of a party," Canning said. The fruit juices succeeded beyond expectation in masking the ghastly taste of the 180-proof alcohol. Many of the men were falling-down drunk. More than a few had to be helped to bed in the huts, while several cars carrying the New Zealand and British guests experienced difficulty driving back to town. "One car, leaving, hit a tree," Barber said. Mitch had a good enough time, mainly watching the heated action on the dance floor from a perch at the bar. "I never thought they could ever dig up that many girls around here, but they did," he said. The good time left him feeling sentimental, though, for Annie Lee and what he was missing by not being with her. "The moon has just been full and it makes me think of you even more and the days we spent together in California. Those were the days," he wrote.

FOUR DAYS LATER, ON JULY 6, 1942, THE JOINT CHIEFS OF STAFF IN Washington, DC, were informed that the Japanese had cleared a strip of land on the north coast of Guadalcanal and were constructing an airfield. The workers were part of a small force that had landed on the island in early June as the Midway sea battle was being fought three thousand miles away. The Japanese plan was to create a single coral runway on a coastal plain near the mouth of the Lunga River as a base for about sixty of Yamamoto's naval aircraft. Planned for completion by August, the base would then be used for further southerly advances.

The intelligence came not from code breakers at Station Hypo in Hawaii or from other cryptology outposts, who were stymied for the time being by the changes in the Japanese code, but from coastal watchers embedded on Guadalcanal. The watchers were mainly civil servants, planters, and farmers who had stayed behind on the various Solomon Islands following the Japanese invasions. They had hidden in the hills and, with the help of native islanders, monitored the movements of enemy planes, ships, and troops. In that instance, they radioed information about the airstrip on Guadalcanal to allies stationed on the east coast of Australia. Their information was verified by aerial surveillance and photo reconnaissance.

The Japanese had already made substantial inroads across the Solomon Islands. They had built a sea base on Tulagi, a tiny island taken in May, just thirty miles north of Guadalcanal. Its harbor was considered the best anchorage in the southern Solomons, and the

Joint Chiefs were already contemplating limited action against the Japanese there. Now news of an enemy airport under construction on Guadalcanal reaffirmed their thinking that something had to be done. Guadalcanal was only 1,200 miles from Australia, meaning that a Japanese occupation of the island could not be tolerated.

Four weeks later, on the night of August 6, US naval gunfire and aerial bombing set afire Japanese barracks and buildings on Guadalcanal. Nearly eleven thousand marines from the 1st Marine Division followed, hustling ashore at Lunga Point, their amphibious landing part of a larger campaign, code-named Operation Watchtower, against Japan's forward bases on the islands of Guadalcanal, Tulagi, and a third island, Florida. The invading Allied fleet, involving some seventy-five war and transport ships, marked a pivot in the Pacific war: the first time the United States had gone on the offense against the Japanese.

Initially the marines faced little resistance while taking control of the half-built airfield. Though the enemy force on Guadalcanal numbered about 2,800 men, only a few hundred were soldiers, the rest being laborers. Surprised and outnumbered, the defenders fled west to the Matanikau River and a peninsula known as Point Cruz, leaving behind plenty of supplies, food, and construction and military vehicles. One early encounter with the enemy, however, was forever seared into the marines' collective memory. Near the Matanikau River, where Japanese forces had fled, patrols spotted a white flag, suggesting that the enemy was prepared to surrender. Further, a captured Japanese sailor seemingly provided confirmation of that, saying during interrogation that the Japanese were starving, suffering from tropical sicknesses, and wanted to give up. Marine Lieutenant Colonel Frank B. Goettge, a senior intelligence officer, led a patrol of two dozen marines to Point Cruz, fully expecting to supervise a major enemy surrender. Only three marines returned. Soon after a nighttime beach landing by the patrol just west of the river, Japanese soldiers perched on a coral plateau began picking them off. Goettge was shot through the head and killed instantly. By morning, another twenty-one marines were dead or wounded. The three survivors, at various times during the night and on orders, ran into the ocean and swam to the 1st Marine Division's stronghold several miles away. But none made it back in time to bring help. The last one making the nocturnal swim through the shark-infested waters was a platoon sergeant, who told a horrifying story. He had witnessed the Japanese make their final charge, firing needlessly into the bodies of dead and wounded marines pinned down at the beach. Even worse, they had mutilated the corpses, hacking at the marines with swords, cutting off the hands of one, cutting out the tongue of another. News of the ambush and the ghoulish massacre spread quickly, setting the tone for a deep and mutual malice in Pacific theater combat. "They were a fanatical enemy," one marine veteran said afterward. "The meanest sonsabitches that ever lived."

"THIS IS THE FIRST OFFENSIVE MOVE WE HAVE MADE," MITCH wrote in his diary. "I'm very glad to see it. Hope and believe it is the first of many." Mitch penned the entry three days after the marine landing on Guadalcanal, words showing that he sensed something big was unfolding, something where he might finally get his chance. "The Solomon battle seems to be going our way," he wrote a few days later. "Sure hope I get in on it."

But Mitch's take on the early going amounted to wishful thinking. The Solomon Islands quickly became the stage of another bloodbath between the two warring nations—with little clarity this time, as opposed to at Midway, as to which side would ultimately prevail. Caught off guard, the Japanese retaliated within hours of the marines' takeover, sending Betty bombers and Zero fighters to begin relentless bombing raids. Yamamoto's Eighth Fleet, including a destroyer and other warships, attacked Allied ships still in the area after transports had dropped off the 1st Marine Division. In Tokyo, military leaders ordered the Seventeenth Army, stationed at Rabaul, to retake the island with support from Yamamoto. And to help direct the effort, Yamamoto moved his naval headquarters to Truk Lagoon to be closer to the fighting.

Under heavy attack, the US ships had cleared the area before unloading the marines' equipment and supplies. Under the command of Major General Alexander Vandegrift, the marines then dug in to establish a defensive perimeter extending about three miles long from the beach at Lunga Point and in an oval shape around the airfield. They began working to complete the airfield, which they named Henderson Field in honor of a marine pilot named Lofton R.

Henderson, who had been killed during the Midway battle. The first plane landed about a week later, when a navy transport plane arrived to remove two wounded marines.

The US commanders were forced to improvise as they went along. "Pretty much a hip-pocket operation," Lieutenant Colonel Strother called the army's scramble to provide aerial support for the marines. Strother's boss, Major General Harmon, was pulling together different fighter units at the army's new base on New Caledonia, but there was no formal Army Air Forces organization. In fact, Harmon and Strother had not intended to send any of their fighters to Guadalcanal until the marines had fully secured the island—late August, at the earliest. But the Japanese counterattack had changed all that. Harmon urgently sent two dozen P-39s and mechanics from his 339th Fighter Squadron to Guadalcanal. "The intensity of Japanese air operations required that they be deployed much earlier," Strother said. "To help stem the frequent air attacks and to bolster the Marines."

In the sky, the two sides went after each other fast and hard. "It's infuriating," complained Yamamoto's chief of staff, Vice Admiral Matome Ugaki. "We shoot them down and we shoot them down, but they only send in more." The constant demand for more planes at Guadalcanal was hampered, however, by the fact that Henderson Field had only a single airstrip. The challenge, said Strother, was "infiltrating a few aircraft and mechanics at a time, as fast as the flying field facilities could handle." By the end of August, though, Mitch's opportunity was at hand. An order went out for a group of pilots from Nadi to report to New Caledonia as a staging area for service at Guadalcanal. "Naturally I volunteered first," Mitch wrote in his diary that night, August 31. "Fourteen of us are to go from here." Mitch made certain that the group of pilots he led, Flight B, was selected—a group including his pal Wallace Dinn, Doug Canning, and a young pilot from San Diego named Julius "Jack" Jacobson. The son of a butcher, Jacobson eventually served as Mitch's wingman. He was not at all surprised by Mitch's advocacy, calling Mitch "very aggressive; he thought Flight B was the best of the three flights." Flight B was chosen. Unhappy about being left behind were Rex Barber and Tom Lanphier, who were in another group and would have to wait.

Late in the morning of September 14, Captain John Mitchell and his fourteen pilots arrived at New Caledonia, transported aboard several heavy bombers—B-17 Flying Fortresses and B-24 Liberators. Within a few days, they were taken to the Army Air Forces base at Tontouta, thirty miles northwest of Noumea, the island's capital. The base was circled by steep mountains rising into moisture-thick clouds. One key problem at Tontouta was too many pilots for too few planes—"Only five planes in commission with about fifty-eight pilots ready to fly them," Mitch observed. He and his men found themselves in limbo until a new shipment of P-39s could be assembled. They settled in to the pyramidal tents housing pilots and mechanics at the foot of one of the tallest mountains. The one comfort was a day room, outfitted with chairs, tables, a magazine rack, a Ping-Pong table, a phonograph, and a GI radio set. One rumor making the rounds was that the much anticipated P-38 Lightning, the twin-engine fighter the army brass had hyped as a speedy match for the Japanese Zero, would soon be making its debut. Mitch listened with mixed feelings. He certainly knew they needed better pursuit planes. "The Zeros have it over us so much there is no comparison," he wrote in his diary while alone one night in his tent. "They out climb and outmaneuver us two to one." But the P-38 had had a troubled history, and Mitch would never forget the horrific death of his friend Ellery Gross. He just hoped and prayed that the army had worked out the P-38's kinks. "I'm not eager to fight in them as you can't bail out," he worried. But he also relished the idea of harnessing the P-38's speed and climbing power. "I really would like to sit up above those damned Japs and let them see how it feels to have someone above you all the time."

As they waited, Mitch and the rest got bits and pieces of war news from the radio in the day room. They learned that while they had been traveling to New Caledonia, the marines at Guadalcanal had weathered a nightmarish forty-eight-hour rampage as wave after wave of Japanese infantry—three thousand soldiers in all—had attacked from a narrow ridge along the Lunga River just south of Henderson Field. In what came to be called the Battle of Bloody Ridge, the Japanese, cursing and screaming, had broken through a marine perimeter and almost reached the marine command post before being repelled. The fighting had been hand-to-hand, as the Japanese poured it on. "The Marines had managed to turn the tide," wrote one war correspondent, "but even when obviously defeated the Japanese kept coming. They went so far as to stage a hopeless bayonet attack in broad daylight." The body count was eight to one in the marines' favor, about one hundred marine losses to the enemy's more than eight hundred.

Mitch and his pilots learned that navy Seabees, the name for navy construction workers, were marking off a grass strip parallel to Henderson Field. The second airstrip, to be called Fighter One, would allow Guadalcanal's plane inventory to grow more quickly. By mid-September, it included a mix of Army Air Forces P-39s, navy F4F Wildcat fighters, navy SBD Dauntless dive-bombers, and TBF Avenger torpedo bombers. They learned that daily atabrine pill treatments were being enforced in earnest, as malaria had erupted after the troop landing in early August and had become a major medical problem, requiring the hospitalization of more than 1,500 men within six weeks.

Most unnerving, they heard from returning pilots about the Japanese fliers' renowned aerial prowess. "The Japs had beat them," Doug Canning said. "And they were scared." Mitch watched his men—rookie fliers who hadn't yet seen any combat—listening to the stories of the Japanese as unbeatable foes. "Wild stories," Mitch observed, "about super planes and supermen, and they told us that if a Zero got on your tail you were dead." As he watched the accounts send chills down his men's spines, the moment became a test of leadership: he had to rebut the morale-busting talk on the eve of their own dispatch to the front line. The men turned to Mitch, and it was no wonder they did, not only because he was their flight leader but because he was a natural leader. The way some men are pure hitters in baseball, or shooters in basketball, John Mitchell knew how to lead. Over their seven months together, they had experienced his selflessness, such as the time at Nadi on Fiji when Mitch had gone off on a rant against a *Life* magazine photo spread showcasing American and British "heroes" from the fighting in Europe against Hitler. Mitch disparaged the gushing publicity as foolish. "There are many who had done much more without any special recognition," he argued. "Seems to me that all they have done is just their duty." He called the adulation unnecessary and over the top. "Maybe it's sour grapes with me, but I do know I would certainly turn down any offer to return to the U.S. as a public hero just because I had been lucky enough to shoot down a couple of Japs." Or there was the time the previous month when a B-25 bomber descending into Nadi had missed the airfield and had been circling helplessly in the sky. While others had watched, Mitch had run to a plane, gone up, and picked up the bomber about twenty miles away. He had led the bomber pilot back, diving toward the field just ahead of him to show him where it was.

One of his superiors, Doc Strother, had always been impressed by Mitch's intensity, his "penetrating eyes," since meeting him at Hamilton Field. His men, at Fiji and now at New Caledonia, admired his humility. His future wingman Jack Jacobson said that Mitch "always lived and flew with us and there was no pretense about him." It was a role Mitch embraced, eschewing the occasional overture to take an administrative position or, as he described it, "to get on the staff and sit back and write papers, stuff like that. And I said, 'No, I don't like that. I like to be responsible. I like for people to depend on me.'"

Now there were the "tall tales from a whipped squadron," as Mitch called them, and he would have none of it. "I took my kids aside and told them this was all crap, that those guys hadn't really fought Zeros and that if any of them believed those tales and thought we couldn't lick the Japs, they could stay there and not go to Guadalcanal." He wasn't about to share his private worries about the enemy's Zeros, stressing instead the stronger firepower and thicker armor that made their own P-39s less vulnerable than Zeros. He knew, too, that during maneuvers the previous months his pilots had watched him to see how he did things, and he had made a point to lead by example—by flying aggressively and preaching that "good fighter pilots never think defensively." That carried over to when they were on the ground and off duty, as he told them he didn't mind if they got into a fistfight or two while carousing, that a streak of wildness was proof that they possessed the daring required of great fighter pilots. So, fielding the fearmongering about the enemy stoked by returning pilots, Mitch's message was clear: "I told 'em that if we had team work we could take 'em. I kept drilling that into them and convinced them they were great pilots.

"Soon they couldn't wait to get into combat."

THE FIRST WEEK OF OCTOBER 1942, AFTER TWO WEEKS OF WAIT-ing, Mitch climbed into the sky from the base at Tontouta on New Caledonia and veered south. He was leading a group of P-39 Airaco-bras on an 850-mile flight. Their destination was Cactus, code name for Guadalcanal. The Pacific islands all had code names, used in every communication, even in letters home. Fiji was Fantan. New Caledonia was Poppy. Tulagi was Ringbolt. And Guadalcanal was Cactus.

The army's top South Pacific commander, Major General Har-mon, had argued for days with his superiors that the US forces on Guadalcanal needed immediate help. He said that the Japanese were preparing another offensive. "It is my personal conviction," he wrote, "that the Jap is capable of retaking Cactus-Ringbolt, and that he will do so in the near future unless it is materially strengthened." Mitch and his men were part of that reinforcement.

There were not enough P-39s to go around, however. Doug Canning, for one, hitched a ride aboard a C-47 transport aircraft. A young *Time* magazine reporter named John Hersey happened to sit nearby. Guadalcanal was the future Pulitzer Prize–winning writer's first assignment in the Pacific. Eager for wartime tidbits, he got little from the army pilot, however. Canning listened politely to questions about the war but shook them off, wanting instead to talk about home. Said Hersey, "Within a few minutes every conver-sation swings around to home." Rather than discuss dive-bombing, Canning talked animatedly "about how he used to drive with girls and boys in his gaudily painted 1936 Chevy from Nebraskan town to town for the fabulous three-day fairs of summertime."

Mitch and eleven others, meanwhile, stopped to refuel and spend the night on another of the Solomon Islands. Then, on Wednesday morning, October 7, they began the last leg of a flight to what, at that moment in time, was the crucible of war in the Pacific: Guadal-canal. The sky was clear, and from a distance Guadalcanal looked welcoming—a tropical island with sun-soaked beaches, coconut groves, green meadows, and inland highlands. Mitch made sure to make a wide turn from shore to steer clear of the Japanese guns he had been warned were embedded west of the Matanikau River. It was then that he spotted Henderson Field, a slice of airfield on a coastal terrain about a mile inland from Lunga Point. Mitch and the pilots following him could see that the base was surrounded by grassy ridges that disappeared into thick jungle, and as they closed in they could make out the defense line the marines had established, the hundreds of foxholes and weapons emplacements armed with Browning .50-caliber heavy machine guns, 37 mm antitank guns, M2 muzzle-loading mortars, and 90 mm heavy antiaircraft guns with fifteen-foot barrels. Looking inside the perimeter, they could see the coconut stumps from the days when the area had been a plantation, and on closer inspection, they noticed bomb craters, tangled plane wrecks, and marine amphibious equipment rusting on the beach next to hulks of Japanese transport ships—the residue of the marine invasion and the fierce fighting since.

The marine base was only about a half mile wide and a mile and a half long. That was the full extent of the US footprint, and from the air Mitch and his men could see how little ground the marines possessed and, worse, how the enemy was all around. The pilots made a final descent through plumes of smoke, one set curling skyward from the cooking fires at the marine base but many more from Japanese camps in the jungle outside the perimeter. It was all there in a single, compact vision, and it looked ghostly. "You could very easily know we were here, and they were there," one of Mitch's pilots observed.

Mitch was the first to touch down, the wheels of the P-39 squeaking as they hit the airfield's glistening Marston matting, long strips of steel rolled out over a base of crushed coral. He had arrived just in time for Yamamoto's next big push to retake Guadalcanal.

PART III

GUADALCANAL

CHAPTER 11

THE FIRST KILL

———

Fighter Two airfield, Guadalcanal
Courtesy of the National Museum of the Pacific War

THEY CAME MAINLY BY TRAIN, A THREE-DAY RIDE DEPARTING
from southern California and covering some 1,800 miles to the eastern side of Arkansas, just shy of the Mississippi border. They ogled
natural wonders along the way as the transport train, filled to the

brim, chugged across Arizona, New Mexico, Texas, and Louisiana. They traveled through the southwest desert, past painted mountain ranges, high plains, the Rio Grande, and the Colorado River. "The states had some pretty sights," wrote one sixteen-year-old girl who'd come from a farm outside Los Angeles. Another teen recalled that some of the passengers had "bought homemade souvenirs from native Indian women when the train stopped in Yuma, Arizona." By contrast, little excitement was felt about the final destination as the hundreds, thousands even, of Japanese Americans began arriving "dusty, hot and tired" to the Rohwer War Relocation Center, a euphemism for concentration camp.

It was early October 1942. Eighty miles away, across the Mississippi River, Noah Mitchell was tending to his pear crop in the tiny hamlet of Enid ("Over one thousand bushels for sale at low prices!") while worrying about the fate of his pilot son stationed somewhere in the South Pacific. Thousands of miles away, Captain John Mitchell was touching down his P-39 on "Cactus," where US forces were bracing for a new Japanese assault overseen by Admiral Isoroku Yamamoto. Meanwhile, at Rohwer, wave upon wave of Japanese Americans arrived at the unfinished facility in the wooded swampland of rural Arkansas. The detainees were of all ages: issei, the Japanese term for first-generation Japanese immigrants, and nisei, their offspring, born in the United States and therefore citizens. "I peered out of the train window," a third teen wrote, "and what I saw was a neatly laid out camp, and this camp was surrounded by barbed wire and there were guard towers at the intervals along the perimeter with search lights." Rohwer was the easternmost of ten internment camps hastily built by the War Relocation Authority, the agency President Roosevelt had created for the purpose of forcibly removing Japanese Americans from the West Coast. Rohwer's population eventually peaked at about 8,400, while nationwide nearly 120,000 people of Japanese ancestry would be incarcerated.

Ten months had passed since the December 7 bombing of Pearl Harbor, the Japanese attack that had changed everything. "Right there and then the world seemed to crumble from under my feet," sixteen-year-old Mary Kobayashi recalled in an essay titled "My Autobiography" that she wrote for English class in the school at

Rohwer. In his essay, Nobuko Hamzawa described how he felt the day his family was taken from their tiny farm near Hayward, California: "I have never felt so helpless, unhappy and frightened in my life as I felt that morning." Mary and Nobuko were among the more than two thousand youths now being held at the detention camp, about which another classmate complained, "This will be my home till this mess is over." Yet another student, Takeo Shibata, was more outspoken than most about the chaotic upheaval in their lives: "There I was, just out of eighth grade, where I was taught that the Constitution guarded all the rights of its citizens. I had to forget all that was taught to me about our democratic country." Takeo had lived in western Los Angeles with his younger brother, older sister, and pet dogs Blackie and Chico. He'd attended Belvedere Junior High School. His father had sold insurance. When word had come that they had to leave, they had sold off all their furniture and other household items. Now in confinement at Rohwer, Takeo was left to long for a better day. "I hope in the near future I will be free to go outside and lead a life like we used to in Los Angeles.

"To lead a life like any other red-blooded American."

THE NIGHT JOHN MITCHELL ARRIVED ON GUADALCANAL, OCTOber 7, Admiral Yamamoto presided over a dinner aboard his flagship *Yamato* at Truk Lagoon that was more business than pleasure. He was hosting an operational conference with ranking naval officers to go over the navy's role in the imminent attack on Guadalcanal involving Yamamoto's Combined Fleet and the Japanese Seventeenth Army. Using darkness as cover, Japanese ships had been passing largely unimpeded through "the Slot" between two chains of the Solomon Islands to unload troops and supplies on the north coast of Guadalcanal. The soldiers scrambled ashore west of the Matanikau River, where Japanese forces were concentrated following the invasion by US marines. Yamamoto's destroyers provided cover by bombing nearby Lunga Point and Henderson Field. The Japanese ground forces on Guadalcanal would soon reach their largest numbers to date, exceeding 20,000 troops. Yamamoto's responsibility was harnessing naval power to lead the way, and he had assembled the strongest sea force since the Battle of Midway. In all, Japanese military leaders had formulated a potent counteroffensive on relatively short notice. The one flaw in their planning—a key one—was the Seventeenth Army's miscalculation of US strength. Contrary to its estimates of 7,500 troops, the total number of US troops on Guadalcanal was actually 19,000 by the end of September and still climbing, reaching 23,000 in early October.

Yamamoto was not his usual self during the dinner meeting, and he hadn't been for some time. Still reeling from the loss at Midway, he revealed a grim fatalism in a private letter to an old friend late that summer: "I fear that I have perhaps one hundred days left to me, and that I must complete my life in their passage." And as much as he tried to maintain an unwavering stoicism outwardly, his closest aides at the dinner could tell otherwise. The cherished admiral's hair had turned grayer in recent months, and he was quiet—almost joyless—at a gathering that, in the past, would easily have turned social once the military business was done. When the group retired to Yamamoto's cabin for an after-dinner whiskey, the best evidence of the admiral's changing demeanor was his stab at gallows humor. When one aide casually asked about his postwar plans, Yamamoto replied, "I imagine I'll be packed off either to the guillotine or to Saint Helena," a reference to the island off the coast of Africa where Napoleon had died in exile in 1821.

THE FIRST THING MITCH DID UPON LANDING AT GUADALCANAL was report to Marine General Vandegrift, the commander who'd led the 1st Marine Division's invasion and was now overseeing the hodgepodge of US forces—marine, navy, army, with a sprinkling of New Zealand pilots—defending their station. Mitch found the general at his dugout headquarters next to the airfield, and the general directed Mitch to join other Army Air Forces pilots at an airstrip known as Fighter One. The pilots were loosely organized into what would eventually become the 339th Fighter Squadron. Mitch already knew some of the men and was introduced to others, including First Lieutenant Besby Holmes, the skinny twenty-four-year-old San Francisco native who combined athleticism with smarts—a junior college swimmer and boxer as well as state champion chess player. Holmes had one of those Pearl Harbor stories that never paled in the retelling—how he'd been at church first thing that Sunday morning nursing a hangover when the surprise raid had started, how he'd fired at an enemy plane with a .45-caliber pistol and then given chase in a beat-up P-36 he'd found abandoned and unscathed on a runway. Until late summer, Holmes had been in Hawaii, and he had just arrived at Guadalcanal via New Caledonia. Like Wallace Dinn, Rex Barber, and Tom Lanphier, Holmes would prove to be Mitch's kind of pilot—bold and fearless—and soon enough the two were flying together in the newly created 339th Fighter Squadron.

The US base was a small coastal pocket on a compact island—only ten square miles—that was otherwise covered in thick jungle. Flies and mosquitoes, millions of them, infested the rain-soaked jungle. "The minute the sun went down you had to get under the mosquito net," one of Mitch's pilots said. "When they landed on you, they would just stand up, drop their thing in there, and you could just see them fill up their bellies with blood." The battle scarring from weeks of warfare was hard to miss. Many more bomb craters dotted the area than Mitch had seen from the sky, some having been converted to dumping pits for empty tin cans and other waste. Bullets had eaten holes in coconut tree stumps and tropical trees. Burnt hulks of Japanese planes lay on the periphery of the airstrip, along with various equipment left behind by exiting Japanese soldiers: shovels, mats, even a steamroller. The Americans were using as their control tower a rickety framed pagoda still standing on the strip the Japanese had built. Mitch quickly learned that to signal incoming Japanese planes a horn was sounded by ground crews and a captured rising sun flag was run up a pole at the pagoda.

The human battle scarring was hard to miss as well. The marines on Guadalcanal were edgy and weary, worn down by combat, sleep deprivation, illness, and the manic slapping at ankles, eyebrows, and elbows to ward off the malaria-carrying mosquitoes. Initially the marines' task had been to take over the airfield, and once that was accomplished, they were to have been replaced by army troops. "But now for almost two months, through ferocious attack after attack by the relentless Japanese, by land, by sea and by air, the landing force had had to fight on," noted John Hersey after arriving on the same transport plane as one of Mitch's pilots, Doug Canning. The day after Mitch got there, Hersey left camp with a marine patrol heading into a jungle valley five miles west of Henderson Field, part of a larger mission to confront the Japanese at the Matanikau River and prevent them from pushing any closer to Henderson Field. Hersey's account of the three-day battle, which ended in a standoff and was known as the Third Battle of Matanikau, ran in *Life* magazine the next month, November 1942. While the Japanese lost upwards of seven hundred men, the marines suffered the most casualties so far on Guadalcanal— sixty dead—and John Hersey used the bloody, up-close combat at the river as a microcosm to inform readers back home "that Americans are not invincible." Hersey saw firsthand "defeat, panic, flight. There had come a moment when the imminence of death simply overpowered a group of men, even though they were members of a hyper-proud service, were well-trained, and were veterans of terrible battles that had proven them brave."

The neophyte war journalist also quickly appreciated the degree to which the marines despised their Asian enemy and would have preferred fighting Nazis over "Nips." "Germans are misled, but at least they react like men," one marine told Hersey. Fighting Germans was a contest between human beings, he said, akin to an athletic performance of matching one's skill against the other's. The Japanese soldiers, by contrast, were "like animals." The marine explained, "Against them you have to learn a whole new set of physical reactions. You have to get used to their animal stubbornness and tenacity. They take to the jungle as if they had been bred there, and like some beasts you never see them until they are dead."

MITCH WOULD NEVER EXPERIENCE THE DIFFERENCE BETWEEN fighting Japanese and Germans, given that he would fight only in the Pacific. What he did know was that as the Third Battle of Matanikau got started, he was going to be right in the thick of things the only way he knew—in the sky. Less than forty-eight hours after his arrival, at dawn on October 9, he was put in charge of eight P-39s taking off from Henderson Field. One of Yamamoto's naval groups, consisting of five destroyers and a cruiser, had been spotted returning up "the Slot" off the coast of New Georgia, an island about 150 miles north of Guadalcanal. The enemy ships had not been able to clear the Slot by daylight, and the Americans saw an opportunity to inflict damage on them. Mitch's fighter planes were escorting navy dive-bombers known as SBDs, short for Scout Bomber Douglas, planes with strong maneuverability and hefty bomb loads. SBDs had played a critical role in sinking Japanese aircraft carriers at Midway.

To provide cover, Mitch took his flight slightly above the SBDs, to an altitude of about 12,000 feet. From there they easily spotted the Japanese ships, and the SBDs went into their dives. But Mitch also spotted at least five Japanese floatplanes—fighter planes equipped with floats for sea landings—accompanying the enemy ships. The floatplanes were at about 8,000 feet, which gave Mitch's planes the advantage, attacking from above. He signaled his fliers to intercept them, and off they went.

This was it, the real deal, and for all the simulated dogfights and intensity he'd brought to bear in his training—such as the time he'd gotten so locked into a dogfight at Hamilton Field that he hadn't even noticed the exercise had been called off—he was now facing live combat. He wasn't feeling panicky, but as he picked the enemy plane to attack, he also had the sobering thought that he "was just a green kid." His P-39 was armed with a 37 mm automatic cannon that fired through the plane's nose, and during exercises on Fiji, Mitch had been unimpressed by its performance. The gun was slow firing, and during drills he could actually hear its sluggish *chong, chong, chong*—really that slow, but it was all he had, so that when he made his move and closed in to where he was only about 300 feet behind the Japanese floatplane, he cut loose.

The third shot was the one that hit, right where the float joined the fuselage. Mitch watched as "that sucker blew, must have blown into a million pieces." Pieces of the shattered plane flew past him in the cockpit. Mitch was pumped up. "Man, this is great" was how he felt, and right away he looked around for a second target. Hungry for more, he suddenly spotted an enemy plane coming at him head-on. He told himself to stay calm, he was fine, not nervous and only thinking, Bring it on. The planes headed toward each other, as in a game of chicken. Mitch waited, and when he felt he was within range, he pulled the trigger again. Nothing happened. He could hear it go *brrrrr*. That was all he got out of the gun. The cannon had jammed. Fortunately, the Japanese fighter missed him as they flew past each other, but when Mitch saw two other floatplanes coming for him, he knew it was no place for him to be, not without a working gun. The SBDs were done anyway, although it wasn't clear what harm their bombs had caused. It turned out that Mitch's other pilots had also had trouble with their guns jamming. Everyone left.

They returned to Henderson Field before midday, everyone accounted for. Mitch had no time to savor the clash, however. He, Dinn, and others were in the air again that afternoon, escorting SBDs on another bombing mission. But that morning's flight was one he would never forget: his first kill. "I won't say you remember it like your first girlfriend," he said afterward. "But it's something like that, the first time you shoot down an airplane."

IN BETWEEN A STEADY PACE OF FLYING, MITCH AND THE PILOTS he'd brought over with him from New Caledonia did their best to get settled in. They moved into tents burrowed into a hillside on the jungle's edge overlooking Fighter Two, another airstrip the Seabees were building several miles west of Henderson Field. The new airfield was parallel to the coast and only about 150 yards off the beach, and it featured a semicircular taxi strip connecting the ends of the runway, as well as a plane revetment, or protected parking area, with blast walls on three sides. The four-man tents came with cots and mosquito nets, and below the tents were a mess hall and a makeshift operations building. Every pilot had access to a shallow foxhole, dug either inside the tent itself or right outside. There were larger foxholes covered with coconut logs and sheet iron nearby.

The army pilots immediately realized the necessity of the foxholes. Within days of their encampment, "Pistol Pete," a nickname given to Japanese field cannons, made their noisy debut. The cannons were positioned on a ridge above the nearby Matanikau River—a location that kept changing to avoid detection—and began intermittently firing shells onto the US base. The shells sailed over the pilots' tents on the hillside. They could hear the reports first, then the shells coming, then the shells landing, *BLAM!* The pilots had about ten seconds from the time they heard the shells to hustle into a foxhole.

A second threat followed: a Japanese pilot nicknamed "Washing

Machine Charlie" commenced nocturnal bombing raids. "It'd go *ga-doink, ga-doink, ga-doink,* that's the way it sounded," Mitch said. "One of those whining, whing-whang engines." The Japanese had intentionally monkeyed with the plane's twin engines to make a racket. Every night the men would be lying in their cots, trying to get a little sleep after flying all day, and then that despised pilot would arrive and drop a bomb. The plane came from Rekata Bay, a base the Imperial Navy had built on the northeast coast of Santa Isabel Island 170 miles away to use for offensive operations against the US forces. The nightly raids began soon after Mitch arrived and continued throughout the fall of 1942. They hardly constituted a full-on assault on the US installation. In fact, it was a bonus for the Japanese if the single bomb—or two—that Charlie dropped on the solo runs smashed anything. The raids were instead designed to torment the Americans, ruin their rest, and drive them to distraction.

The strategy wasn't a bad one. The men, hunkered down in tents, would hear the grinding engine off in the distance. Mitch would lie there, telling himself "I'm not going to get up. I gotta sleep." He'd turn over and try to sleep until "finally when that damn bomber is getting pretty close, you'd get up and run for a foxhole and jump in the hole, an open pit thing, you'd hit it on the run, squat there for a minute or so, and he'd go on." Not only was a night's rest ruined, but Washing Machine Charlie had succeeded in getting the Americans out from under their mosquito nets, exposing them to the malaria-carrying insects. "When we'd go for the foxholes in skivvies they'd chew on us." The marines returned fire, but that only made matters worse; the antiaircraft guns were located on a hill not far from where Mitch and the pilots slept, and the guns shook the tents whenever a round was shot. Further, even though searchlights roamed the dark sky and sometimes even held a light on Charlie at 25,000 feet, the noisy guns never succeeded in hitting their mark. Among Pistol Pete, Washing Machine Charlie, and the antiaircraft guns, the wear and tear was evident on the men's faces at breakfast. Doug Canning was one pilot who got fed up with all of the jumping from cot to foxhole. "I just moved my cot down and slept in the fox hole from then on and until I left Guadalcanal. That was my home, a foxhole up on a hillside."

The waking hours gave Mitch plenty of time to think about the first kill. He'd felt a surge of adrenaline, that's for sure, and afterward he wanted a repeat of the rush that had come from squaring off in the sky. Of his enemy he said, "He's a flying machine armed and capable of killing you, just the same as you are capable of killing him." He recognized, though, that it wasn't like the ground fighting, which was marked by a butchery that had become commonplace, as wave after wave of Japanese rushed the marines who were dug in and holding the perimeter. Corpses littered the areas surrounding Henderson Field, body limbs tossed about, faces mangled, soldiers who'd been cut in two by machine guns. Battle-weary marines, stunned and exhausted by all the bloodshed, took on the dark, vacant look that came to be tagged "the Guadalcanal, or 1,000-yard stare." The disposal of the dead near the Matanikau River became such a problem that engineers tried blasting an overhanging cliff to bury a pile of bodies at the bottom. A few weeks later, there was a night when Mitch tossed fitfully on his cot, unable to sleep because of the all-night gunfire between the marines and the charging enemy. In the morning he was in the mess tent getting breakfast when one of the marines he'd befriended waved him over. "Mitch, you want to see something?" The soldier took Mitch to the site of the previous night's fighting, right down to the battle line, and showed him the Japanese bodies stacked up after a banzai charge. Mitch winced. The sun was hot, and the Japanese corpses had begun to rot. He turned away.

"I couldn't eat breakfast."

Ground combat was almost personal: a soldier took aim at a guy in his sight and shot him through the head. It wasn't like that in the sky. One pilot never really saw the other when he hit his "flying machine." The shooter saw the plane burst into flames or a wing come off, or maybe the plane started to spin and he saw it crash. But it wasn't as though the shooter sat there in his cockpit and followed his target all the way down. Because if he did, he was going to get shot. The shooter was usually past the target by that time anyway, looking around for other planes that he could be attacking. Mitch could see that shooting down airplanes wasn't anywhere near as intimate as killing on the ground and was more businesslike. There was a certain anonymity to it. But even with that clinical distance,

Mitch made no bones about his single-minded focus on destroying his enemy, especially this one. If he did see a Japanese pilot bail from a plane he'd blasted, he wouldn't hesitate to finish the job. "I'd shoot him, too, if I had the time." It was kill or be killed, that simple. No mercy.

"War is not an ethical thing," he said. "It's a horrible thing."

CHAPTER 12

NIGHTS TO REMEMBER

—

Rex Barber with a sheared P-38 wingtip
Courtesy of the National Museum of the Pacific War

THE JAPANESE BID TO RETAKE GUADALCANAL ISLAND THAT BE-
gan in earnest on October 13, 1942, came in two waves: the first,
smaller, one aimed at setting the stage for a massive second one. It
started that Tuesday with a major convoy of Japanese troopships
arriving to the coast west of Henderson Field, where the Japanese

ground forces had relocated. The convoy's task of depositing several thousand more infantry was the latest in a steady buildup of troops for the climactic assault, a buildup that would reach about 25,000 men. What made this delivery different from previous ones was the cover Yamamoto was providing. From his naval headquarters at Truk Lagoon, 1,300 miles away in the central Pacific, the admiral had sent a task force of two mighty battleships, *Kongo* and *Haruna*, along with nine destroyers, to attack Henderson Field. Intended to freeze the Americans in place, the shelling would be the fiercest and most concentrated since the marines had taken over the airfield in August.

The naval task force, aided by a plane dropping flares to illuminate targets, unleashed its firepower shortly before midnight October 13, firing hundreds of rounds of armor-piercing high explosives and blasting Henderson Field for nearly two hours. In addition, twenty-two Japanese bombers, escorted by Zero fighters, bombarded the base. Joining the action was "Pistol Pete," the long-range 150-mm howitzers, lobbing shells at the Fighter Two airfield from their position near the Matanikau River. The ground artillery assault began on October 13 and continued into the next day and the one after that.

The seventy-two hours that followed October 13 came to be known as Three Nights to Remember. Shells from Yamamoto's battleships systematically tore apart the main runway, with exploding gasoline tanks lighting up the night. The bombing pattern moved from the airstrip to campsites in the palm grove and on the surrounding hillside. There were moments when the flares and the ammunition and gas dumps that burst into flames combined to make it seem as bright as midday. Seabees of the 6th Construction Battalion scurried to try to keep the runway in operation, unloading dirt from dump trucks and slapping precut Marston matting over the bomb craters. They'd duck and run to fill a crater, tamping down dirt, laying down the mat, and then scattering as more bombs whined overhead. But as soon as they filled one crater, another bomb created a new one. They could not keep up. The bombardment was relentless as the Japanese bombers continued to attack Henderson Field during the stretches when Yamamoto's battleship assault from the ocean was halted.

The US forces sprinted, jumped, and threw themselves into fox-holes, where they spent the nights trying to stay out of harm's way. Mitch jumped into a hole with the rest. "It was very hot. Lots of men puked." In the morning, the men emerged to find shells from the battleship guns—their jagged noses measuring fourteen inches in diameter—littering the ground, unexploded. Tents had been either sliced by shrapnel or crumpled. Forty-one men were dead. The day before the attack, Henderson Field had had ninety working aircraft, a combination of navy SBD dive-bombers, navy F4F Wildcat fighters, and army air forces P-39 and P-40 fighters. Mitch and the others checked on their planes and found them in shambles. Together they pushed most of them off to the side. They counted forty-two planes still able to fly. Henderson Field was so pockmarked with bomb craters, however, that heavy bombers didn't have enough room to take off. The nearby grassy strip known as Fighter One—though rough and short at 2,000 feet—proved usable, although only for the lighter planes that had survived the attacks. But the US forces also faced a fuel crisis that would not be resolved anytime soon; cargo ships carrying fuel could not be expected to get through to the island, given all the Japanese warships and fighter planes still in the area. "We don't know whether we'll be able to hold the field or not," one marine colonel reported.

Their prospects only got worse. During the afternoon of October 14, two waves of Japanese bombers and fighters descended virtually unimpeded, freely inflicting further damage and more US casualties. Nighttime saw continued bombardment from the warships, and at dawn on October 15, the Japanese convoy reached the waters off Lunga Point. In plain sight of the US forces at the crippled Henderson Field, the five Japanese transport ships unloaded supplies and troops, protected by escorting warships. The message couldn't have been any clearer: the enemy was readying for a ground offensive. Japan's Seventeenth Army leaders were emboldened, at once believing that the US troops' morale had been crushed by the bombing and supremely confident that their plan to encircle Henderson Field from several directions with their beefed-up infantry divisions would bring them victory.

The marines could only dig in and await the anticipated ground attack. In the meantime, aerial assets were used as best as possible, with flights cobbled together to challenge the Japanese buildup. With little fuel left, the ground crews scrounged around the jungle along the runway for gasoline drums that hadn't been hit, while others siphoned fuel from disabled planes. Pilots with their parachutes strapped on bolted from foxholes during breaks in the bombing raids and ran in a zigzag to planes loaded with 100-pound to 500-pound bombs. Firing up the engines, they looked like drunk drivers weaving down the pockmarked runway that the Seabees had treated with their first aid of dirt and mats. "The Seabees were marvelous," pilot Doug Canning said, "the way they ran out with tampers after raids to fill holes, repairing the field fast and then putting the Marston mat back down." Several of the flights managed to strike the oncoming troopships using the 100-pound bombs, but the damage did not slow the convoy. Another group of planes went looking for "Pistol Pete," hoping to silence the nearly constant artillery fire, but they couldn't locate the jungle-concealed cannons sharing the nickname and, running low on fuel, had to give up. On the third day of the Japanese push, October 15, the fuel crisis eased a bit when C-47 transport planes arrived at intervals from the US base at Espiritu Santo, six hundred miles to the south. Each plane could carry only twelve drums of fuel, however, just enough to keep twelve planes in the air for an hour. But everyone—pilots, mechanics, and ground crew alike—was now pitching in to get the fighters into the air, patching up and fueling aircraft, loading bombs by hand, and belting ammunition into the guns.

Throughout the day and night of October 15, Mitch and Wallace Dinn were part of a group of army fliers dispatched to disrupt the enemy's unloading of troops and supplies. During one of their flights, the P-39s fired on a Japanese transport while an accompanying navy SBD bomber hit and sank the ship with a 500-pound bomb. One P-39 pilot was shot down during the attack but was immediately picked up unharmed. The next attack was even more intense and dramatic: the fliers ran into a bunch of Zero fighters. The notoriously fast planes were covering for the ships and barges unloading troops, but the army fliers still managed to get through. Dinn made a direct midship hit on a transport. The bomb blew the vessel apart and sent wounded and dead enemy soldiers flying into the ocean. The water turned red from blood, and Mitch watched sharks dart in to grab flesh. He stared in horror: "The sharks were having a heyday." The next morning, Mitch, Dinn, and the others were back at it again, this time coming upon beached transport ships unloading men. "Troops went into the water but we strafed the hell out of them," Mitch said.

For a day and a half, the pilots had flown as often as they could, returning at dusk or even later. Landing at night was difficult, with only a handful of makeshift lights dimly illuminating an airstrip often drenched in tropical rain. They managed during that time to sink two transport ships, set several others on fire, and repeatedly strafe the beach landings, losing some of their own planes and men in the process. But their effort was a nuisance at best, for by the time the Japanese convoy departed Guadalcanal on October 16, it had succeeded in landing thousands more troops and most of its cargo and ammunition.

THEN CAME A LULL, SAVE FOR SPORADIC SKIRMISHES. YAMAMO-to's main task force cleared the area and set sail for the naval base at Truk. Meanwhile, the 2nd Division of Japan's Seventeenth Army left its stronghold in the village of Kokumbona west of Henderson Field and began hacking its way through the jungle. The battle plan called for Lieutenant General Masao Maruyama of the 2nd Division, leading nine infantry battalions totaling about 5,600 men, to loop around the field and set up for a surprise attack from the south. Another Japanese unit, the Seventeenth Army artillery, headed for the ridges overlooking Henderson Field, from which it would shell the airfields, simultaneously diverting the marines and providing cover for troops charging from the west across the Matanikau River as well as General Maruyama's men from the south. Completing the three-pronged assault, a coastal force consisting of about 2,900 men, a tank company, and extensive heavy artillery—150 mm howitzers, field artillery pieces, and other guns—departed along the beach for Henderson Field. The Japanese commanders enveloping the airfield from the south and west and the coastline were instructed to annihilate the enemy and not let up until Marine General Vandegrift himself and his staff emerged carrying a US flag and a white flag to surrender. Then, once Henderson Field was retaken, the forces were to report their victory to headquarters using the code word "Banzai," an expression translated as "Ten thousand years," or "Long live."

Changes in US leadership were afoot. Admiral Chester Nimitz, commander in chief of the US Pacific Fleet, decided that a more forceful frontline leader was needed to tackle the perilous situation on Guadalcanal. He chose Admiral William F. "Bull" Halsey, Jr., and on October 15, the order was delivered to Halsey as he arrived at the island of New Caledonia: "You will take command of the South Pacific area and South Pacific forces immediately." The crusty, tough-talking Halsey was known widely for his go-get-'em aggressiveness. Following the Pearl Harbor attack, he'd told the press, "Before we're through with them, the Japanese language will only be spoken in hell." News of his appointment spread quickly in the South Pacific, providing a morale jump on Guadalcanal.

On October 16, the US forces emerged from their foxholes and ate their first hot meal in days, having survived on an underground diet of cold hash and hardtack, a bland, dry biscuit. Some of the men took the opportunity to scrub their damp, sour clothes to try "to keep down some of the stink," as Mitch described it. Mitch used the respite to write to Annie Lee. "You can see now that I do get around. Here I am in Guadalcanal," he announced in the October 22 missive, ignoring the custom of using code words for the islands. It was his first letter since arriving at Henderson Field two weeks before. He mainly chitchatted, asking Annie Lee for news from home, passing along what little he'd heard from his father, Noah, and telling her that he had relished reading a bunch of her letters, some dating back to May, that had been forwarded from Fiji, where he'd been stationed previously. "I enjoy them so much," he said. "Since things are well blacked out here at night and since I've been going pretty hard and fast during the daytime I haven't even had a chance to read all of them. But that gives me something to look forward to—very much so!" He was hoping, too, that she'd get his letter in time for her birthday—she would turn twenty-six on November 11—and poured on the love and affection: "I know I have the best, sweetest and most beautiful wife in all the world. It seems that my being apart from you just makes me realize more and more how much I love you and how much I need you to make my life complete."

Mitch included a peek at the war but cast the combat in a rah-rah way, bragging that he was "going to town with the Japs and loving every minute of it." He told her about the enemy plane he'd destroyed—his first—and boasted that more kills were coming, as he'd "only gotten started." The pounding Japanese attack that had just ceased? He downplayed it. "We are being bombed every day and have an occasional shelling," he said, "but it's really not too bad as we usually manage to get in a well-protected spot." Overall, he said, if not for an agitator at night—an oblique reference to Washing Machine Charlie—things were going great. "I have never felt better in my life," he said. "We work hard but have no way of stepping out of line. Haven't had a drink in weeks. If these yellow devils would leave us alone at night so we could sleep it would be about perfect."

The truth was far less rosy. His letter contained no hint of the worry and the mounting tension felt by the US forces waiting for the renewed Japanese effort to overrun Henderson Field. The best that Secretary of the Navy Frank Knox could muster when reporters in Washington, DC, pressed him on whether Guadalcanal could be held was "I certainly hope so." Editorial writers were not reassured. "Secretary Knox, in a game effort to be optimistic, simply stated that he hopes our boys will win," wrote the *Los Angeles Times*. "There is not the slightest doubt the enemy has succeeded in landing considerable reinforcements, not only in men, but also in mechanized equipment and field artillery."

Annie Lee Mitchell would have to rely on the press coverage, not her husband, for the broader reality—that US control of Henderson Field hung in the balance. And Mitch's epistolary spin clearly did not have a calming effect. "Johnny is in the Solomons," Annie Lee fretted to her Aunt Ludma, "and I can hardly stand the thought of that."

THE JAPANESE DRIVE BEGAN THE DAY AFTER MITCH WROTE HIS letter, and it started with thousands of fresh troops and tanks moving eastward across the Matanikau River toward Henderson Field. The new commander in the South Pacific, Admiral Halsey, monitored events from New Caledonia, where a chaplain from Guadalcanal conveyed the concern, even despair, felt by the US defenders bracing for the attack. Halsey would have none of it. "We're going to win," he said, "and you and I will see Yamamoto in hell."

Four times that Friday night and into the next morning, Japanese soldiers and tanks emerged from the jungle and attempted to cross the river, hordes of soldiers rushing the entrenched marines. Each time, the marines managed to stop them, with enemy bodies stacking up on the sand bar. One reason the marines were able to repulse the attack was that their attention was undivided. The second Japanese force, led by General Maruyama, had not yet reached the south side of Henderson Field, slowed by the unexpectedly grueling march through the gnarly, hot, rainy jungle, meaning that the grand plan for a coordinated assault fell apart. Two more days passed before Maruyama's battalions began multiple frontal attacks; nighttime charges continuing for the next two nights by infantrymen screaming battle cries, throwing grenades, and firing weapons. Equipped with rifles, machine guns, antitank guns, mortars, and other artillery, marine and army soldiers at the perimeter time and again mowed down the storming Japanese troops. Some gunners peed into the machine guns' water jackets to keep them firing and not overheating. The US defenders bent but did not break, even though the Japanese soldiers managed several times to close within a few hundred yards of the airfield. But a US counterattack succeeded in killing them and clearing the field. The carnage was jaw-dropping. Whereas the United States lost about 60 men during the raids, the enemy death toll exceeded 1,500. The Japanese had grossly underestimated the US troops' strength, and the roughly 23,000 Americans they faced overall, though stretched thin, were more than three times the preassault estimates.

The fliers pitched in from the sky: Mitch, Wallace Dinn, Doug Canning, Mitch's wingman, Jack Jacobson, and others from the Army Air Forces, Marines, and Navy. Mitch led several raids on the Japanese base at Rekata Bay on Santa Isabel Island, about 170 miles from Guadalcanal. In one, he, Dinn, and Jacobson dropped 500-pound bombs and then joined marine fighters, dive-bombing and destroying eight floatplanes. Mitch strafed a gasoline dump of about three hundred barrels and set it afire. During a return raid that afternoon, Mitch and Jacobson fired on two planes sitting on the beach, and then Mitch, on a final pass, suddenly felt his P-39 rock. Bullets from a ground gun had riddled his plane. He was not wounded, but his right wing was hit, severing the cables to the ailerons, the panels at the tips of the wings that go up and down and enable a plane to bank and roll. Mitch compensated and headed back to Guadalcanal. Both ailerons had dropped an inch or so, and he had to fly home holding his stick all the way to the right side of the cockpit.

The marine commanders also put Mitch and his pilots to work assisting the ground forces, dive-bombing the enemy troops mobilizing around Henderson Field. They'd attack a specific location from the air and then watch as the marines mopped up. Mitch said, "The Marines would find dead Japanese, their shoes blown off and no visible wounds. They were already dead or almost dead. The Marines finished the job." In one instance the pilots were directed to hit a hillside near the airfield where the enemy had dug in. Mitch burned the grass off the hill with incendiary bullets. Several Japanese ran out of foxholes into nearby woods. The pilots dive-bombed and strafed just in front of the marine advance. Other times they dropped depth charges into the jungle to flush out the enemy, whereupon marine shooters picked them off. Mitch later learned that the infantry had rounded up a cache of souvenirs—samurai swords and the like—and was irked when the soldiers wanted to peddle the combat mementos for money or a precious bottle of whiskey. "That made me sore since I knew we'd killed lots of the enemy for them." The souvenir grab during the vicious fighting in the Pacific devolved into macabre barbarism on both sides. Japanese beheaded dead marines and mutilated their corpses, while marines made a point of rifling through the packs and pockets of the enemy dead. "Helmet headbands were checked for flags, packs and pockets emptied, and good teeth extracted," one marine wrote later. "Sabers, pistols, and hari-kari knives were highly prized." Some of the infantrymen took to wearing necklaces made of the teeth of Japanese soldiers or affixing severed ears to their belts. Admiral Nimitz, appalled by the gruesome trophy taking, issued an order that "no part of the enemy's body may be used as a souvenir," but the abuses continued nonetheless.

Mitch got his second kill during one of the many dogfights over Henderson Field the first day, October 23, that the Japanese resumed all-out fighting. He and navy pilots flying Grumman F4F fighters were up against sixteen Japanese bombers and twenty escorting

Zeros. "Grummans got 19 Zeros and I shot down one," he wrote matter-of-factly in that day's combat report. Very quickly, in three weeks of intense fighting, Mitch had received his combat baptism by fire—harrowing encounters he survived while absorbing lessons that he then noted in his diary, combat reports, and letters. "There are all sorts of things to be learned that one learns only from experience," he wrote. The takeaways covered the gamut, from the strengths and weaknesses of the P-39 ("Very good for strafing and dive-bombing," he said, "not too suitable for combat at high altitudes.") to the role luck played in whether he returned from a mission or not. "We're not out there playing tiddlywinks," he wrote. It was kill or be killed, and "luck is part of it. But you make your own luck. I believe that. I think if you're prepared and you're ready, when the time comes, you're going to be able to handle the situation." He also discovered something else during October's pivotal Battle of Guadalcanal, something he hadn't necessarily expected. "I really like it out here," he wrote. Sure, the conditions were tough, and he missed "the nice things I could have if I were in the states." But there was also this: "Great personal satisfaction in being up on the front lines—of getting in there and really scrapping and of knowing you are doing your very best to lick hell out of 'em." He continued, "It's a satisfaction nothing else can replace. It lets a man know how much guts he really has got, and allows him to revel in that fact."

Mitch was on the fast track to becoming an ace fighter pilot—a status defined by five or more kills—and got his third on "Dugout Sunday," the name the soldiers and press gave to the unrelenting Japanese bombing assault on October 25. "Today marked the most persistent series of air blows in any single day of the bitter struggle for Guadalcanal, to say nothing of the machine gunning and shelling by warships and shore batteries," wrote Robert Cromie, a war correspondent for the *Chicago Tribune*. The heavy fire started shortly after midnight, startling the troops and sending them into their foxholes. Breakfast, wisecracked Cromie, was butter, eggs, "and Zeros." The continuous bombing was supplemented by a rare daytime appearance of a cluster of Yamamoto's destroyers off Lunga Point and a steady diet of shelling by "Pistol Pete," the enemy's land battery. The Japanese pilots at first focused on some planes at one end of Henderson Field that were parked in formation, blowing them apart, but they were actually disabled, battered hulks that the Seabees had positioned as decoys while camouflaging working aircraft along the edges of the airstrip. The skies were a clear blue, but heavy overnight rains had turned Fighter One into a wet bog, grounding US aircraft and preventing any immediate response. "Field was muddy, and there were Zeros over the field all day," Mitch complained. "Five dive-bombers bombed us, a Japanese destroyer sneaked in and destroyed three of our Guadalcanal to Tulagi boats."

By early afternoon, the hot sun finally dried the strip, and orders were given to go after the warships using all available P-39 fighter planes equipped with 500-pound bombs. Only four were ready, however. Mitch took one up, along with Dinn, Jacobson, and another pilot from Texas, named Fred Purnell. Mitch's engines faltered and he lost speed, requiring him to turn back. The others dropped their payloads but made no hits—misfires that were frustrating but not all that unexpected. It was difficult to dive at a ship that was moving, turning, and twisting. "Once you commit to a dive you don't have the ability to twist and turn like they can, so we didn't have much success," Mitch's wingman, Jacobson, said. The four tried again later in the day, and that time Jacobson landed a bomb directly on the bow of one of Yamamoto's battleships. Several B-17 Flying Fortresses followed with additional hits, and Mitch had a perfect view from his cockpit as the battleship sank. Then, just before dark, it was Mitch's turn. He and the others were sent north of Guadalcanal to attack one of Yamamoto's naval support groups, made up of a single cruiser and fourteen destroyers. But on the way they spotted a number of Zero floatplanes and got sidetracked. Taking advantage of the cloud cover, they attacked the Zeros before they themselves were seen. "I got one on the first pass, Lieutenant Purnell got one on his first pass and Lieutenant Jacobson got one," Mitch reported later. They ended up not going after the naval force, having dropped most of their bombs.

The Japanese losses at sea, in the air, and on the ground were continuing to mount. By nightfall, US fliers had shot down twenty-two enemy planes, and antiaircraft artillery had picked off another five. Similarly, on the ground, the US infantry followed the pattern of previous nights, under siege but holding tough and slaughtering a dwindling number of soldiers from General Maruyama's 2nd Division. Several thousand corpses had piled up on the two main fronts, south of Henderson Field and along the Matanikau River to the west. Wrote war correspondent Cromie, "Japs in all the curious postures of death lay awaiting burial, some sprawled across barbed wire and still clutching their useless weapons."

The next day, October 26, the Japanese commanders decided to call off further attacks, and their hopes of a resurgence at sea were dashed that same morning when a fleet of carriers and battleships that Yamamoto had ordered to an area east of Guadalcanal was stymied by a smaller US force scraped together by Admiral Halsey. Technically speaking, the Japanese won the two-day Battle of Santa Cruz, sinking one US carrier and damaging another while losing none of their own. But the outnumbered Americans inflicted significant damage to two Japanese carriers and destroyed so many Japanese planes—nearly a hundred—and, just as crucially, their skilled crews, that Yamamoto ordered the fleet's withdrawal. The bid to retake Henderson Field, begun so confidently, had failed.

On Guadalcanal, the US forces felt their situation improving considerably, with time now to repair the two airstrips and finish another, Fighter Two. Their fuel, ammunition, and food stocks were resupplied, and they even initiated offensive operations against the Japanese headquartered in the village of Kokumbona across the Matanikau River. Army air forces, navy, and marine pilots flew daily missions, and early on the morning of October 28, Mitch took off in a P-39 Airacobra for yet another attack on the Japanese seaplane base at Rekata Bay on Santa Isabel Island. Pilots Jack Jacobson and Wallace Dinn accompanied him, each in a P-39. Their job was to fly cover for four navy SBD Dauntlesses, strafing and dive-bombing while the SBDs released their bombs, and then to accompany the SBDs safely back to Guadalcanal. The flight began routinely enough. The SBDs did their part and left, while Mitch and his two pilots strafed enemy floatplanes sitting in the bay and blasted a few ground installations. There wasn't even any enemy fire during their several runs.

Then, just as they were about to head back, a new opportunity arose. Wallace Dinn called Mitch's attention to a camouflaged gasoline dump on the beach. Mitch signaled him to go ahead and strafe it. Within seconds, though, he wished he hadn't. Dinn dove at the dump and fired his .30-caliber wing guns. He hit the target, and one end of the fuel dump exploded into a small fire. But when Dinn circled back for a second run, his plane was hit by ground fire, temporarily stunning him. Smoke filled the cockpit, while leaking antifreeze pooled at his feet. He managed to veer left, away from the Japanese base.

Mitch could not raise Dinn on the radio and watched as the plane shook and trailed smoke. Dinn had to bail out. The P-39 crashed into the sea, and the last Mitch saw was his friend's parachute disappearing into the jungle below. Worried and shaken, he and Jacobson hurried back to Henderson Field. They returned to Rekata Bay that afternoon on another bombing mission and also searched for their missing friend. They flew up and down the beach several times looking for Dinn but saw no sign of him or his parachute. Days passed, with more missions and still no indication whether Dinn was alive or dead. Then, eight days later, on November 4, the suspense finally ended when a small boat docked at Henderson Field. Wallace Dinn stepped ashore along with some other servicemen and, of all things, a Japanese prisoner. The weakened Dinn had made it back with a jungle survival story about natives who had hidden him from the enemy and then transported him by canoe to the safety of a British boat on the other side of Santa Isabel Island. The pilots shook their heads, impressed that Dinn and his native protectors had even taken a prisoner. It was all pretty amazing, for sure, but, drama notwithstanding, Mitch was mainly relieved and happy to have his best combat buddy back.

AT THAT POINT, MITCH AND THE BAND OF PILOTS WHO MADE UP his flight merited a break. "They say we are due for some rest," he reported to Annie Lee. "Personally, I would just about as soon stay on and fight these yellow so and so's right here." But orders were orders, he said. "The powers that be say we have earned a vacation." Mitch, Dinn, Jacobson, Canning, and a few others departed Guadalcanal on November 10, first stopping at "Poppy," the code name for New Caledonia, and three days later, arriving in Australia. The break was scheduled for a week, but Mitch ended up being gone a month. When it came time for him to board a plane for Henderson Field, he was shaking with a bad case of the chills. He'd quit taking atabrine during his leave, which wasn't an uncommon thing to do. Soldiers disliked the antimalaria drug, whose side effects included sweating, shaking hands, and, over time, a yellow discoloration of the skin. There was also a false rumor that atabrine caused impotence. Mitch was not allowed to depart with the others and, instead of returning to the front line, was ushered to a hospital, where blood tests confirmed that he had malaria.

Mitch's hospitalization meant that he was not around for the ongoing combat on Guadalcanal, fierce fighting that steadily but stubbornly tilted in favor of the Americans. The Japanese commanders desperately sought to reload for one final try to dislodge US forces from Henderson Field. Yamamoto once again ordered his destroyers to bombard the air base while transport ships scooted down the Tokyo Express from Rabaul to drop off more reinforcements and supplies. The fighting that month, including another sea battle at midmonth, was chaotic and costly for both sides, although more so for the Japanese, whose overall morale plummeted and whose battered infantry was hit hard by exhaustion, tropical illness, and malnourishment. "Their uniforms, little more than rags, hung from emaciated limbs," observed one historian. "Their hair had grown long and crawled with lice; their skin was dirty and pocked with open sores." By month's end, the Americans succeeded in restricting the Japanese resupply efforts to submarines, which hardly sufficed to overcome the severe shortages in food, ammunition, and fuel. For the Japanese army, Guadalcanal had become known as "Starvation Island."

Mitch didn't cotton to being sidelined, but the hospital stay did provide him with an opportunity to surprise Annie Lee. From the hospital in Sydney he was able to send her a telegram via the Postal Telegraph-Cable Company early on the day before Thanksgiving. Annie Lee received it after nightfall in her tiny apartment at 147½ East Woodland Street in San Antonio. She'd fallen asleep on the couch when the ringing phone awakened her at 10:30 p.m. It was Western Union calling to say it had a cable from Captain John Mitchell. Annie Lee sat bolt upright. She'd been anxious ever since learning that Mitch was on Guadalcanal, and some of her relatives hadn't helped matters by telling her that if Mitch was in the Solomon Islands, she probably shouldn't count on seeing him again. The husband of one cousin had said it would take all of Annie Lee's prayers and then some to keep him safe—hardly the kind of support she needed. Annie Lee had snapped back, "If they couldn't say anything better just keep quiet." Now she held the receiver and waited for the Western Union worker to read her the cable: WEEK LEAVE EVERYTHING FINE ALL MY LOVE = JOHNNIE MITCHELL. It was a brief, simple message in which Mitch, not surprisingly, made no mention of his illness; he would get into that in a later letter. But it was just what Annie Lee needed. There was no way she could fall back to sleep, and she later wrote her Aunt Ludma about the telegram and especially the part about Mitch saying he was safe. "That cable, of course, was after the big naval battle in the Solomons on the 14th and 15th," she said, "so I felt much better. I'm sure you realize how eager I am to have word from him."

Mitch returned to Guadalcanal on December 5, "itching to get going again," he wrote his wife, this time disclosing his illness. "I'm completely cured of my malaria," he continued, and "gonna run that three [kills] up to ten this time, give myself a couple of yellow bellies for a Xmas present." He joined the aerial assault against a last-ditch Japanese effort to resupply troops using the "drum method," whereby empty fuel drums were sterilized and stuffed with provisions and ammunition. The drums, roped together, were then dropped from ships into the ocean waters close to the Japanese encampment at Kokumbona, west of Henderson Field, so that small craft and Japanese soldiers could pull them ashore. The drops were made at night to avoid detection, and hundreds of drums would still be adrift at daylight, sitting targets for Mitch and other pilots to strafe and destroy on a daily basis. "I have been going from 3:30 a.m. until about 10 p.m.," Mitch said.

Two anniversaries came up in quick fashion, one of international significance, the other personal. December 7 marked one year since Yamamoto's attack on Pearl Harbor, and to honor the day the

Japanese government-controlled media and propagandists published roundups of the war that were celebratory, full of false data about US losses, and exulting in the nation's inevitable victory. The truth—that Japan's prospects had taken a dark turn at Midway and continued to worsen at Guadalcanal—was a closely held secret within government and military circles. Yamamoto, for one, was clear-eyed, writing to a friend, "Here we are, one year since the commencement of hostilities, and I feel sad to see the handicap [advantage] we were given at the beginning being gradually whittled away." To another friend he included a short poem: "Looking back over the year, I feel myself grow tense, at the number of comrades, who are no more." Meanwhile, in the United States, a clear lift in the nation's spirit was evident. An interview with the new commander in the Solomons, the navy's Bull Halsey, appeared on the front pages of newspapers all over the country. "We are now definitely passing to the offensive; we're just starting!" said the admiral, who just six weeks earlier had said that the South Pacific war was touch and go. Halsey made certain to single out the country's archenemy of Pearl Harbor: "Tell Yamamoto that peace will be dictated in the White House but not as he envisaged," he said. Meanwhile, advertising campaigns promoting war bonds continued to exploit hatred toward Japan's naval hero. "Remember Pearl Harbor," exhorted one ad published in the *Washington Post* on December 7. "One year ago—exactly—Admiral Yamamoto and his talented knifemen exposed the true savagery of their nature." With that, urged the ad, "Buy War Bonds."

The other milestone was personal: Mitch and Annie Lee's first wedding anniversary on December 13. It had been a year since the couple had hustled into City Hall in San Antonio to get married during Mitch's stopover en route to California. Mitch made a point of writing his wife a love letter dated on their special day, and he was able to tell her that he was going to be promoted to the rank of major. December was the month, too, when Vic Viccellio, Mitch's boss, told him he'd soon be taking over as commander of the 339th Fighter Squadron, a unit composed of fliers previously attached to several other squadrons, including the 70th Fighter Squadron. His flying buddies Wallace Dinn, Doug Canning, and Jack Jacobson were already in the fold, but by the end of the year he would be welcoming men from the 70th he'd trained with on Fiji who were finally getting to join the war. "It's going to be fun to see ole Vic and some of the boys," he said. "Of course, we will have some tall tales to tell them and will naturally have some big bull sessions."

Reunited were pilots whose last big gathering had been the July Fourth Texas-style barbecue at Nadi on Fiji's western coast. Rex Barber and Tom Lanphier, two standouts, were among the new-comers, as were Joe Moore, Robert Petit, George Topoll, A. J. Buck, Roger Ames, and Delton Goerke. The fliers were in for a different fireworks display than that of the previous Independence Day; they were greeted on arrival at Henderson Field by one of the frequent Japanese bombing runs. And joining the unit officially was another crack pilot who was already at Henderson. Besby Holmes had ar-rived in October, in time to participate in the intense fighting over control of the base. He'd yet to record any kills but had blown up a beached Japanese ammunition ship.

Mitch and his squadron took the new men to the pyramidal tents burrowed in the hill overlooking the new Fighter Two airstrip, where they would live and work together. They'd be sharing planes, flying in rotation. "Our routine was mostly eat, sleep, fly," A. J. Buck said. From K-rations and wormy Japanese rice the menu expanded marginally to include Spam. "Spam spam, spam—a dozen different ways," Buck said. "Got mighty old, but still better than 'K' rations." Rex Barber in particular was happy to see Mitch and to join the fray. "I was always eager to get into combat just like I would be eager to get into a football game from the sidelines," he said. "I was out there to fight." Barber got his chance—and his first kill—in a matter of days. The Japanese were discovered building a new airfield at Munda on the southern tip of New Georgia Island, and Barber was sent to conduct surveillance. Though under orders not to attack, he couldn't help himself when he spotted a Japanese twin-engine "Nell" bomber making its final ap-proach. "It was too good to miss," he decided. From an altitude of 7,000 feet, he dived in behind the bomber. He fired away, hitting the right engine, which burst into flames. He watched the bomber crash into the sea and then spotted enemy Zeros to his left. He fled, lucky to have cloud cover to aid his escape. He later won a Silver Star for the action.

Tom Lanphier likewise didn't take long to make his mark—and, like Barber's, it happened over the Munda airfield. Lanphier was in a P-39 assigned to defend nine bombers as they attacked the strip being carved out of the jungle. Shortly after the raid began, a Japanese Zero zipped across his line of sight, racing after one of the B-17s. Lanphier gave chase as the speedy Zero curved toward the bomber, which was about a mile in front and unaware, preoccupied with flying its line to the target below. The Zero settled into its firing position to the rear of the bomber, and Lanphier, frantic but locked in, dropped in behind the Zero and turned on his gun switch. "I found myself close enough behind the Zero, who was ignorant of my lethal presence, to have him filling my whole gunsight." He pressed the trigger atop the P-39's control stick, and the cockpit shuddered as the .50-caliber guns fired. Lanphier missed, the tracers falling behind and below the enemy plane. He made a correction by lifting his plane's nose to raise the stream of gunfire. Then it seemed as though he'd overcompensated—it looked as though he was firing above the Zero. But just as the Zero was about to disappear below his line of sight, it exploded. The plane disintegrated before his eyes, bits and pieces flying off in all directions. Lanphier realized, "The same light-weight aluminum aircraft frame which lent the Zero its superior maneuverability also instantaneously converted it into a firecracker." He had gotten his first kill, and, to his surprise, his second came moments later, when he and other P-39 pilots took on two Zeros climbing after the B-17 Flying Fortresses as they turned to head for home. Lanphier later won a citation for his success during the Munda sweep.

Both pilots, Barber and Lanphier, possessed the mix of self-confidence, gung-ho attitude, and accountability that Mitch looked for in pilots. "One thing mattered to Mitch—were you there when you were needed?" Barber said. The two were different in other ways, though. Barber was gregarious and a team player who turned to his football-playing days to describe taking down an opposing plane. "Just like on a football field making a tackle," he said. "You have made an open field tackle, and it's a little difficult against a good runner, but you nailed him hard and you put him on the turf. And you get the same feeling when you nail this man on the airplane and you put him down. It's elation." Lanphier was not the backslapping type—he was more detached—but he was just as ambitious, if in a way that seemed self-centered to some pilots. There was the time, for instance, during the previous summer at Nadi when Lanphier had hustled his way onto a B-17 bombing mission, which had annoyed Captain Viccellio. That unauthorized escapade had left other pilots scratching their heads, but Lanphier had defended it to Barber as being part of his plan to build a war record that would help launch a postwar political career. Whether it was aloofness or the fact that Lanphier, as the son of a decorated colonel, had friends in high places, some of the other pilots grew wary of him. "Selfish motives did begin to appear," one of the other pilots in the 339th, Lieutenant A. J. Buck, said. "Petty jealousies." Fortunately, Buck said, the feelings did not impact their performance, even as pilot gossip "did point the finger at Tom Lanphier as being a 'Glory Hound.'"

THE MEN OF THE COMPOSITE FIGHTER SQUADRONS SPENT CHRISTmas 1942 together, joining the marines, sailors, and soldiers at Henderson Field for a special holiday meal that had been shipped in for the occasion. While their Japanese counterparts a few miles away at Kokumbona starved on rations of rice and coconuts, the men enjoyed roast turkey, potatoes, cranberry sauce, and mince pie. The mess hall was festooned with red and green rope, Christmas tree balls, and spangles, and under the hot tropical sun church services for different faiths were held. There was even a Santa Claus—a soldier in a red coat and shorts—riding around in a truck with a bunch of musicians playing Christmas carols and other tunes.

Military-wise, an early gift of the holiday season—one that benefited Mitch and his army fighter pilots specifically, as well as the US cause in the South Pacific generally—was the November arrival of the much-hyped Lockheed P-38 Lightning. Hundreds of men turned out to greet the new fighters, lining both sides of the runway and waving wildly as a dozen of them landed. More P-38s arrived on a ferry from New Caledonia. Mitch wasn't around when the speedy twin-engine fighter planes first showed up—he was still in a Sydney hospital recovering from malaria—and despite all the fanfare that the new plane could climb higher and fly faster than the Zero, some pilots at Fighter Two initially seemed skittish. Their caution was due

partly to the P-38s' troubles during tests in California, where difficulties bailing out had some calling it a "mankiller." Given its high-
altitude flying capability, the cockpit heaters were inadequate, and
during steep dives the plane tended to buffet. Finally, training, if it
could even be called that, was a hurried, on-the-job affair. "The transition to the P-38 was a chore," Marine Major General John P. Condon observed. "Twin engines, a wheel instead of a stick. It was a lot
of airplane." The marine pilots had so far been accustomed to having
army pilots in P-39s providing cover during their bombing missions,
but Condon noticed that during the initial deployment of the P-38s,
his marine pilots sometimes couldn't find the new planes. The P-38s
were going so high up, said Condon, that "they often failed to see
things happen. They were up at 35,000 feet and didn't see any Zeros,
didn't see any action." The marines needled the army pilots, complaining that they were in the clouds to avoid combat. They started
calling the P-38 a "high altitude foxhole."

In early December, when Mitch returned to Guadalcanal and was
joined by the likes of Barber, Lanphier, and Besby Holmes, concern
about the P-38s was soon "in a total turnaround," Major General
Condon noted. Mitch and the others, Condon said, flew "with a kind
of aggressiveness at any altitude. Down on the deck, or up at 35,000
feet. It was another ballgame." The chiding ended, and Condon was
assured that his marine pilots would now be accompanied by just
about "the best airplanes and best pilots."

For Mitch, the death of his friend Ellery Gross in a P-38 training
crash certainly gave him pause, but when presented with an actual
plane, he did not hesitate. Right off he loved the P-38, the superb tone
of the two big Allison engines humming on each side of him. "Very
smooth, like a Cadillac," he said. Most planes had guns in the wings,
requiring pilots to shoot from three hundred or more yards out in
order to have the angle needed to hit the target. The P-38 had four
.50-caliber machine guns in its nose, pointing straight ahead, along
with a 20 mm cannon—a design creating a cone of fire that enabled
pilots to aim straight ahead with an unobstructed view and unleash
a relentless stream of lead. In addition, the plane's twin engines provided a top speed of 395 miles per hour at 25,000 feet and a rate of
climb that made the Lightning the fastest aircraft in the Pacific.

"We'll clobber 'em now," Mitch said.

The frustration he'd felt in October about the P-39 not being able to fly as high as enemy bombers was over. To prove the point, he practically begged for permission to go after Washing Machine Charlie in late December after the enemy bomber had buzzed the base and ruined yet another night's sleep. "Let me go up in a P-38," he urged, "see if I can shoot him down that way. If you hold him in the searchlight, I'll get him." His army superiors gave him a cold-eyed no, saying they preferred using ground artillery to chase Charlie away. Mitch groused, "Ground officers knew nothing about air."

But he, Dinn, Barber, Lanphier, and the others, flying in what was the first twin-engine squadron in the Pacific, got plenty of other chances to demonstrate the prowess of the P-38. They devised a new aerial defense for Henderson Field against Japanese Zeros and Betty bombers that were accustomed to flying higher and beyond their reach. The new P-38s could now meet the enemy planes at 30,000 to 32,000 feet. "We'd wait for them," said Mitch, "and it ruined their tactics." They went on patrols up the Slot, the chain of islands north of Guadalcanal, looking for enemy bombers or fighters, and they learned to box in a Zero when they found one alone. "If we could get a guy on either side of him, we had him, no matter which way he broke," Mitch said. They learned that the lightly built Zeros were "easy to burn, and poorly protected." The P-38s were a lot tougher and could absorb punishment. If a Japanese Zero did get on their tail, they could lose him with a high-speed climb. With new confidence, the men accompanied B-17 Flying Fortresses on bombing missions. During one mission, on January 4, 1943, Mitch, Wallace Dinn, Besby Holmes, and others successfully escorted B-17s to the Japanese stronghold on Bougainville, an island in the north Solomons.

"We felt cocky," Mitch said. The rush felt good—until it didn't. The very next day, January 5, Mitch, Wallace Dinn, Besby Holmes, and three other P-38 pilots were again escorting five B-17s to Bougainville. The bombing raid went off without trouble. On the return to Guadalcanal, however, they spotted a swarm of Zeros—twenty-five of them, in fact. Mitch and his men, outnumbered four to one, could have flown away. But they did not, and, as the flight leader, Mitch had to make the decision to engage or not. "We went after 'em," he said, "even though we didn't have to." The fighting was over quickly, with Mitch's fighter pilots managing to hold their own. They had taken out three Zeros, and maybe more, with Mitch accounting for one. But getting his fourth kill was cold comfort for the fact that First Lieutenant Wallace Dinn was blown out of the sky—the burning engine on the left side falling off, the plane crashing into the ocean. Dinn would not find his way home this time.

Back on the ground, Mitch was shaken. He was met by Lieutenant Colonel D. C. "Doc" Strother. Mitch had known Strother since his training days at Hamilton Field in California, and Strother was now overseeing all army pilots on the island as the flight officer for United States Army Forces in the South Pacific Area. Strother got onto Mitch about losing a pilot, and Mitch blew up. He was exhausted, had gone all day without eating, and Wallace Dinn was dead. He yelled at Strother to back off, sore that Strother had said nothing about the good job they'd done, only attacked him about losing Dinn—which Mitch was already down about. Six months into the Guadalcanal Campaign, and some four hundred Allied pilots and crew members had been killed; this one was personal. He wrote Annie Lee the next day, "I lost the very best friend I have. His name was Wallace Dinn and he came from Corpus Christi. I got the dirty so and so that got him, but that is small consolation."

Mitch made sure to apologize to Strother for talking back to him the way he had. Strother cut him some slack, told him it was okay, that he knew Mitch had been upset and he probably shouldn't have broached Dinn's death so soon. Privately, Strother even admired the way Mitch had stood up to him. But for Mitch, the outburst was undisciplined and emotional, which was not good; it was as if he became even more determined as a pilot and squadron commander to channel his anger where it belonged: toward the enemy.

Which he did—channel his anger—a few weeks later. The Betty bomber known as Washing Machine Charlie made a disruptive, maddening bombing and strafing pass over Henderson Field early one morning, and Mitch was fed up. While the warning siren blared, he marched out of his tent, down the hill, and toward the airstrip. Without anyone's permission, he got into a P-38 and took off. He called the control tower and said, "This is Mitchell, I'm airborne."

He told the tower to notify antiaircraft not to fire at him but then, after a pause, wisecracked, "Well, I don't care if they shoot at me or not, they can't hit anything." Mitch climbed as quickly as he could to about 30,000 feet, scouring the sky for Charlie and hoping that the searchlights would find the enemy aircraft in the predawn darkness. But he saw nothing. Then the tower called, saying the bomber had just flown low across Fighter Two, strafing at about fifty feet off the deck. Mitch dived and flew across the airstrip a few minutes later, also at about fifty feet "looking for this guy, but by then he's gone." The tower had spotted him, though, flying over the water just north of the airfield, just "stooging around out there." Mitch turned in that direction. He looked hard, and, sure enough, he saw Charlie silhouetted against the morning sky. He flew out to sea toward the enemy nuisance.

By that time, many of the men had gathered to watch the show, standing outside tents or foxholes or near the mess tent. A combat reporter for the army magazine *Yank*, Mack Morriss, was among them, scanning the sky for the hunter and the hunted. Someone yelled and pointed offshore. "At first we could see only the 38 in the distance, but then, beyond it, was another plane," Morriss wrote later. Mitch, racing to catch up to the Betty bomber, could feel the full power of the plane's twin engines. Meanwhile, onlookers on the ground were following the action, mesmerized. "The plane in front was going hell for leather," said Morriss. Even so, the P-38 closed in, and Morriss watched as the Betty bomber suddenly exploded, "a terrific burst of flame—just like the sudden flare of a match, same orange color. The ball of fire kept going and then dropped into the drink."

Mitch had pulled in right behind him. "I burned him. And he crashed."

Mitch turned back toward Fighter Two and saw that he had an audience. He steered the P-38 in a slow, lazy roll across the airfield, what was called a victory roll—"To shine my butt," he joked.

The mission had been unauthorized, for sure, but no one seemed to care. "When the Jap went to blazes we cheered like bastards," Morriss said. Rex Barber and the other pilots in Mitch's squadron were beside themselves, jumping up and down and hollering. One ranking officer said contentedly, "Bagged one before breakfast!"

Mitch was greeted like a folk hero. He'd gotten rid of Washing Machine Charlie, which surely provided a measure of revenge, a tonic of sorts for the loss of his pal Wallace Dinn. The kill did something else, too: it displayed the self-assurance at the core of Captain Mitchell's flying spirit, a reminder that for a special mission, he was the one.

CHAPTER 13

MOON OVER GUADALCANAL

——

Navy Secretary Knox and Admirals Nimitz and
Halsey on Guadalcanal *WWII Database/public domain*

A LIGHT SNOW FELL IN WASHINGTON, DC, AS PRESIDENT FRANK-
lin D. Roosevelt entered the House of Representatives chamber to
deliver a wartime State of the Union address aimed at shoring up
the nation's spirit. "The coming year will be filled with violent conflicts," he told the joint session of Congress, "yet with high promise
of better things." Looking back at 1942, he honored the bravery of
the 1.5 million soldiers, sailors, and marines fighting overseas: "the

heroes, living and dead, of Wake and Bataan and Guadalcanal, of the Java Sea and Midway and the North Atlantic convoys." He cheered the fact that in Europe the Nazis' superior airpower, which at the war's start had enabled the relentless bombing of London, Warsaw, Rotterdam, and Coventry, was no more. Said FDR, "the Nazis and the Fascists have asked for it—and they are going to get it." In the Pacific Theater, the president singled out the Battle of Midway as a critical turning point, a victory halting Japan's expanding dominance in the region: "We know that as each day goes by, Japanese strength in ships and planes goes down and down, and American strength in ships and planes is going up and up," he said in a message that was rebroadcast in twenty-six languages.

Eight thousand miles away, Admiral Isoroku Yamamoto celebrated the changing year aboard the flagship *Yamato*, anchored at the main naval base at Truk Lagoon in the central Pacific. He dined with his officers on a New Year's soup called *ozoni*, but his mood could hardly be called celebratory. Gloomy was more like it, as he considered the prospects for the upcoming winter and beyond. "I am distressed," he wrote a friend in early January. "Things here look about as bad as they could." In another letter written a few weeks later to an old navy comrade, Yamamoto was just as blue, if not more so: "I have acquaintances and beloved subordinates both in this world and the next. Part of me feels that it wants to go and meet them again, and another part feels that there are things it still wants to do here."

Reluctantly, Japan's military leaders, including Yamamoto, had by year's end concluded that Guadalcanal was a lost cause. They had unanimously recommended to Emperor Hirohito that the navy devise plans for removing army soldiers from the island. The propagandists and media in Japan, taking their cue from the government, would spin the evacuation not as a retreat but as a tactical move to deploy military assets elsewhere in the Pacific. Indeed, while assenting to the withdrawal, the emperor had also insisted that plans for a new operation against New Guinea be conceived so that Japan would still be seen as being on the move. But Yamamoto knew the truth: the abandonment of Guadalcanal was a major blow. The numbers told the story: more than 14,700 Japanese soldiers had been killed or gone missing during the six months of fighting on the tiny

tropical island, with another 9,000 men dying from disease or starvation. The death toll of nearly 25,000 did not include another 4,300
men who had died at sea aboard transport vessels while trying to
reach Guadalcanal. By contrast, the US marine and army combat
fatalities numbered 1,600, with another 4,300 men felled by tropical disease—a total of not quite 6,000. Both sides had lost about
the same number of planes—more than six hundred—but the Japanese had lost many more pilots and crewmen than the United States
and its allies had. Yamamoto, the advocate of carriers and airpower
who'd guided the navy to dominance in that area, was acutely
aware of the difficulty of replacing the 1,260 pilots and crew killed
in the South Pacific. "The enemy's replacement rate is three times
ours," he noted in a letter written during the last raid on Henderson
Field. "The gap between our strengths is increasing every day."

To accomplish the new year's evacuation, called Operation KE,
the Japanese bluffed the Americans into thinking that their navy and
army were marshaling resources for yet another attack on Henderson Field. Bombing raids were ordered to make it seem as though
something major was in the works. False and misleading radio messages were sent to hoodwink US commanders into thinking that the
mobilization of warships was for an offensive operation when, in
fact, it was to shuttle troops off the island. The gambit worked; US
Army and Marine soldiers on Guadalcanal, now numbering more
than 50,000, made ready to defend against new troop landings—an
invasion that never came. The remaining Japanese, instead of advancing, hustled to the northwest corner of the island, where, beginning on the night of February 1 and continuing for two more
nights, they departed in a series of evacuation runs. By February 8,
an estimated 13,000 men—less than a third of the 36,000 soldiers
who'd once made up the Japanese ground force—had been saved
from "Starvation Island." There wasn't much left to them; they were
so emaciated, undernourished, and sickly that they resembled prisoners of war. But they had gotten away, and Yamamoto was relieved.
"You did it very well, indeed," he told the admiral in charge of the
large-scale removal. "The Army will be pleased to know we can send
back its soldiers in great mass."

The mass evacuation was one of the few successes in Japan's

otherwise humiliating failure at Guadalcanal. The pullout not only marked Japan's first land defeat but also relinquished control of the sea and air south of the Solomon Islands to the Americans. "Guadalcanal is now ours," the *Washington Post* editorialized on February 11. "The Japanese Imperial headquarters insist that the withdrawal of Japanese forces was only carried out 'after their missions have been fulfilled.' But this is obviously a verbal smoke screen put up in order to hide a major disaster." Bull Halsey, the admiral recently put in charge of US forces in the Solomons, had come to view the South Pacific islands as rungs in a ladder ascending to Tokyo. For him, Guadalcanal was the first rung, and, having climbed it, the only direction to go now was up. On Guadalcanal itself, *Yank* magazine reporter Mack Morriss captured the elation of the men at Henderson Field when they realized that the Japanese, instead of attacking, had fled. "People are going around with grins on their faces," he scribbled in his diary as he was sorting out the significance of the moment. "I wonder if this is really the end of organized resistance," he continued. "On its face it could be nothing else. Where could any more Japs be—further around the coast on the south side? It doesn't seem likely. In the hills? Possibly a few." Morriss could only arrive at a single, momentous conclusion: "that Guadalcanal, scene of the first U.S. offensive blow, is at last really ours after six months and two days."

WITHIN DAYS OF THE EVACUATION, YAMAMOTO OFFICIALLY moved from the *Yamato* to a new ship anchored at Truk, the recently commissioned *Musashi*. Like the former flagship, the *Musashi* was a colossal battleship, weighing 71,000 tons. The difference was that the new ship's bridge and admiral's suite were customized, enlarged, and air-conditioned to befit its service as the flagship of the commander in chief of Japan's Combined Fleet. Yamamoto's arrival was a formal affair, with a military band playing Japan's national anthem as he boarded the ship and, outfitted in white uniform and gloves, reviewed the ship's crew. Joining him was his valued confidant, Vice Admiral Matome Ugaki, and in the weeks to come they began haggling with army and other military leaders about the new offensive mission that, in part, had been mandated by the emperor as the next step after abandoning Guadalcanal.

Ugaki and others continued to notice the shadow, first detected in the fall, over their beloved admiral's mood. His physical well-being was becoming an added concern. It wasn't just that his hair had turned gray and he seemed to have aged quickly; to be sure, he was about to turn fifty-nine, on April 4, 1943. Rather, he was experiencing swelling in his legs and numbness in his hands. When he turned to one of his favorite pastimes, composing poetry or letters with a writing brush, his hand sometimes shook. Long gone were the days of surprising his crew by performing a headstand on a ship's railing. He was also keeping to his cabin for larger chunks of time, far more than his staff officers were accustomed to, and he rarely joined in the lighter moments, such as a game of ring toss on deck. Yamamoto himself noted that he'd barely left the ship since the start of the fight over Guadalcanal. "I've been ashore four times since August to visit the sick and wounded, to attend services for the dead, and so on," he wrote a friend. "But apart from that I've been stuck on board."

The commander in chief seemed depressed, as if he were privately bearing the full weight of Japan's reversal of fortune in the Pacific war. His staff wondered what to do, and there was talk of attending to his personal needs. Not in terms of his wife and family— there is little evidence that Yamamoto stayed consistently in touch with them after the outbreak of war. They were thinking instead of his lover, Chiyoko Kawai. Yamamoto missed her; he hadn't seen her since just before the Midway disaster the previous May. They corresponded regularly, and Yamamoto learned that Chiyoko had made a major decision; at the end of 1942, she had stopped managing a brothel, severed her ties with the geishas she had employed, and settled into her home in the Kamiyacho neighborhood in eastern Tokyo. Though Yamamoto had not returned to the mainland during the past year, several staff officers had, and they had visited with Chiyoko and brought news of her back to their admiral. What if, his officers pondered, they flew Chiyoko in from Tokyo to boost his spirits?

The reunion never panned out. Instead, for the remainder of February and into March plans for the next military operation were hashed out, and, with the prospect of a new battle, Yamamoto's manner, at least outwardly, seemed to pick up. The internal machinations of Japan's military establishment were rarely harmonious, meaning that Yamamoto, Ugaki, and their staff tangled with the army's leaders about how best to go after New Guinea. Yamamoto argued that prior to an invasion they would need to ensure the security of their existing bases in the Solomon Islands—at Rabaul on New Britain Island and on Bougainville Island, for example. The best way to accomplish that, he said, would be to again attack the Americans at Guadalcanal, not to retake the island but to contain the US forces there and hinder their interference with the New Guinea operation. But the army wanted to go full bore, and its leaders complained that Yamamoto was not showing enough support for the notion of taking New Guinea without delay. In the debate, the army, which generally possessed more influence than the navy at Imperial General Headquarters in Tokyo, was not helped by its failure in early March to capture an airport on New Guinea. US planes had intercepted and destroyed a convoy of eight Japanese ships loaded with troops, as well as a number of accompanying destroyers. Three thousand soldiers had been killed. Yamamoto was then given the green light for his plan to cripple Henderson Field from the air.

With his aides, Yamamoto devised what they called Operation I-Go. Given the goal of immobilizing Henderson Field, as opposed to invading it, no ground forces were involved, only airpower. Yamamoto ordered five aircraft carriers to Rabaul, and by late March the number of Japanese planes that had been assembled constituted the largest Japanese aerial force to date in the South Pacific. The start was set for early April, and Yamamoto decided that he and Ugaki would temporarily move the Imperial Navy's headquarters from Truk Lagoon to Rabaul in order for him to initiate and oversee the operation personally. On April 2, the night before his departure, he wrote Chiyoko one of his longest letters ever, thanking her for her two letters of March 27 and March 28 and also for hosting several members of his staff during their recent trip to Tokyo. "Their stories made me feel that I myself had returned home for a while and enjoyed a visit with Chiyoko." He insisted that he'd begun to feel much better and was now in "excellent condition with the blood pressure equivalent to persons in their thirties." The numbness he'd suffered in his fingers, he said, "was completely cured after I was given forty injections of vitamins B and C combined." She need not worry any more about his health or about the fact that he was leaving for the front line the next day. He was feeling energized, he said, in large part because of her endless love. "I will go in high spirits since I have heard about you," he wrote. Moreover, he was animated about Operation I-Go: "I am delighted to be going to launch attacks on the enemy a bit." With that, he cautioned that he would not be able to write again for two weeks, and he enclosed a freshly cut lock of his hair along with a two-line poem he had composed for her: "If I think of you as ordinary passion dictates / Could I have had a dream of only you every night?"

The next morning, April 3, Yamamoto, his staff, and Vice Admiral Ugaki left Truk for Rabaul in two seaplanes. The split into separate planes was for security purposes, so that the two high-ranking commanders would not be flying together. Yamamoto arrived at Rabaul Field that same afternoon, in time for the final preparations for the aerial attacks on Henderson Field, where the Army Air Forces' John W. Mitchell and other fliers, along with more than 50,000 troops, had solidified their occupation. As for the letter Yamamoto had written to Chiyoko, it would turn out to be his last to her.

DURING THE WINTER, THE US ARMY AIR FORCES ON GUADALCAnal saw their operational structure reorganized. The bomber and fighter squadrons on "Cactus" were blended into the newly activated Thirteenth Air Force, whose main base of operations was set up on the island of Espiritu Santo six hundred miles south of Guadalcanal. The Thirteenth would report to the Navy's Bull Halsey, the commander of all South Pacific forces, or COMSOPAC, who was stationed further south at New Caledonia, while Halsey would report to Admiral Chester Nimitz, commander in chief, US Pacific Fleet, or CINCPAC, who was at Pearl Harbor.

Halsey created an advance command team on Guadalcanal to oversee the air units for the three services there: Army, Navy, and Marines. The team's offices were several tents just west of Henderson Field toward the Lunga River, over a slippery, muddy hill that offered protection from enemy shelling and was reached by jeep slithering through the ever-present muck. The command team would run the three airfields, each now with a distinct use: Henderson Field mainly for navy SBD bombers and transport planes; Fighter One, on the interior side of Henderson Field, for navy fighters and marine Corsair fighter-bombers; and Fighter Two, nearly two miles away and parallel to the coast, for the Navy's Grumman F4F Wildcat fighters and the Army Air Forces' P-39 fighters and new P-38 Lightnings. Right next to each strip was an operations office—a dank, smoke-filled dugout lined with coconut logs. Under the new structure, each plane grouping was assigned an operational commander. The bombers, for example, got a bomber commander, while the fighter planes were overseen by a fighter commander. Taking up the latter duty during the winter was an officer Mitch and the other army fighter pilots knew well, Lieutenant Colonel "Doc" Strother. Meanwhile, their old squadron commander, Vic Viccellio, was made fighter operations officer, working alongside Strother. Vic also won a promotion to lieutenant colonel. Though on paper the airfields at Guadalcanal seemed to be bulging with aircraft, the fact was, as one officer later said, that "there were rarely more than 138 operational aircraft available on any one day, including search, photo and other non-combat types."

To acknowledge the changing balance of power in the region, Secretary of the Navy Frank Knox made a much publicized inspection tour at the end of January, covering some 20,000 miles and including stopovers at Pearl Harbor, Midway, and Fiji. Escorted by dozens of fighter planes and accompanied by Nimitz and Halsey, he arrived on Guadalcanal in a twin-engine amphibian patrol bomber the morning of January 21. He immediately began chatting up the men and reporters surrounding him, describing how the previous night at "Buttons," code name for Espiritu Santo, there had been an enemy air attack—the first against the Thirteenth Air Force's new headquarters. Somehow the Japanese had known that the high-ranking cabinet official from Washington, DC, was there. "How the Japs got the information we don't know," Knox said. "But they didn't do us any harm." That night at Guadalcanal, there was a second, heavier bombing, and although few of the men got any sleep, no casualties resulted. Knox toured the base with Halsey and Nimitz, visiting troops on the perimeter, where he announced what the men already knew. "We have dissipated the threat of the Japanese ground troops," he said. "We are now dominating the island."

By winter's end, Halsey sent a career naval officer to Guadalcanal to take over the team there, his official title being Commander of Air Forces in the Solomon Islands, or COMAIRSOLS. He was Marc "Pete" Mitscher, a fifty-six-year-old, battle-tested rear admiral who was all business and known for the long-billed cap he always wore to block the tropical sun. Halsey championed Mitscher's tenacity. With much of the fighting in that region turning into an air war, he wanted Mitscher in charge. "I knew we'd catch hell from the Japs in the air," Halsey said. "That's why I sent Pete Mitscher up there. Pete was a fighting fool."

IMPLEMENTING A COMMAND STRUCTURE THAT HAD BEEN LACK-ing during much of the Guadalcanal campaign was a good and necessary move, but bureaucratic matters weren't the concern of fighter pilots like Mitch and his men. It was above their pay grade, for one thing, but mainly they simply wanted to get on with the war. Mitch got his eighth kill on February 2, when he shot down a Zero that tried to interfere with bombers he and three other P-38 pilots were escorting. He was already over the five kills required of an ace pilot; he was the first ace in his fighter group. "Boy, am I gonna have a sore arm from patting myself on the back!" he told his wife a few days later. "Anyhow, ol' darling, I'm in there giving them all I've got and every time I burn one of them I say there's one for Annie Lee and 'Junior.'" His record earned him a special honor, a Distinguished Service Cross, awarded on March 9 for "having shot down a total of eight confirmed, destroyed an undetermined number on the ground and water during strafing strikes at Rekata Bay and Munda Point, and led numerous other missions in addition to those specifically mentioned, in over 150 hours of combat flying time as Flight Commander." He penciled the medal onto stationery so Annie Lee could see it. "I'm no artist," he told her, but "you get a general idea of how it looks. The 'bird' in the middle of course is an eagle." His promotion to major finally became effective, too, and it took getting used to. "When someone says major this or that I look around to see who they are talking to, to find it's me!" he said. But it was true: John W.

Mitchell, from the tiny hamlet of Enid, Mississippi, was now twenty-eight-years old, a major in the Army Air Forces, and an ace pilot with a Distinguished Service Cross. He marveled to Annie Lee what a difference a year had made: "Guess it's going to seem funny to you seeing me as a major when I was only a 1st lieutenant when I left."

Other fliers in the group were also showing off their stuff, eager to engage the enemy as they mastered the new twin-engine P-38 Lightnings and strapped "belly tanks" onto their planes for added fuel on longer missions. Rex Barber, Tom Lanphier, and Besby Holmes all racked up kills throughout the winter months, although Holmes became sidelined with an awful case of jungle rot. To stay airborne, he sat in the dispensary tent between flights soaking his feet in a bucket of potassium permanganate, a chemical compound used as a fungicide. The rot got so bad, however, that he was finally sent to a hospital in New Caledonia in March to cure it. The pilots were also pleased with their handling of the quirks of their designated airstrip. Because Fighter Two ran parallel to the beach, offshore winds blew constantly onto the strip at a right angle. The army pilots managed the crosswinds in the P-38 Lightnings handily enough and were bemused by the fact that the Navy's F4F Wildcat fighters were not as adaptable. They took to razzing their navy counterparts for awkward landings featuring wavering tail skids and for sometimes having to pull up for a second try. In all, Mitch and his cohort of pilots from the 339th Fighter Squadron and the former 70th Squadron were making their mark at Guadalcanal, "as outstanding as any combat airmen I ever saw anywhere," their former commander, Vic Viccellio, said. "There were no bars, no women and the Japs were moving everywhere and so we had plenty of incentive." Indeed, when a transport plane landed one morning in early March and a woman happened to step out—a blond army nurse named Mae Olson—the men at Henderson Field were agog. "There's a woman aboard!" shouted one soldier at the sight of the twenty-six-year-old lieutenant from Little Falls, Minnesota. Mae Olson was the first woman to have arrived on Guadalcanal since the war had begun, there to help remove the wounded and sick. Even though she was there for only about a half hour, she drew a crowd, the soldiers rubbernecking her every move, with one flippantly saying later that "two generals and several colonels went scurrying for cocktails."

Mitch kept Annie Lee up to date as best he could about that and other news, as their letters to each other grew into the hundreds. Annie Lee continued to send Mitch piles of magazines, described Hollywood movies she'd watched, and said she was learning to play bridge. Mitch sent Annie Lee a photo of himself sporting the mustache he'd grown, which he shaved before winter was out, and he regularly sent money to her and also to his father and stepmother in Mississippi. "I hope to someday fix them up with water works there in Enid," which was his way of saying that Noah Mitchell's house lacked running water and toilets. When Mitch got hold of a new watch, he acted as if he'd won the jackpot. "Have just drawn a beautiful wrist watch from supply," he exclaimed about the critical piece of equipment for any pilot but especially for a flight leader tasked with keeping everyone in sync. The maker was the Elgin National Watch Company outside Chicago, one of the oldest watchmakers in the United States, dating back to the 1860s, when Mitch's forebears were first homesteading in Mississippi. "It's an Elgin," he said, "and keeps *perfect* time!" The one favor he asked of Annie Lee was that she send him a new wristband, given that bands rotted quickly from constant perspiration in the tropical heat. "Cloth or leather would be fine," he said.

Mitch talked about movies, too, as they now had the luxury to screen films since the Japanese evacuation and the ground combat around Henderson Field had ended. They'd gather in a palm grove, swatting mosquitoes and keeping one eye on the screen while keeping another toward the sky just in case there was an unexpected raid. Mitch watched *Mrs. Miniver*, a romantic drama set in England during the war, directed by William Wyler (Wyler being one of five prominent Hollywood moviemakers, including Frank Capra, John Huston, George Stevens, and John Ford the others, who had begun making war films as a way to help the cause). Mitch also saw lighter fare, such as *Sullivan's Travels*, a comedy directed by Preston Sturges and featuring Veronica Lake in one of her first major roles. "I thought it very good," offered Mitch as amateur film critic. "I see Veronica has what it takes," he said coyly, adding "She can't hold a candle to you though!" And of course he wrote about the bright full moons over Guadalcanal. "Gosh, it's pretty tonight!" he said one winter night when, unable to sleep, he had left his tent. "I sat and looked at the moon for a while, naturally my thoughts were of you." No matter how far apart he and Annie Lee were, a shared moon in the big nighttime sky served as his way to connect their love, as if they were actually seated side by side. "Reminds me of several occasions where in the moonlight I have held you close in my arms," he said, adding, "There will be another day when I shall do just that again."

Therein lay the rub. Annie Lee was always joyous and relieved to hear from him, reading his news and his proclamations of love, but the one theme running through their letters of late 1942 and early 1943 was Annie Lee asking her husband when he was coming home. Most every letter involved that question one way or another, and Mitch tried to reassure her as he juggled his desire to be with her and his duty to fight. "If it were not for you I would try to get them to let me stay on here the rest of the war," he said at one point about the impossible tug of love and war. Though he loved her without a doubt, he was also a fully committed army fighter pilot. He sometimes tried humor to counter her complaint that the length of their separation kept growing. "Don't you worry a minute about us feeling like strangers," he said. "Just remember Mrs. Mitchell we've been married nine months already, so we'll be old married folks!" Joking was all well and good, Annie Lee replied, but the fact was that after their hurried marriage they'd been together for only a few weeks, and she still wanted to know when he'd get to come home.

"You keep asking about my return," he said late in the fall. "I still know no more than before." Then, to his surprise, there was suddenly talk at headquarters that he might be able to return to Texas during the holidays. But just as quickly word came down that there was no way he'd be home by Christmas. Trying to remain hopeful, he offered, "I *might* be home around March 1—or a little sooner." Just before 1942 expired, he further raised her expectations, writing that his homecoming hopefully "won't be long after the first of the year." Annie Lee was ecstatic. "He stated everybody tells him he should be home in January!" she told her Aunt Ludma. If that was true, Annie Lee said, "that means I'll only have three more Sundays to spend alone." She could barely contain herself. "I'm so proud of him I hardly know what to do, and I'm all set to spoil him good when he comes home—I feel that anyone who has done what he has and has been away so long deserves to be spoiled."

But once again hopes of a reunion were dashed, as January came and went and Mitch remained stationed at Guadalcanal. It got to the point where Mitch wrote a midwinter letter six pages long that was part pep talk—asserting that his love for her was boundless—and part admonition—that she needed to quit obsessing over his return.

"You seem to be living with only one thought in mind," he said, "and while it is flattering to me I do not think it is by any means the best thing for you." He encouraged her to get out—visit Aunt Golda, go bowling, go to dinner or to the movies with her friends. For one week he wanted her to try not thinking about him. "I'm afraid you are getting in a rut and just thinking that until I come home nothing else matters."

He understood that there had been moments when it had seemed as though he might be on his way, and then he wasn't—and now his latest update was simply going to add to that roller coaster. "Things have changed here," he said, "and I see March 1st fast arriving and leaving without me being on my way home." Pilots were being sent on short leaves to New Zealand and Australia and then returning for another round of combat, and that was the way it was. "You mustn't forget there are some who have been here longer than have I—also, and most important, this is a war we are in and not just something to inconvenience us. It's much bigger than you or me, much bigger than all of us." There was a new factor, too, he needed to mention— one that was good for him careerwise but that would postpone his homecoming. "My promotion and recent acquisition of a squadron are going to slow my return," he said, and he implored Annie Lee not to fret and to understand that he "must and should stay" as long as he was needed. He restated what had emerged as a mantra of sorts to deal with his dual devotions: "Frankly there's only one reason that I want to come home—that of course is you. Otherwise I would just as soon stay here until the war is over."

WITH THE ISLAND UNDER US CONTROL AND THE MEN AT HENDER-
son Field no longer under threat of constant attack, the winter months
saw a more balanced mix of flying and frolicking. That was especially
true in early March, when a lull in the air war lasted several weeks.
It was summer in the South Pacific even though the calendar said
winter, and the men dressed more casually than ever, trying to keep
cool—in loose-fitting khaki trousers or shorts, T-shirts or fatigue jackets, with canteens attached to their belts like handguns, as they had
to drink up to a gallon of water daily to stay hydrated. Mitch set up a
radio inside his tent, and when the reception was clear he could pick
up favorite songs along with new releases, the latter often reminding him how out of touch and far away they were from the United
States. "The announcers occasionally come out with 'one of the most
popular tunes of the year' and we won't even have heard it before,"
he said. There were now nearly a hundred army air forces pilots on
hand, and they converted one tent into a club of sorts, featuring a record player. Their record collection wasn't extensive, so they got the
most out of what they had, playing Glenn Miller's "Harbor Lights"
and "Serenade" over and over again, as well as a tune distinctly ap-
ropos for flyboys, "He Wears a Pair of Silver Wings," by Kay Kyser
and his orchestra, which had been the hottest hit for four weeks the
previous year. Dinah Shore recorded a version, crooning "Although
some people say he's just a crazy guy/To me he means a million other
things/For he's the one who taught this happy heart of mine to fly/
He wears a pair of silver wings." Some of the songs were more risqué,
including Johnny Mercer's "Strip Polka," with its line "Take it off."
The Andrews Sisters sang a version, and, according to Rex Barber,
"We played it till it wore out." The men sat around late into the night
listening to music, talking, and drinking—preferably American beer,
if available, or, on rare occasions, cheap whiskey mixed with grape-
fruit juice. Some started card games, but the gambling was never as
constant as it had been on Fiji, when they had been idle so much of
the time that poker games had lasted for days. The later the hour
the more likely they'd break out into song themselves. For one sing-
along, they adapted the old folksy tune "Put on Your Old Grey Bonnet" to their own circumstances, composing new lyrics that, like the
Kay Kyser piece, were about the glory of being a fighter pilot:

Tell the U.S. Marines and the fighting Philippines;
And the boys in Manila Bay;
That the Air Corps' comin', with their big bombers hummin';
And we'll fight all the harder;
When we think of Pearl Harbor;
And we'll make those yellow bastards pay.

Even with the March lull, however, the brutality of the recent Guadalcanal campaign was always at hand—literally over their shoulders. No one could miss the cemetery the men unofficially called Flanders Field, after the World War I burial ground in Belgium. It was created in the center of a grove of trees near Lunga Point, and it kept expanding as more dead marines were brought back from the jungle, some exhumed from the hills, for a proper burial at the US base. The graves were dug in rows, each with a cross and a palm leaf resting on the earthen mound. The markers on more than fifty graves read UNIDENTIFIED, while the rest bore the names of the fallen. Some had mess kits, bullet-ridden helmets, and even propeller blades cemented into the ground. Inscriptions ranged from a few words—KILLED IN ACTION or OUR BUDDY—GONE BUT NOT FORGOTTEN—to more ambitious remembrances. One marine honored a fellow leatherneck with a well-known war epitaph: WHEN HE GOES TO HEAVEN / ST. PETER HE WILL TELL / "ANOTHER MARINE REPORTING, SIR / I'VE SERVED MY TIME IN HELL." The marines had paid their respects en masse with a special service on the morning of New Year's Eve day; under a blistering sun two columns of soldiers, each carrying a rifle and wearing a helmet, had marched in silence to the graveyard. Following a bugle's call, a band had played sorrowful music and a priest had said a memorial Mass, chanting in Latin. A makeshift altar had been constructed using spent Japanese shell casings for supports. The music carried throughout the base, and near the end of the service one marine bugler playing "Taps" was answered softly by a second bugler farther away. The service for the dead ended with the band playing the "Marines' Hymn." Robert Cromie wrote, "I wish every person back home could walk between the crosses, see the names and read the inscriptions."

Fighting picked up again at the end of March, with bombing raids aimed at disrupting and disabling the Japanese bases on the islands north of Guadalcanal. The army air forces pilots worked as tag teams, one group piloting the new P-38 Lightnings while the other took leave. Lanphier and Barber had traveled mid-March to Auckland, New Zealand, and Sydney, Australia, where, Lanphier said, the men "sampled the entire menu of delights available, each to his own device—or vice. For myself, I got out to the beach at Bondi for a couple of days and blew two months' pay on a photo finish at Royal Randwick race course. I hit a few bars and knocked back a gallon or two of that muscular ale they use for sustenance down there."

The more combat hours they flew, the savvier Mitch and the others became about piloting the P-38s, continuing a shift in the balance of airpower that Admiral Yamamoto worried about privately in letters to friends. The shift, which accelerated as Japan struggled to replace its flying corps, was something Mitch tracked as well. "Early they had good Navy carrier pilots," he said, crediting the fliers trained under Yamamoto's command. "They were hot then, but when those had thinned out, it was easier for us. We got better as they got poorer, and our scores went up." The enemy pilots of the previously dominant Zeros had taken to calling the twin-engine P-38 "the fork-tailed devil."

Lanphier and Barber, having just returned from leave in mid-March, went on a particularly successful raid that did justice to the plane's nickname. They were tasked with attacking thirty enemy flying boats and Zeros that a US reconnaissance plane had spotted anchored in the Faisi lagoon, a channel running between the Shortland Islands and Faisi Island. Lanphier, now a captain, led eight P-38s, with eight marine F4U Corsair fighter planes assigned to fly with them. They took off in the dark from Fighter Two on March 29 for a surprise attack at sunrise, but severe weather complicated matters. When they reached the Faisi lagoon, Lanphier and Barber looked around to see that the original attack force of sixteen planes was down to six. Several P-38s and all but one marine Corsair had returned to Guadalcanal, due to either the bad weather or engine troubles.

Lanphier organized the remaining planes into formation, and the P-38s made several strafing runs that set fire to at least seven

planes while managing to avoid enemy antiaircraft guns. They exited ahead of a counterattack and soon spotted what looked like a
Japanese destroyer. "We got some ammunition left," Lanphier radioed the others. "We got some gas—let's strafe him a little." The
six planes circled and made diving runs, strafing the stern, the midship, and the bow with machine-gun fire. Heavy smoke began bil-
lowing from the ship. The army pilots could see sailors starting to
jump overboard. They lined up for a final run, and as Barber began
his dive, he got so locked in that he experienced "target fixation," in
which a pilot's focus becomes fanatical. He kept firing, unaware of
how rapidly the distance was shrinking between him and the target,
until suddenly he realized that he was so close that the side of the
ship was in front of him. He pulled back on his controls as hard as he
could. The Lightning slid across just above the deck, and although
he managed to avoid full impact he did not come away unscathed.
His left wing clipped the ship's radio mast, tearing about forty-four
inches off the tip of the wing. The impact almost threw him into the
ocean, but he managed to right the airplane, and after he trimmed
it up, it flew okay. He offered no excuses: "This was being a little bit
stupid, but I did get back."

Word of the successful raid, and especially of Barber's near crash,
spread quickly, and all six pilots were later awarded the Silver Star.
Bull Halsey sent an airmail-gram to the men featuring the admiral's
"attaboy" bluntness: "Congratulations on a nice Faisi roast." The
coverage back in Washington, DC, took a different slant, however.
"Capt. Lanphier Cited in Fight on Jap Vessel," read the headline in
the *Washington Post*. The article was all about Tom, starting with
the opening sentence: "One of the fighting and flying Lanphiers of
Washington and Detroit has just had his name inscribed high on
the Army Air Forces' roll of honor by crippling and probably sinking a Japanese ship by aerial gunfire." The raid was cast as practically Lanphier's alone. And in the retelling, an inflated tally of his
overall Pacific record was included, reporting that to date Lanphier
had destroyed "17 Nipponese planes—nine shot down in air combat
and eight wrecked on the ground." It was an article that had his
father's fingerprints all over it. Colonel Thomas Lanphier, Sr., was
stationed in Washington and was known to chat up reporters. In all,

the article's one-sidedness served to highlight a tendency to buff up Lanphier's exploits at the expense of others. No mention was made of Rex Barber or any other P-38 pilot; rather the newspaper reported that Tom Lanphier was "one of the most effective fighter pilots in the United States Army Air Forces."

THE RAID OF MARCH 29 CAME JUST A FEW DAYS BEFORE YAMA-moto, stationed at the base on Rabaul, intended to commence Operation I-Go, the all-out attack to cripple Henderson Field. On the morning of April 4, Yamamoto's birthday, he and his vice admiral, Matome Ugaki, appeared at Rabaul Field to see the departing fighter squadrons off. "Now we are approaching the difficult battle, a sequel to the last one," he told his men from a platform erected at the airfield. "However difficult a time we are having, the enemy also has to be suffering." Fierce tropical squalls and relentless rain forced the planes to return that day, and continued bad weather over the next several days led to similar false starts. Each day, however, Yamamoto appeared at the airfield, dressed impeccably in a snow-white uniform with gold braid, Ugaki at his side, waving his cap to send them off or, when the pilots were told to turn back, to greet their return in pouring rain. The admiral's repeated appearances at the airfield and his commitment to his men were characteristic, the very things that over the course of a career had fostered deep loyalty—a devotion that remained steadfast even after the devastating loss at Midway and the abandonment of Guadalcanal. He was "the personification of the Navy," one of the pilots at Rabaul that day wrote later. Whether in rain or intense tropical heat, "Yamamoto was every inch the perfect military figure, and conducted himself on occasions with military reserve and aplomb."

The weather finally cleared on April 7, allowing squadrons of Japanese fighters and bombers—nearly two hundred planes in all—to attack Henderson Field and other US airfields on nearby islands, as well as Allied ships gathered around Guadalcanal. By that time, with help from Allied code breakers who'd picked off bits of enemy intelligence, the US forces at Henderson Field were alert to an imminent Japanese attack. "Condition is red," *Yank* magazine writer Mack Morriss wrote in his diary that day. "Anticipation of some 100 Bogies coming down. Seems like everything we've got is in the air. This should be a dilly." The initial wave of Yamamoto's attack squadrons was indeed confronted in the skies over Guadalcanal by every plane that Pete Mitscher, the newly arrived air commander at Henderson Field, had on hand: seventy-six fighters from the three services, Wildcats, Corsairs, Warhawks, and, most notably, twelve P-38 Lightnings. Rex Barber, Tom Lanphier, and their army air forces comrades once again distinguished themselves. That first day, they rode their P-38s to about 30,000 feet, higher than the Zeros could fly, and waited for the enemy fighters escorting the bombers to show up. When eleven of them appeared, the army pilots paired up and dived in formation, firing and picking them off. In minutes, they shot down seven Zeros, with Lanphier accounting for three and Barber for two. Marine pilots, meanwhile, targeted the Japanese Val bombers, with one marine pilot downing seven of them.

The fighting continued for nearly a week. Yamamoto remained at Rabaul to see off waves of attackers and then kept busy while awaiting results. During the day he met regularly with Ugaki and other officers to go over matters concerning the fleet or to enjoy a game of *shogi*. He walked the sprawling base so that his presence was known and made a point of going to the hospital to visit with the wounded and sick. To relax, he returned to the cottage where he was staying, up on a hill overlooking the base. Day after day, Yamamoto was given updates from returning pilots that made it seem they were achieving fantastic results—first at Guadalcanal and then in raids on Port Moresby and on New Guinea. Japanese bombers did indeed sink a US destroyer, the *Aaron Ward*, in the waters off Guadalcanal, but the fliers' reports of shooting down hundreds of US planes and sinking dozens of ships were gross exaggerations. In truth, the United States' losses were far less than Japan's. Besides the *Aaron Ward*, the tally was two transport ships, one tanker, and twenty-five planes, while Yamamoto's naval air force lost at least forty aircraft. But believing that Operation I-Go had achieved its goals, and with congratulatory messages coming in from the emperor and Imperial General Headquarters in Tokyo, Yamamoto was ready to cease offensive operations.

It was at that moment that Yamamoto decided to head south to see his men at various forward bases before heading north to the flagship, *Musashi*, at Truk Lagoon. He figured they deserved to see him and hear directly how proud he was, how valiantly they'd fought, and he saw an efficient one-day inspection of bases at Ballale, the Shortland Islands, and Buin on the southern tip of Bougainville as a morale booster. He was especially interested in making a first stop at Ballale, a small island off Bougainville, where the depleted troops under Lieutenant General Masao Maruyama's command were recovering after fleeing Guadalcanal; they were soldiers who had survived the slog through the virtually impenetrable jungle during the failed November attack on Henderson Field. Yamamoto wanted to thank them personally. No surprise, his closest aides were taken aback and voiced concern about his last-minute change of plans. Being at Rabaul, weren't they already close enough to the actual fighting? Why take any chances, possibly moving into harm's way? But Yamamoto had made up his mind, and he set Sunday, April 18, 1943, as the day for his departure. His staff hurried to put together an itinerary and choose the aircraft and crews that would make up the admiral's flight. Once everything was arranged, his detailed schedule was transmitted to the appropriate officials at the various destinations. But the moment it was sent, the radio transmission was intercepted by unintended recipients, meaning that the minute-by-minute description of the trip and the flight routes to be taken by the commander in chief of the Japanese Imperial Navy fell smack into enemy hands.

PART IV

VENGEANCE

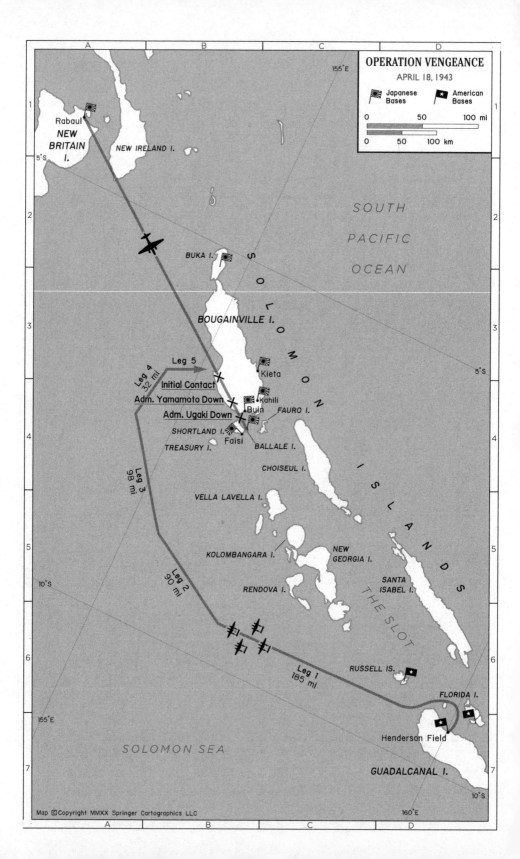

CHAPTER 14

FIVE DAYS AND COUNTING

———

Yamamoto at Rabaul, April 1943
WWII Database/public domain

WHEN THE CODED MESSAGE CONTAINING ISOROKU YAMAMOTO'S
travel plans was transmitted late in the afternoon of April 13, 1943,
the signal was picked up right away by a US radio station in Wahiawa, just north of Pearl Harbor. The interception began as routine
fare—just one message among hundreds gathered in the radio net by
technicians working around the clock. But after being forwarded to
the Dungeon, the basement office where a band of code breakers la-
bored in secrecy, the message stood out from the heap. It concerned
Yamamoto. Deciphering the message's contents became top priority,

and the transmission was fed into the Dungeon's IBM computing machines on punch cards—a first step in converting into text the five-digit number groups making up the latest version of the enemy's JN-25 code. That revealed the second significant feature: the number of recipients. The classified message was not a communiqué between Yamamoto and a single colleague or underling but had gone to naval commanders at the several Japanese bases located in the northern Solomon Islands.

Taking the lead at that point was linguist Marine Lieutenant Colonel Alva B. "Red" Lasswell, the intense, lanky night duty officer who was not only a translator but a skilled cryptanalyst. Nearly a year had passed since he and fellow Japanese-language expert Joe Finnegan had played crucial roles in unraveling Yamamoto's grand plan to crush the US Pacific Fleet at Midway, a code-breaking coup that had set the stage for the stunning US victory there in May 1942. Since then Lasswell, Finnegan, and a core group had continued to anchor the Station Hypo decryption unit at Pearl Harbor, officially called Fleet Radio Unit Pacific (FRUPAC). But there had been some changes in personnel. Most notably, Navy Captain Joe Rochefort, the gung-ho, strong-willed officer in charge of Station Hypo during its Midway heyday, was gone, reassigned to command a floating dry dock in San Francisco. It was the humiliating price for having been on the losing end of political infighting with the officials in Washington, DC, who ran the rival naval intelligence decrypt unit Op-20-G. The boss's fall from grace had dismayed Lasswell and the loyal crew, most of whom Rochefort had handpicked to serve at Station Hypo. Lasswell called Rochefort "a very valuable man" and later lamented the captain's transfer to a floating dry dock as "a waste of talent." Other turnover among Hypo's officers and enlisted men was headline grabbing only in history's future tense: a young navy lieutenant who would eventually become a justice on the US Supreme Court had joined the unit. John Paul Stevens, about to turn twenty-three years old, was not one of the cryptanalysts but served as a traffic analyst. The analysts, also known as scanners, were the ones at Hypo who, by monitoring the source and destination of enemy transmissions on Tuesday, April 13, had caught notice of the multiple endpoints in the message involving Yamamoto.

Following the wildly successful and impactful Midway decryption the previous spring, the Japanese had altered their JN-25 code in a more consistent and timely fashion. In late summer 1942, they had switched from JN-25(c) to JN-25(d), and by year's end they had made another alteration. The changes had slowed the code breakers at Station Hypo and elsewhere, creating blackout periods in cracking Japanese transmissions. But between traffic analysis and a network of coastal watchers on islands in the South Pacific, the United States continued its clear advantage in gathering intelligence and providing Nimitz, Halsey, and other commanders a leg up in plotting strategy and combat operations against the Japanese. The retrieval of the Yamamoto message from the pile now carried the potential for yet another huge "get."

Holding a raw, encrypted version of the message, Red Lasswell pulled his chair up to a gray metal desk and, adjusting his signature green eyeshade to fend off the fluorescent ceiling lighting, readied himself to pick up where the IBM tabulating machine had left off. He was obsessively organized. "Lasswell approached cryptanalysis like a chess player maneuvering relentlessly to untangle his problems," said the unit's information officer, Lieutenant Commander Wilfred J. "Jasper" Holmes. With input from fellow cryptanalysts, he extracted the meaning of coded geographic symbols essential to understanding the message. "RR" meant Rabaul, the major Japanese base where Yamamoto was stationed. "RXP" was Buin, a Japanese base on the southern tip of the island of Bougainville. "RXE" meant Shortland Island, just south of Buin, a six-minute ride by plane. "RXZ" meant Ballale, a tiny island next to Shortland, with a small airfield.

Toiling through the night and into daybreak on Wednesday, April 14, Lasswell decoded numerical values into plain Japanese text and worked up an English translation. And just as film image goes from latent to visible as a negative soaks in chemical developer, the full meaning of the message was eventually revealed: on the morning of April 18, 1943, the commander in chief of the Imperial Japanese Navy was going to fly from his Rabaul stronghold for a quick inspection of various southern bases.

"Lasswell had worked it out in every minute detail before he showed it to me," Holmes said, and the contents that Lasswell shared began:

SOUTHEAST AREA FLEET/TOP SECRET.

The Commander in Chief Combined Fleet will inspect Ballale, Shortland, and Buin in accordance with the following:

0600 depart Rabaul on board medium attack plane (escorted by 6 fighters).

0800 arrive Ballale. Immediately depart for Shortland on board subchaser (1st Base Force to ready one boat), arriving at 0840.

0945 depart Shortland aboard said subchaser, arriving Ballale at 1030. (For transportation purposes, have ready an assault boat at Shortland and a motor launch at Ballale.)

1100 depart Ballale on board medium attack plane, arriving Buin at 1110. Lunch at 1st Base Force Headquarters (Senior Staff Officer of Air Flotilla 26 to be present).

1400 depart Buin aboard medium attack plane.

Arrive Rabaul at 1540.

The dispatch was relatively brief but shockingly loaded with detail, practically an hour-by-hour itinerary of Yamamoto's travels.

"We've hit the jackpot!" Lasswell announced.

EVERYONE ON DUTY INSTANTLY UNDERSTOOD THE PROMISE OF the decryption: the chance to ambush Japan's top admiral. But after the initial adrenaline surge, the idea actually gave Lasswell pause. Most, if not all, of the key intercepts he had handled had involved ship or troop movements. The Midway intercept, for example, had yielded an entire battle plan. That, generally speaking, was the fruit of the code-breaking labors, uncovering information about Japanese navies and armies, not individuals. But intelligence revealing the movement of a single person was exactly what they had in hand—intelligence that, if acted upon, would be tantamount to Yamamoto's death warrant. Lasswell couldn't put his finger exactly on the reason for his discomfort. He didn't think there was necessarily anything wrong with going after Yamamoto, but doing so, targeting an enemy leader, felt strangely personal. It felt more complicated than the clean and clear-cut euphoria following the Midway intercept, and it was why he always favored the Midway accomplishment. "I got greater satisfaction out of that than any other thing," he said later, whereas with the Yamamoto intercept, "I didn't feel, somehow or other, the joy in that one." The best way he could explain why was that he "felt more of a snooper in the latter."

Whether to use lethal force—morally or legally—to target an enemy leader was certainly not Lasswell's responsibility, but already during the course of World War II, its application had gained traction. The British were at the forefront, having created highly trained units within their armed forces to carry out special operations, and Winston Churchill had given the green light in late 1941 for a commando unit to kill Nazi general Erwin Rommel, "the Desert Fox." The mission had failed, but the notion of a "leadership decapitation" or "targeted kill," as it was variously called, was increasingly seen as part of the wartime playbook. Moreover, the Hague Convention of 1907—the body of international law applicable during World War II—did not forbid a directed killing of an enemy leader.

Even so, Lasswell and his colleagues at Station Hypo were more than relieved to run the Yamamoto message up the chain of command. Holmes, the information officer, used a secured line to call the fleet intelligence official at Pearl Harbor, Commander Edwin T. Layton. Holding Lasswell's translation, he read the explosive contents over the phone. Then, so that Layton would have an actual hard copy to deliver to Admiral Nimitz, he and Lasswell raced out of the Dungeon. They passed the security guards at the door, climbed the stairs, stepped into the bright light of dawn, and hurried across the grounds to CINCPAC headquarters.

"Lasswell and I were glad to have it out of our hands," Holmes said.

Lasswell, for one, certainly had full confidence in Nimitz. The admiral had backed Station Hypo's decryption saying that Yamamoto was planning to attack Midway when code breakers in Washington had adamantly insisted that the targets were different. Said Lasswell, "He laid it on the line when that controversy came up regarding Midway, and he risked everything on his judgment of me." But just as important, Lasswell had gotten to know Nimitz personally in the twelve months since. The two men were housed not far from one another and on many mornings walked to work together. "I saw a great deal of him," Lasswell said. "He was a great man." It was that simple for the star code breaker. Nimitz would know what to do with the Yamamoto intercept. But Lasswell also knew that the admiral did not have much time. The message of April 13 said that Yamamoto would be on the move on April 18, which meant in five days.

BY THE SECOND WEEK OF APRIL, WHILE ADMIRAL YAMAMOTO WAS deciding to fly south on his inspection tour, Major John Mitchell of the 339th Fighter Squadron was back at Fighter Two on Guadalcanal. Mitch had been on leave the first part of the month, enjoying a respite in Auckland, New Zealand, which had become the leading "leave town" for pilots rotating off Guadalcanal. He'd gone swimming in a cold mountain stream, its bed solid rock, its water as clear as crystal. He had found peace and beauty in the country's volcanic landscape and had thought of Annie Lee. "I will have to find a nice quiet spot like this when I get home," he wrote, "where we can spend a week or ten days by ourselves." It was quite a contrast to the pouring rain on Guadalcanal in early April, where day after day began with a downpour, up to three inches some days, the men sloshing to the mess tent through mud.

Mitch returned to "Cactus" rested and had lots to catch up on. He'd missed plenty of action—the entirety of Yamamoto's concerted but unsuccessful bid to disable Henderson Field. The men told him stories, none more dramatic than the combat trifecta featuring either Tom Lanphier or Rex Barber or both, showcasing their mastery of the P-38 Lightning. There had been the one on March 29, when Lanphier, Barber, and several other pilots had relentlessly strafed a Japanese ship, during which attack Rex had experienced "target fixation" and nearly crashed into the ship. He'd come back alive but without three feet of one wing that had been sheared off by the ship's mast as he had swerved out of harm's way. There had been another on April 7, when Lanphier, Barber, and every other available pilot had taken on Yamamoto's aerial raid, a day that had seen Lanphier increase his kill tally by downing three Zeros and Barber increase his by two. The third would have made an aerial combat highlight reel, if such a thing had been possible. It had occurred on April 2, the day after Rear Admiral Pete Mitscher's arrival at Henderson Field, and what a way to impress the new boss. Lanphier and two other pilots—not Rex Barber this time but Doug Canning and Delton Goerke—had been returning from a flight near Japanese-occupied Vella Lavella, northwest of Munda Point, when they had spotted an enemy freighter along the shore, partly camouflaged by trees. With plenty of fuel left in their belly tanks, they had decided to try out Vic Viccellio's idea of deploying the tanks as skip bombs, like skipping flat stones across the water's surface. Lanphier and Canning had zeroed in on the freighter, releasing their tanks as close to the vessel's hull as possible. The tanks had burst on impact. Right behind them was Goerke, who had opened fire, the tracer bullets from the plane's .50-caliber machine guns igniting the gasoline. Flames had roared, a spectacular sight, as the ship had begun burning instantly. Back at the base, the usually stone-faced Mitscher had been ecstatic about the pilots' ingenuity. He had dashed off a note to Bull Halsey, who had been equally impressed. VERY NEAT USE OF HEAT, Halsey wrote back. YOUR TREATMENT IS HARD TO BEAT. The three pilots enjoyed the attention, none more than Lanphier. The belly-tank trick became another instance when his war record was embellished. In Washington, DC, Lanphier's father spoke to a *Time* magazine reporter, and a few weeks later the resulting account was all about young Tom, with no mention of either Canning or Goerke. Lanphier had catapulted to the "top crust of fighting airmen," *Time* correspondent Jim Shepley wrote in a memorandum to his editor summarizing his interview with the senior Lanphier.

Mitch, who was glad to hear all about their aerial exploits, made it no secret that he was antsy to "get back in the scrap." He was reinvigorated, not just physically as a result of his break but emotionally as well, the result of his last letter from Annie Lee. They'd fallen into a rut all winter long when Annie Lee had seemed to be hounding him about his returning to Texas, to which in early March he'd finally told her that enough was enough. Then, while he had been on leave, he had received her most recent letter, and when he had opened it and begun reading, a calm had washed over him: she had written to tell him to do what he had to do, that she now understood. It was a stunner. She said that nothing had changed—she still wanted him home as much as ever—but that if he "had not had enough fighting" he should stay and fight; she would be "waiting and carrying on" until he did come home. "That's exactly what I needed to hear," Mitch replied. "You know that I'm coming home when I can, and when I feel that I have done enough." But if he were to return before then, he said, or "before war weariness or pilot fatigue started then I would not be content at home and would very soon be wishing I were back here or somewhere where our men are spilling their blood for America." He didn't think it would be much longer, perhaps after the next rotation, though "Uncle Sam may have different ideas." But no matter when it was, her words had bolstered him, and he was so appreciative. "To have you backing me makes it 100%."

COMMANDER LAYTON, MANILA FOLDER IN HAND, HURRIED DOWN the hallway in the headquarters building at Pearl Harbor toward the first-floor office of Admiral Nimitz. It was 8:00 a.m. on Thursday, April 14, 1943, and Layton, the fleet intelligence officer, was arriving for a morning briefing with the admiral. The briefing was part of their daily routine, but there was nothing routine about this one. Inside Layton's folder were the sheets of paper Red Lasswell and Jasper Holmes had handed to him just minutes before—sheets with information about Isoroku Yamamoto on them, the detailed plan for the admiral's island-hopping scheduled to begin in four days.

Though code breaker Lasswell had never met Yamamoto during his tour of duty in Japan prior to the war, Layton had. As an assistant naval attaché in Tokyo in the late 1930s, Layton had socialized with Yamamoto, then a navy vice minister. He had played bridge against him. He had gone on a duck hunt at the Japanese emperor's hunting preserve with naval officers from several other countries, all of whom Yamamoto had hosted with memorable charm and courtesy. Yamamoto had even given every officer a duck to take home at the end of the hunt. But their personal connection did not matter. That was then, and this was war, and the file held urgent intelligence about the Japanese naval commander responsible for the Pearl Harbor attack— the enemy leader most loathed in all of the United States, save for

Adolf Hitler. They were still picking up the pieces at Pearl Harbor sixteen months after his sneak attack, sifting through the wreckage and
rebuilding the base. In fact, below Nimitz's office at naval headquarters the mangled, rusting upper hull of the USS *Arizona* still jutted
out of the deep harbor waters. The wreckage served as a memorial to
the 1,100 sailors and marines who had died in the flames and explosions after a 1,765-pound bomb had hit and destroyed the 29,000-ton
battleship.

Layton handed Nimitz the folder and took his customary seat in
a split bamboo chair to the admiral's left. Nimitz opened the file.
"Our old friend Yamamoto," he said. He began to read, and Layton
waited. It was so quiet that Layton could hear a marine sentry's
footstep on the linoleum floor in the hallway outside. Hanging on
one wall of the plain, cream-colored office was a prized keepsake
from Yamamoto's raid—a samurai sword that had been recovered
off Pearl Harbor from the body of the commanding officer of a Japanese midget submarine. If Nimitz's heart began racing as he read
the enemy dispatch, he didn't show it, his body characteristically
erect, his demeanor poker faced. He would have noted, though, that
the date of the beginning of Yamamoto's travels—Sunday, April
18—fell on Palm Sunday. In addition, it was the first anniversary of
the Doolittle Raid, and although the bombing of Tokyo had been
at once a morale booster for the United States and a shocker to the
Japanese, who'd thought their homeland was invulnerable, reports
afterward of the torture of eight captured US pilots and the Chinese
villagers who'd tried to assist them had further deepened the animosity Americans already felt toward their Japanese foes. To act on
the deciphered message would add to the wartime significance of
the eighteenth day of April.

"What do you say?" Nimitz asked.

Layton detected a slight smile on the admiral's face.

Nimitz continued, "Do we try to get him?"

The discussion that followed touched briefly on the propriety
of such action in terms of wartime customs and practices: plotting
to kill a specific enemy leader was highly unusual in US military
history. Nimitz then notified naval headquarters in Washington,
where code breakers were also going over the message and further

discussion about rules of war and possible legal precedents were being held. Whether President Roosevelt, who was not in Washington at the time, and Secretary of the Navy Knox were directly involved in discussions remains unclear. No paper trail exists confirming that Roosevelt was ever made aware of the fast-breaking events, while some officers stationed on Guadalcanal later recalled seeing classified paperwork with Knox's name attached to it. What was clear was that Nimitz, as the commander of the US South Pacific Fleet, had operational authority to make the call. He and Layton spent that Wednesday talking through a range of what-ifs—as in, if they did go after Yamamoto, what would be the upside? Layton was quick to reply. "It would stun the nation," he said of Yamamoto's home country. Layton and the admiral well knew that Yamamoto was probably second only to Emperor Hirohito in popularity in Japan, idolized by naval officers, sailors, and civilians alike. The elimination of Yamamoto would be a morale crusher. Equally important, his death would remove from battle Japan's top naval strategist, a leader, as Layton said, "unique among their people. He's the one Jap who thinks in bold strategic terms." But what if his replacement proved to be an even more effective commander? Nimitz wondered. Wasn't the devil you knew better than the one you didn't? They came up with the names of the top-tier naval leaders in line to replace Yamamoto, reviewing each one's strengths and weaknesses, assessing how they measured up. Layton said, "Yamamoto is head and shoulders above them all." Layton, though not addressing legal precedents per se, then brought up the nineteenth-century military theorist Carl von Clausewitz, the Prussian general whose classic work *Vom Kriege* [*On War*] was familiar to any aspiring military officer. Layton paraphrased Clausewitz's view that "direct annihilation" of enemy forces was paramount, reminding Nimitz that in his book the revered strategist had once written "something to the effect that in war the object is to direct the attack at the heart of the enemy." Layton said that a strike against Yamamoto would be a "practical and direct application of that principle, for Yamamoto represented the heart of the Japanese Navy."

Then there was code breaking, the secret and powerful advantage the United States had enjoyed for so long. If US fighter planes were to suddenly appear at one of Yamamoto's scheduled destinations at the same moment as his plane, how would that look? Would it be seen as more than mere coincidence, leading the Japanese to conclude that their code had been breached? That would be a horrible setback, potentially catastrophic to the Pacific campaign. The cracking of the Japanese code, though not perfect, had yielded so much helpful intelligence during the Guadalcanal campaign, and it had given them Midway. Had they not had advance warning of Yamamoto's massive assault at Midway Island, the Japanese could well have succeeded in crippling the US Pacific Fleet once and for all. Instead, the US victory had been a pivotal turn in the balance of power in the South Pacific Theater. Keeping the source of the intelligence secret was essential. Layton suggested a cover story that coastal watchers had come through once again and spotted Yamamoto's movements. "We could say it came from Australian coastal watchers around Rabaul," he said. "Everybody in the Pacific thinks they're miracle men."

Layton and Nimitz then considered the mission logistics, starting with planes—the enemy's and their own. Yamamoto would be flying as a passenger in the high-performance Mitsubishi bomber the Allies had nicknamed "Betty." That was actually fine by Nimitz. The Japanese might value the Betty for its long-range capability, but the plane's lack of armor and lightweight materials, plus all the fuel it carried to achieve distance, could be viewed as a gift, rendering it vulnerable to aircraft fire. The key question was whether Yamamoto's trip to the various northern Solomon Islands would bring him within range of any of the US aircraft at Guadalcanal, the lone base within striking distance. Nimitz studied the map of the South Pacific on the wall of his office, locating Rabaul, Bougainville, and other Japanese outposts in the Solomons and then Guadalcanal. It would have to be an aircraft that could handle well in excess of eight hundred miles of flying, round-trip. It was probably too far for the Navy's F4F Wildcat fighter or the Marines' F4U Corsair fighter, but maybe not for the new fighter plane that had been in service for less than six months: the Army Air Forces' P-38 Lightning.

But Nimitz would leave the basic operational details and challenges to commanders on the ground. He had made up his mind.

In the end, authorizing the special operation against a prominent, high-valued target was likely not difficult. Nimitz had no doubt that eliminating Yamamoto would carry enormous advantages, demoralizing Japan's military and its people. That Yamamoto was so deeply despised in the United States would help inoculate against any moral second-guessing after the fact. Besides, wasn't going after Yamamoto while he was in a plane akin to bombing his flagship while he was aboard—a combat action rather than some kind of a violation of wartime ethics?

"We'll try it," Nimitz said.

His next move was to notify the admiral he'd installed in late fall to take on Yamamoto and the Japanese in the South Pacific. "Let's leave the details to Halsey," Nimitz said. Layton began work on the order that Nimitz would send to Halsey's headquarters at Noumea on the island of New Caledonia, where it was already Thursday morning, April 15. In it, Halsey, the outspoken hater of all things Yamamoto, would be informed about the prized intercept and told to begin planning an aerial attack, with this proviso: that "all personnel concerned, particularly the pilots, are to be briefed that the information comes from Australian Coastwatchers near Rabaul." Nimitz did not want to have any pilot or pilots shot down and interrogated spilling the beans about the deciphered intercept. Nimitz looked over the typewritten dispatch Layton had prepared, initialed it for release to Bull Halsey, and then added a personal note: "Best of luck and good hunting."

It was payback time, vengeance for the December 7, 1941, surprise raid on Pearl Harbor.

THE SAME EVENING, APRIL 13, THAT THE DISPATCH WAS TRANS-mitted announcing his inspection tour in five days, Admiral Yamamoto crashed a party at the Rabaul base. He'd caught wind that a handful of underlings, classmates years earlier at the Naval Cadet School on the island of Etajima in Hiroshima Harbor, were planning a small reunion. Yamamoto showed up and, to the officers' delight, brought along two bottles of Johnnie Walker Black, the high-quality blended Scotch whisky that was a favorite not only among officers of the Imperial Japanese Navy but also of Great Britain's prime minister, Winston Churchill. The men traded stories from days gone by, enjoying the kind of socializing—drinking and smoking—that had been strictly banned when they were fledgling naval warriors at the austere academy. The occasion brought back to Yamamoto memories from the turn of the century when, just sixteen years old, he had left his home in remote Nagaoka to enter the academy. His final year, spent like that of every midshipman sailing the world's seas on the academy's square-rigged ship, was one of his fondest memories. By night's end, the officers were making up greetings on paper with their signatures to send to several more classmates stationed elsewhere, with Yamamoto proudly writing their words in his skillful calligraphy.

The evening spent in the soft glow of nostalgia served as a relaxing break from wartime matters. The next morning, April 14, Yamamoto was back at it, monitoring planes departing Rabaul for a strike on Milne Bay and nearby US ships, the final aerial attack of Operation I-Go, after which he ordered planes to return to bases and carriers to head back to Japan. He held a series of meetings that day and the next with staff and commanders from the various air corps, congratulating them on their hard work, warning against contentment despite all of the positive combat reports coming in, and cheerleading about sea and air battles soon to come. "Every man who attended these special meetings could not help but be impressed by the admiral's sincerity," said one Zero squadron leader. But despite the outward appearances, Yamamoto had to know in his heart of hearts that conditions were hardly as promising as the exaggerated victory reports trickling in from the war front suggested. For one thing, Japan had suffered serious losses in pilots and planes—an attrition that was becoming increasingly problematic. And its series of attacks against US bases, even if causing substantial harm, had not ultimately mattered. Henderson Field had not been ruined but remained fully operational. That meant the change of fortunes of the United States that had begun with the Japanese loss at Midway and had then been followed by the failed Japanese effort to retake Guadalcanal would continue, with Allied forces pushing northward from one island to the next or, as Allied military leaders liked to put it, climbing the rungs of a ladder toward Japan.

Yamamoto's key aides, too, continued to fret over his upcoming tour. His loyal, longtime staff officer and frequent *shogi* opponent, Captain Yasuji Watanabe, was distressed upon learning that the itinerary had been sent by radio rather than the secure alternative of delivering orders to each base commander by hand. Watanabe tried to enlist Yamamoto's chief of staff and confidant, Matome Ugaki, on his side, but the vice admiral agreed with Yamamoto about the value of the trip. Even if Ugaki had been opposed, he was in no position to say so; he was hospitalized with fever, diarrhea, and overall muscle pain from a bad case of a mosquito-borne illness, dengue fever. Army Lieutenant General Hitoshi Imamura, on the other hand, tried to persuade Yamamoto to reconsider by relating his own near miss at Buin. In February, when Imamura had flown to the base on southern Bougainville to visit his soldiers, his plane had unexpectedly run into a flight of enemy planes. His quick-acting pilot had turned straight into storm clouds and been able to escape undetected. But Yamamoto was untroubled by the close call and simply praised the pilot's evasive action. Then there was the commander at one of the three bases Yamamoto was planning to visit on Sunday. Rear Admiral Takaji Joshima, the commander of the Eleventh Air Fleet on Shortland Island, seemed to speak for all commanders when he was handed the radio message. "What a damn fool thing to do, to send such a long and detailed message about the activities of the C. in C. so near the front!" he had told his staff officers. "This kind of thing must stop."

BULL HALSEY WAS NOT AT NOUMEA, NEW CALEDONIA, WHEN ADmiral Nimitz's initial dispatch arrived on the morning of Thursday, April 15, but his executive officer was, and he immediately gave a heads-up to the freshly installed commander of combined air forces on Guadalcanal, Rear Admiral Pete Mitscher. The chain-smoking, thirty-seven-year veteran of the Navy was given the message at the tent near Henderson Field he used for a command office. Hardly posh, the tent was at least large enough to hold his desk in one corner, three other desks for his top officers, and folding chairs for visitors. Mitscher studied the message, occasionally using a handkerchief to wipe the moist humidity from his gold-framed glasses. The previous

April 18, he had been the captain of the USS *Hornet*, the aircraft carrier from which Lieutenant Colonel Jimmy Doolittle and his Tokyo raiders had taken off, and Mitscher, like so many others, had raged afterward upon hearing of the torture of captured US pilots. "I hate those yellow bastards worse than if I were a Marine," he had been heard to say more than once. The chance, one year later to the day, to go after Admiral Yamamoto was irresistible, and Mitscher assembled his senior staff.

Halsey, once back in the loop, contacted Mitscher the next day, Friday, April 16, and with characteristic wit wrote, "It appears the Peacock will be on time. Fan his tail." By then Mitscher had begun brainstorming with his navy and marine staff. They considered basic requirements for the high-stakes mission, including making sure that pilots kept radio silence and stayed away from Japanese-held islands to avoid detection. They began studying maps and making more precise distance calculations. Even though a straight flight from Guadalcanal northwest to Bougainville was about 325 miles, the mission's total round-trip would be closer to 1,000 miles. That was because the fliers would have to take a circuitous route to avoid being spotted by any of the Japanese outposts along the direct line between the two points. The numbers confirmed Nimitz's hunch that P-38 Lightnings were the only fighter planes suitable for the long-range assignment, as Mitscher notified Nimitz in one of the secured communications now making the rounds between Pearl Harbor, New Caledonia, and Guadalcanal. The navy and marine officers working with Mitscher were not thrilled; they had been angling for their own fliers to lead the charge in F4F Wildcats and F4U Corsairs. Instead, the Army Air Forces' P-38s would carry the day. But even for the gas-guzzling P-38s, the distance was a stretch; they would need belly tanks to increase their fuel capacity. Mitscher's office notified a base on New Guinea to deliver all the drop tanks on hand, preferably large three-hundred-gallon tanks, as quickly as possible.

To man the mission, Mitscher already had in mind a couple of P-38 pilots he figured had the right stuff: Tom Lanphier and Rex Barber, the pair who had made such a strong first impression as he was taking command at Guadalcanal. Their bold flying had certainly debunked the early grumblings of the navy and marine pilots who, when the P-38 had first arrived, had complained the Army's fliers were soft and sometimes hard to find, hiding high in the clouds. When he was told that Lanphier, Barber, and other pilots in their flight were scheduled to go on leave in a matter of days, Mitscher ordered the leave canceled. To get a fuller take on the personnel flying P-38s, Mitscher then turned to the most obvious person, Army Air Forces Lieutenant Colonel Vic Viccellio, who before his recent promotion had commanded nearly every pilot now manning "the fork-tailed devils." In his new position as fighter operations officer, Viccellio often hopped around between Henderson Field and Halsey's headquarters on New Caledonia, but this particular week ended in Mitscher's tent, where he was briefed on the mission, with Mitscher sharing the message from Halsey that ended, "Fan his tail." It had been a race against time from the moment code breaker Red Lasswell first deciphered the Yamamoto message, and now it was even more so. Yamamoto was scheduled to depart Rabaul in less than forty-eight hours. The mission planning had gotten started but was an ad hoc work in progress, still without a hard-and-fast flight route and decisions on the number of planes to use and who would fly them. There was still a lot that needed sorting out, with Mitscher and his advisers agreeing that over such a long distance, the only chance of success would be to develop a precise flight plan that the fighter pilots would then need to execute to near perfection.

Mitscher wanted to know: Who should lead the strike?

Viccellio did not hesitate: Major John W. Mitchell, commander of the 339th Fighter Squadron, he said. Mitch had been back from leave only a few days, and he was easily the most logical choice. Vic had known Mitch since before the war, going back to Hamilton Field outside San Francisco and those early training days. He'd always wanted Mitch, a natural born leader admired by the rest of the men, nearby. Moreover, Mitch was an ace pilot, the first in the fighter group, and over the winter, during multiple bombing raids and dogfights, he had proven that the P-38 Lightnings were the best thing the Allies had going in the Pacific skies.

Viccellio got no argument from anyone. Hands down, Mitch was the one.

CHAPTER 15

ONE DAY MORE

——

P-38 Lightnings in formation
WWII Database/public domain

WHILE REAR ADMIRAL MITSCHER, LIEUTENANT COLONEL VICCEL-
lio, and others were talking about him at Henderson Field headquarters, Major John Mitchell was not in any of the usual places—his tent,
the pilots' "club," or the Fighter Two operations dugout. He'd gone

off by himself the afternoon of Friday, April 16, to find some privacy up on a hill above the pilots' tents overlooking the Fighter Two airstrip. There he settled in and began a long letter to Annie Lee in San Antonio, still in the glow of her last letter in which she had seemed calmer and been reassuring about his determination to keep fighting until he felt he'd done his part. The thing was, he said, "It's *awfully* quiet here now and not near so much fun." He didn't get into specifics, but he was referring to Yamamoto's Operation I-Go, which was now over—and which, to his chagrin, he'd missed. He wrote instead about this and that—about his good luck at dominoes, winning $435—and about some rearranging he'd done to his tent: "I even chiseled me a mattress and though it's not too good a one it sure beats nothing and feels good to me." He rambled on about finances, money orders he'd sent home, and the frustrating fact that he'd been with her only briefly in the sixteen months since their wedding. Marriage by mail was wearing thin. He then tiptoed back to the topic of his return, although prefacing his remarks by acknowledging that to do so was probably risky: "I know darn well I shouldn't even mention this as you will begin to get ideas at once." He speculated that he actually might be heading back to the States, suggesting that it might finally be his turn to leave for good. "That doesn't mean tomorrow or even next month, but I am hoping that I may see you and have a drink with you on my birthday." His birthday was June 14, and he teased her about the date in a way that was actually another reminder of their long separation. "If you've forgotten when that is you won't know when to expect me home," he deadpanned.

Then, as if he was worried that he'd said too much and Annie Lee would get worked up about a possible homecoming, he tempered his remarks. "I shouldn't mention anything," he repeated. "Every time I have ever planned anything along that line it has fallen through."

Mitch was right. The next afternoon he found himself in a jeep hurriedly rattling along the rough coral road from Fighter Two to the navy command tent at Henderson Field nearly two miles away. He'd been summoned by Vic Viccellio and been told to bring Tom Lanphier along. "Some mission or other" was about all Vic would say on the telephone, although he added, "You'll like it."

It wasn't the first time Vic had dropped a surprise on Mitch out of the blue. Two years earlier, in April 1941, when Mitch had been a mere second lieutenant training at Hamilton Field outside San Francisco, Vic had ruined Mitch's Easter plans to visit Annie Lee in Texas with the news he would be heading overseas on a special mission. Mitch had then spent the rest of that spring and the early summer in bombed-out London, part of a small contingent of talented young pilots selected to soak up what they could about combat flying from the Royal Air Force. He'd gotten his first taste of war while there, when German antiaircraft guns had fired on him after he'd flown across the English Channel in a Spitfire. What did Vic have in mind this time?

Mitch and Lanphier arrived at the operations headquarters at Henderson Field, nicknamed "the Opium Den" because of the constant fog of cigarette smoke trapped inside. Entering the tent, they were greeted not just by Viccellio but also by Rear Admiral Mitscher and other senior staff. Mitch hadn't yet met Mitscher, who was still only two weeks into his new role as Commander of Air Forces in the Solomon Islands, or COMAIRSOLS. He eyed the admiral, who, standing in the dank bunker, looked small, almost emaciated. Steely eyed and leather faced, he wore a sun helmet—a long-billed brown cap—even indoors. And he chain-smoked; when Mitscher went to light a new smoke, Mitch saw what others already knew: that the reserved navy veteran, chin jutting, curled his hand to protect the cigarette, wind or no wind, the way cowboys did in Hollywood westerns.

ADMIRAL ISOROKU YAMAMOTO SPENT MUCH OF SATURDAY IN CONferences with his commanders at naval headquarters at Rabaul. His chief of staff, Matome Ugaki, now discharged from the hospital, was feeling strong enough to preside. The morning session was spent reviewing the unit reports from Operation I-Go's aerial strikes. Yamamoto, who was unaware how exaggerated they were, expressed his approval. Ugaki seconded the praise while adding a note of caution. Even with the operation's success, the rear admiral reminded everyone of the "present plight" of the navy's diminishing airpower. Moreover, he said, given the strain on manufacturers and materials back home, they could not count on "swift reinforcement." The group took a break for lunch and reconvened for a second session in the afternoon, that one focused on, as Ugaki explained, "policies and principles." It was essentially a forward-looking discussion about the navy's operational needs, whether in equipment or the repair and upkeep of the various airfields and bases. Some bickering erupted among several commanders from forward bases over priorities and also about who bore responsibility for the declining morale detected in the squadrons. The tension was troubling, and Ugaki's discomfort was only made worse when a rash broke out all over his body, related to his debilitating bout of dengue fever. Yamamoto, meanwhile, seemed in fine spirits. He maintained a steady hand and quelled the internal wrangling with reminders that they must all work together to stop the enemy advance up the chain of islands from Guadalcanal.

Throughout, and in between discussion of the main items on the day's agenda, the officers' chatter ranged from the humdrum—what to wear on the base inspection—to the weighty—the admiral's safety. Ugaki raised the topic of attire with Yamamoto and other staff officers. The issue was whether a green khaki outfit, rather than Yamamoto's formal white uniform, would be suitable and more comfortable, given the tropical heat, for the one-day tour. There was even the suggestion that because of the heat the travelers be allowed to wear open-neck shirts. Though some staffers favored a relaxed sartorial approach, Ugaki did not like the idea. He did not think it would be as hot as some were saying, and, besides, they would be in a plane much of the time. No open-collared shirts, he insisted. "It wouldn't be proper for the commander in chief of the Combined Fleet and his staff officers to visit officers and men at the front wearing unofficial uniforms," he said. That seemed to settle the matter. Buttoned-down khaki uniforms would be the dress *du jour*. But then came disagreement about footwear. Ugaki assumed that they would wear their standard boots with matching leather leggings. But Yamamoto and the others favored airmen's boots as much lighter and sensible for getting into and out of the planes multiple times. Ugaki did not protest too much, but he was reluctant to wear the airmen's boots himself.

More significantly, Yamamoto's staff continued to fret about mission security. One minor adjustment had been made to his itinerary, but the change had been more about logistics than anything else. The original travel plan had called for Yamamoto to fly first to the island of Ballale, then ride a subchaser to island-hop to various bases, and then end up at the airfield at Buin, which was called Kahili, on Bougainville Island. The revised itinerary called for him to fly first to Buin and continue on from there. To his closest aides the last-minute tweaking mattered little; they wanted the mission aborted altogether. Longtime senior aide Yasuji Watanabe, who had overseen the itinerary down to the smallest detail and had been alarmed when it had been sent out by radio rather than delivered by hand, was openly opposed to the admiral's making the trip. The commander at Shortland Island, Rear Admiral Takoji Joshima, who had been distressed earlier in the week when he'd received the admiral's schedule by radio, had actually flown to Rabaul on Saturday afternoon to make his case in person. Joshima, insisting that he knew the conditions at the front better than anyone else, argued, "Please, sir, this is dangerous." Do not go, he implored.

Yamamoto appreciated the concern, but neither he nor Ugaki considered the trip particularly risky; it was a mission that was, as far as they were concerned, a closely held secret. To be sure, enemy planes, particularly the new US P-38 Lightnings, had made an occasional flight as far north as Bougainville. And Yamamoto's army comrade Lieutenant General Hitoshi Imamura had already tried to dissuade him with his story about a dangerous close encounter with the enemy near Bougainville. But Yamamoto and Ugaki knew that P-38s had never made a full-scale raid to that area, the main obstacle being the extra fuel the gas-guzzling fighters needed to reach Bougainville from Guadalcanal. (Yamamoto did not know that at least one US flier had been eager to attempt such a raid: Major John Mitchell. Following intelligence reports in late winter of increased Japanese shipping activity, Mitch had been asking to lead a fighter sweep up to the southern tip of Bougainville. But his requests had always been denied, he said, "as it was felt the risk of losing some of our precious P-38s was too great.") Moreover, Yamamoto felt quite comfortable traveling with a cover escort of six Zero fighters, one of which would be flown by one of the navy's best pilots, Shoichi Sugita, otherwise known as the "Shoot-Down King." "This is no cause for concern,"

Yamamoto reassured Joshima. It would be a quick, one-day trip—departing first thing the next morning, returning late in the after-noon. "You must have dinner with me tomorrow night," Yamamoto said cheerily.

STANDING INSIDE THE POORLY LIT COMMAND DUGOUT, MITCH took stock of the group assembled with Mitscher. It was a gaggle of high-ranking officers—a clear sign that something big was in the works. Though every US military service was represented, Mitch noticed that the group was top heavy with navy personnel. Besides Mitscher, there was his assistant chief of staff, Vice Admiral William A. Read, as well as Navy Commander Stanhope C. Ring. The Ma-rines had General Field Harris, Colonel Edwin L. Pugh, and Major John P. Condon. Mitch was unfamiliar with most of them, but he knew Condon. The major had been around since the first of the year and was in charge of the Marines' fighter pilots; interservice rivalry aside, Condon was one who had readily acknowledged the positive impact the Army's P-38s had made since arriving at Guadalcanal the previous year.

Someone handed Mitch a piece of paper. Mitscher and the others began explaining that a message concerning the whereabouts of Admiral Yamamoto had been intercepted. In short order they told Mitch that he had been chosen to lead a flight of P-38s to go after the Japanese naval commander—and to do so at dawn, meaning in eighteen hours, give or take. There would later be a dispute as to whether Mitscher or anyone else had ever discussed the source of the intelligence—the breaking of Japanese codes. Some insisted not, but even if US code-breaking know-how was not part of the conversation—the decryption being top secret—Mitch came to realize during the briefing that the information had come from intercepted communications. For him, though, the issue was not how they'd come to learn of Yamamoto's travel plans, it was the fact that there was little time to hammer out logistics for a mis-sion that, at best, seemed a long shot. Moreover, it was clear that no matter how poor the odds, the assignment was of paramount importance—that he and the pilots chosen for the mission were to make every effort to take out Japan's naval genius, even if it meant, as several of the officers in the dugout asserted, "ramming Yamamoto's plane."

The paper he was handed delineated the timetable of Yamamoto's trip. Mitch took it all in, not saying much at first, mostly listening. He noticed that a lot of people had a lot to say who weren't going to do anything on the actual mission, and the thought came to him: "As usual, in a crowd like this, there's always a bunch of big mouths." No one seemed interested in his perspective, and the way the navy officers were talking, Mitch could tell they had basically decided on a way to get Yamamoto. In their view, the time to attack Yamamoto was after he'd made his first stop at Ballale and boarded a subchaser for the short ride to Shortland Island. Blast him then, when he's on the boat, they said. It seemed a done deal, and some of the naval staff had even begun studying the ocean currents around Ballale for the next morning. That was when Mitch finally had enough. "I don't want to do that," he said. The moment he broke his silence, the tent quieted—albeit briefly. The others began questioning him: Why not? The standoff escalated, as the navy officers persisted, their voices getting loud as they argued with him. But Mitch would have none of it. He told them he was an army fighter pilot and didn't know a subchaser from a sub, a wisecrack no one thought humorous. The point was, he wouldn't be able to identify for certain which boat carried Yamamoto. And even if he hit the right one, he continued, Yamamoto could very easily jump into the water and survive. "And that wasn't what we're up there for. We're up there to get him." The navy contingent gave Mitch hard looks, and the debate went on for precious minutes as they tried to get him to go along. He would not budge. He insisted on getting Yamamoto in the air and refused to consider any other plan.

Mitscher, a former aviator himself, broke the logjam by making a critical observation: that Mitch was the one leading the flight. "Since Mitchell has to do the job," he said, "I think it should be done his way." The moment was pivotal. Mitch, by speaking up and asserting himself, had made it clear to all the navy and marine brass in the room that he was in charge—he was the mission leader upon whose shoulders the flight would succeed or not. Moreover, had Mitch stayed silent and had the others devised a plan to attack Yamamoto in the subchaser, an interception would not have been possible. Yamamoto's itinerary had been slightly modified. He would not be flying first to Ballale, as the code breakers had decrypted. Instead, his first stop would be Buin, on Bougainville Island. The men in Mitscher's tent did not know that; they were working from the travel plan they'd been given, and if Mitch had headed to Ballale for an attack on a subchaser, the timing would have been way off; Yamamoto would not have been there. In that regard, it was providential that Mitch insisted on a plan to intercept Yamamoto in the sky as he approached Bougainville, no matter the location of his first or even second stop.

Mitscher's ruling settled matters, and the discussion turned toward the mission's operational needs. Pilots' names came up. Mitscher mentioned how impressed he'd been with the recent flying feats of Tom Lanphier's flight—a reference to the four pilots who'd strafed an enemy ship: Lanphier, Rex Barber, and two others, Joe Moore and Jim McLanahan. Though not an order, Mitscher's comments carried weight; and it was as if he were reading Mitch's thoughts, because as he assessed who in the 339th Fighter Squadron had the experience and the ability for a mission like this, Lanphier and Barber, as well as Besby Holmes, along with some others, were the ones who quickly came to mind.

The staff then informed Mitch that auxiliary belly tanks would need to be attached to the planes because of the amount of fuel required for the trip. There were 165-gallon tanks at the base already, which the pilots had used in the past to increase their planes' fuel capacity, but this trip's distance would call for even bigger tanks. Mitscher's staff had therefore ordered that 300-gallon tanks be sent over from a US base on New Guinea as quickly as possible. Mitscher asked Mitch if he could think of anything else, and Mitch considered the P-38s' flight instruments. He thought of his new Elgin wristwatch and appreciated its precision, but in flight the optimal value of a watch—any watch—came in combination with a compass, and therein lay the problem: the compass in the P-38 cockpit was one of the plane's few shortcomings. Mitch considered it unreliable, giving erroneous readings. The navy's fighter planes, meanwhile, had excellent compasses—ones that Mitch had eyed and envied—and so in reply to the rear admiral's question he said, "A compass—one of those big Navy jobs." Mitscher nodded, ensuring that the request was as good as fulfilled.

The meeting had now run deep into the afternoon. Mitch began to feel an urgency to get back to Fighter Two to get ready. He began, as well, to feel the magnitude of the mission. He'd flown more than two hundred missions during the eight months he'd spent on Guadalcanal. He'd shot down eight enemy planes and become the Army Air Forces' first ace on Cactus. But neither he nor anyone else, for that matter, had ever had an assignment like this one. "The longest planned intercept ever," he realized. Not only the longest but an aerial intervention with huge unknowns. They had Yamamoto's basic itinerary but not his route or altitude, nor did they know whether he'd be on time. Furthermore, there was no way for them to know his speed—or whether *they'd* get there at the right time. Everyone in the dugout had concurred on certain constraints to minimize forewarning Yamamoto in any way: radio silence, low-altitude flying, and, to avoid enemy coastal watchers, flying no closer than twenty miles to shore. It meant he'd be leading P-38 Lightnings for more than four hundred miles, hugging the ocean's surface the entire way and with no land references in sight. No checkpoints at all. None. They'd be dependent on his watch, his compass, and his airspeed calculations. Dating back to his boyhood in Mississippi, Mitch had been comfortable in the wild, finding his way home by relying on the moon and the stars—Johnnie Bill and the moon. But this was way different. The high-stakes mission, he concluded, would rely on dead reckoning: "By guess and by God."

The danger, too, was self-evident. Although according to the intelligence, six Zero fighter planes were to escort Yamamoto's Betty bomber, everyone in the tent expected more enemy firepower than that. Much more. On the ground at Bougainville were another seventy-five or so Zeros, and a substantial number of those planes would likely greet Yamamoto and bring him into the airfield. That was what the Americans had done for Navy Secretary Knox in January to protect him during his island-hopping tour. As Mitch recalled, they'd taken all the available fighters on the base and flown them to their extreme range in order to pick up Knox's aircraft and escort him back to Guadalcanal. With that in mind, they had every reason to think, as Mitch realized, that "the Japanese would come up with possibly as many as 50 of these 75 fighters." The Army Air Force had eighteen P-38 Lightnings at Fighter Two, which meant Mitch could expect to face three to four times as many Zeros. No one had to say anything. The math was obvious; Mitch and his fliers would be outnumbered. The consequences were just as obvious: there was a high probability that some of them would not return.

"We expected a hell of a battle" was how Vic Viccellio later framed it.

ON RABAUL, WHEN THE ALL-DAY MEETINGS RUN BY CHIEF OF staff Ugaki had concluded, senior aide Watanabe could see that Admiral Yamamoto was worn out. Watanabe accompanied Yamamoto on the ride back to his quarters atop the hill overlooking the harbor. The weather was quiet and clear, and it was cooler on the hilltop than on the sprawling base below. The two men dined alone and afterward moved to another room to resume their long-standing pleasure of opposing each other at *shogi*. The pastime was a relaxing distraction. Watanabe, still worrying about the trip, wanted to go along, but Yamamoto had a different idea. He interrupted the play in a way that revealed he was always mulling over his naval responsibilities even as he competed in the board game. The tension among his base commanders during the day had nagged at him, and he wanted his trusted aide to address the matter the next day. "I want you to stay here tomorrow and finish the conference," he told Watanabe. "If we don't pull them together now, we're in trouble. You know what to tell them, and make things very clear." The commander in chief then retired for the night.

WHEN MITCH AND TOM LANPHIER RETURNED TO FIGHTER TWO early that evening, Mitch headed right away to the mess tent, which had a large table he could use as a desk. Given the stakes, part of him wanted to assign himself the task of shooting down Yamamoto. He was an ace, after all, perhaps the best P-38 pilot on hand. But as mission leader he knew he couldn't let ego and glory guide his thinking; he needed to make selfless, not selfish, decisions, meaning that he would need to devise a plan without favoritism that would give the squadron the best shot at success. He decided that the best use of his own skills would be to protect the pilots targeting Yamamoto. He would assign others to the "killer flight" while he flew over them at 20,000 feet, a high-altitude position to handle the fifty-odd Zeros they expected to come up from Kahili, a large fighter base, to provide an honor escort for the admiral. Neutralizing the escort would hardly be a minor matter and, thought Mitch, provide plenty of action: "I wanted to get after those Zeros."

Mitch rounded up maps of the Solomon Islands, pencils, and a slide rule and set up in the tent to work on a flight plan. Major Condon had prepared one possible route, but Mitch rejected it. Timing was perhaps the most crucial factor in plotting a path, especially given the flying distances and fuel demands, and Mitch knew that Condon had no firsthand experience flying a P-38. Condon did not know Mitch's airspeed nor his ground speed, nor did he know where he intended to make the interception, and without knowing those factors, Mitch concluded, there was no way Condon could put him into the right place at the right time. Besides, Mitch was still smarting from the way the officers at the briefing with Mitscher had acted, as if they knew the best course and he should simply fall into line. "It was my squadron, my neck and my responsibility and I think it ludicrous that anyone on the ground would have given me detailed instructions how to make the flight," he fumed.

Given the many unknowns, the plan would have to be built upon assumptions, and Mitch began by making several: that the punctilious Yamamoto would arrive at his first stop on time; that his Betty bomber's airspeed would be 180 miles per hour; and that the optimal point of interception would be about thirty miles out from Yamamoto's destination. The base had antiaircraft guns and aircraft, and Mitch wanted to strike and be gone before the ground personnel knew what was going on. Studying the map, he plotted the best point to be just south of Empress Augusta Bay in central Bougainville. Then, working by lantern and flashlight, he took up the tasks of projecting Yamamoto's expected route, imagining one for himself and determining how the twain would meet. He did so by backing everything up, meaning he looked at the point of interception over Bougainville and plotted a route working back to the Fighter Two airfield. He drew up five legs that, after takeoff, would take his flight west of Guadalcanal, then north, and then east to Bougainville and the point of interception. In the event that they didn't see Yamamoto, they'd continue across the island to its east coast in the hope of finding the admiral there. Mitch incorporated estimated wind speeds and weather forecasts, which called for clear, cloudless skies, provided by two intelligence officers so that he could calculate compass settings, distances, and the timing of each of the five legs. In all, it was a looping route covering about 435 miles and avoiding the string of Japanese-held islands between Guadalcanal and Bougainville. He'd also keep his planes flying only about fifty feet above the water. The goal was a low-altitude course that would minimize the chance of being picked up by the Japanese—whether by their coastal watchers, their radar installations, or the ever-present small boats circling off the Japanese islands. For if they were spotted and Yamamoto was notified, not only would the admiral turn back, but Mitch and his men would likely fly headlong into far more than the fifty Zeros they were expecting.

While Mitch worked, curious pilots wandered by, with Doug Canning asking what was up. Mitch quickly explained, and Canning's jaw practically dropped. Everyone knew about Yamamoto, thought Canning, "How he had planned Pearl Harbor, and was planning to ride up Pennsylvania Avenue to the White House to dictate peace terms." Word began to spread, and Mitch had to wave them off, telling Canning and the others he'd brief them shortly. That everybody wanted to go did not surprise Mitch. "If there was going to be a big show, they all wanted in on it." It was well after dark before Mitch completed his calculations. He checked and double-checked the numbers with the help of the two intelligence officers, Navy Lieutenant Joseph E. McGuigan and Army Captain William Morrison. He made "strip maps"—strips of paper about a foot long and eight inches wide to rest on a pilot's knee—marked with headings for the 435-mile course they'd be flying. When he examined the maps and assessed the challenges, he put the odds at "about a thousand to one that we could make a successful intercept at that distance." But long odds were not something he could dwell on; he had his orders and would need to tackle the job as just another mission. One thing going for him: "I was full of confidence in those days."

Eighteen P-38 Lightnings were in service, meaning that Mitch could choose seventeen fliers from about forty pilots assigned to the several squadrons that shared the P-38s. He conferred with Major Louis Kittel, the commander of another army air forces squadron, and together they settled on a pilot roster for the mission. In the cover flight were Kittel and lieutenants Roger J. Ames, Everett H. Anglin, Doug Canning, D. C. Goerke, Lawrence A. Graebner, Raymond K. Hine, Besby Holmes, Jack Jacobson, Albert R. Long, William E. Smith, Eldon E. Stratton, and Gordon Whitaker. Mitch designated four pilots as "the hunters" in the attack, or killer, flight, Captain Tom Lanphier and First Lieutenant Rex Barber as one pair; First Lieutenant Jim McLanahan and First Lieutenant Joe Moore as the other. Mitch, as mission leader, would fly the first Lightning in formation and be in charge of the cover group as well.

Close to midnight, the men got the word to gather on the hill behind the pilots' tents, the same hill where, higher up the prior afternoon, Mitch had written his letter to Annie Lee completely unaware of what was in store for him. The forty-odd fighter pilots, along with some of the mechanics and ground crew from Fighter Two, found Mitch waiting. He'd hauled out a blackboard and on it had written the names of the sixteen pilots he and Kittel had selected. Then, in businesslike fashion, he told them about the order to intercept Admiral Yamamoto and described the dogleg route around some islands to Bougainville Island, a route that was water all the way and not a rock in sight. As Mitch continued talking, he distributed the strip maps so that each pilot had one. He'd lead the formation, he said, and the rest were to follow him—at a low altitude and in radio silence. "No one was to touch that mic button from the time we took off until we engaged the enemy planes." The outbound route of about 435 miles would take 150 minutes, according to his calculations, and the return—a straight line back to Guadalcanal covering about 300 miles—would go faster. They would need every ounce of fuel to complete the more-than-700-mile round-trip and would have little time at Bougainville to find their target or to linger. It was get in and get out.

Mitch said that Jack Jacobson would be his wingman, meaning that Jacobson would fly his plane slightly behind Mitch's, guarding Mitch's flank like a second set of eyes. He read out the names of the fourteen pilots in his cover flight, and he identified the four

pilots in the killer unit—Lanphier with Barber as his wingman and
Moore with McLanahan on his wing. If something went awry with
any of those four, Besby Holmes and Ray Hine were told, they'd
be the ones to drop out of the cover group to join the attack unit.
Mitch stressed flight discipline repeatedly: follow him, and fly low
but not too low. "Watch yourselves," he cautioned. "If you drop just
a few feet lower, you're in trouble. That drink will hypnotize you
and you'll be in it before you know it." They were to use hand signals only, no radio, and stick to their assigned roles. He did not want
planes from his cover group going after Yamamoto, flying around
helter-skelter and possibly shooting at one another. He didn't want
eighteen airplanes trying to knock down one lone bomber. He told
his cover group its job was "to watch the attack flight and cover it
until the bomber is down. Nothing more."

When Mitch was done, every pilot on the hill recognized that the
mission was a crapshoot, with Mitch shaking the dice and letting
them rip. So many things could go wrong, given all the mission's
moving parts: a mechanical breakdown along the way; unexpected
winds that would either slow or quicken the flight; unexpected tropical storms they'd have to fly around to avoid; even a last-minute
decision by Yamamoto to alter his itinerary or call the trip off altogether. Mitch would have to do a perfect job of navigation and
timing over trackless ocean through winds of unknown velocity to
reach the appointed spot at the correct second. Even if everything
fell into place and Mitch's timetable held up, they'd be straining to
spot a dot in the bright sky upon their arrival at Bougainville. Nonetheless, Lanphier was all for it, appreciative that Mitch had put him
and Barber into the attack unit. Barber was likewise afire. Other
pilots asked questions, as if looking for ways to allay their skepticism. Why did they think Yamamoto would be on time? What about
the weather? There'd been an evening shower; the air was heavy
and humid. The two intelligence officers were reassuring: the admiral was notoriously punctual and the forecast was for clear, hot,
and windless weather. The briefing ended with Vic Viccellio, who'd
commanded nearly every pilot on hand, stepping up to give a brief
talk. He stressed the mission's importance and their instructions "to
get the admiral at all cost."

While the hillside at Fighter Two was a pocket of intensity, else-where on the base men were welcoming the return to Henderson Field of a bomber pilot who had gone missing for two months. Captain Thomas J. Classen and his crew had been flying a B-17 Flying Fortress named *My Lovin' Dove* in February when they had been attacked by eight Japanese Zeros. During the fight Classen and his eight crewmen had all suffered wounds, and Classen had been forced into crash-landing at sea. Everyone had somehow survived, and for sixteen days they had paddled in inflatable rafts before reaching an island where they were taken in by friendly natives, despite being in Japanese-occupied territory. They had slowly made their way from island to island and eventually, sixty-six days later, had been picked up and taken to Guadalcanal.

Mitch's pilots were not part of the homecoming; their minds were on Yamamoto as they headed for their tents to try to catch a few hours of sleep. Mitch went to his and fell onto the mattress he'd made, the new creature comfort he'd written about to Annie Lee the day before. He could hear music coming from Doug Canning's tent—Glenn Miller's "Serenade in Blue," with its easy-listening lyrics, "When I hear a serenade in blue/I'm somewhere in another world,/alone with you"—and he actually fell asleep.

MITCH WAS AWAKE BEFORE DAWN, TAKING A SEAT AT THE LONG table in the mess tent he'd converted into an office the prior evening. He ate with his men—powdered scrambled eggs, milk, Spam, and coffee. Members of the Fighter Two ground crew were there, too, having worked through the night attaching 300-gallon belly tanks under one wing of the P-38s and 165-gallon tanks under the other. The auxiliary tanks affected a plane's maneuverability, which was why the plan called for Mitch's fliers to drop them upon reaching the Bougainville coast. The ground crew hurried as best it could to get things right, connecting fuel lines and electrical circuits so that the pilots would be able to open, close, and drop the tanks. Crews also loaded ammunition containers for the five guns on each plane: four .50-caliber machine guns and a 20 mm cannon. The machine guns, the nose pointing straight ahead, provided the Lightnings' "cone of fire" that Japanese pilots had learned to fear in a few short months.

The mechanics and armorers worked in a steady rain, but the morning of Palm Sunday, April 18, 1943, was breaking clear and blue—and hot. And the eighteen P-38s were checked and ready. Takeoff time for the more-than-four-hundred-mile trip was set for a little after 7:00 a.m. Their target was scheduled to fly from Rabaul about an hour after they were airborne, at 8:00 a.m. Guadalcanal time.

Mitch hustled down to the airfield and climbed onto the wing of his plane and into the cockpit. He was pleased to see that the navy compass he'd asked for had been installed. He wore a lightweight khaki flying suit and, like most pilots, the snug-fitting rawhide marine boots that, in the event of a parachute drop, would not slip off. He adjusted his helmet, goggles, radio headset, and throat mic. He checked the instrument panel, the fuel mixture control, the prop and rudder settings, and the control yoke. Once settled in, he positioned the strip map he'd drawn on lined white paper securely on his knee. It was his flight plan, written in extreme detail and upon which the Yamamoto mission depended, so if his calculations were off and the pilots failed, that failure would be on him and him alone.

Mitch fired up the twin engines and taxied across the pierced-steel matting that made up the runway's surface, past the maintenance crew standing by with a truckload of coral. The steel spikes in the matting came loose despite regular inspections, and coral was a ready repair. He was the first to take off at 7:10 a.m., as scheduled. Rear Admiral Pete Mitscher was stationed at the end of the runway to see them off; it was a year to the day since he'd stood on the deck of an aircraft carrier to watch Jimmy Doolittle and his bombers take off for their raid on Tokyo. Mitscher wasn't the only one on hand, though. Despite repeated reminders about the mission's secrecy, word had spread, so that navy and marine fliers were gathered in pockets along the runway, rooting for Mitch and his men to take out the United States' deeply loathed enemy, Isoroku Yamamoto.

Next to take off was Jack Jacobson, Mitch's wingman. Fighter Two was only 3,500 feet long, meaning that with the heavy fuel load the pilots cut it close and for additional lift had to use their dive flaps. Mitch and Jacobson circled the airfield at an altitude of about 4,000 feet, waiting for the others to join them and form up. Doug Canning and D. C. Goerke followed. Next in line were the four pilots in the killer flight. Tom Lanphier, Rex Barber, and Joe Moore all got off fine, but as Jim McLanahan lumbered down the runway and was about to take off, one tire caught a loose spike and blew. The tire shredded rapidly, and the P-38 veered off the runway to a stop. McLanahan's plane was disabled, unable to go. McLanahan was out just like that, and for the moment the killer unit was down to three planes.

The remaining ten planes in the cover flight then took off and fell into line without incident. Mitch checked his wristwatch and saw that, just as he had planned, the takeoffs and formation had taken fifteen minutes. It was 7:25 a.m. Everyone was ready to go. Except, suddenly, Joe Moore. Once airborne, all the pilots switched to their external fuel tanks for the trip to Bougainville, saving their internal gas for the interception and flight back. But Moore couldn't get his cross-feed valves to work, and the belly tanks weren't feeding. His engines sputtered each time he turned the switch, and no matter how many times he tried, he could not correct the problem. Mitch looked and noticed Moore gesturing at him. It took Mitch a minute to understand that Moore was signaling he couldn't get his belly tanks to feed. Mitch saw no alternative: Moore was done, and Mitch had to get him out of there. He signaled the bad news, and the dejected Joe Moore banked to the left to turn back.

They were not even twenty minutes into the mission, and already Mitch had had to resort to a contingency; he waved to Besby Holmes and Ray Hine to leave the cover group and join the killer unit. He then turned the flight—now sixteen P-38 Lightnings—in a westerly direction to commence the first leg, which he'd calculated would take fifty-five minutes. He kicked the plane's rudders so that they fishtailed, his way of signaling the others to spread out. He'd said beforehand that he wanted them strung out on the way up to avoid the strain of flying in close formation all the way. He also took them down to about fifty feet above the water's surface, where within minutes another design flaw in the otherwise gem of a fighter plane was revealed. The Plexiglas bubble over the cockpit was locked tight. Ordinarily that was a nonissue. The P-38 was a high-altitude twin-engine fighter usually flown at 20,000 feet and above. But now, hugging the ocean, the bubble was like a magnifying glass for the sun's unfiltered rays, and there was no way to open it. The temperature inside soared. Mitch read the gauge: 95 degrees in the cockpit. Everyone was soon soaked in sweat.

THIRTY MINUTES LATER AND 650 MILES TO THE NORTHEAST, YA-mamoto and his chief of staff, Matome Ugaki, arrived by car at the east airfield at Rabaul. Ugaki was feeling upbeat: "The sky was quite clear and the early birds sang pleasantly in the trees." Because Yamamoto and Ugaki were returning by day's end, they traveled light, taking only items that fit into their pockets: cigarettes, eye-glasses, a small diary, and handkerchiefs. Ugaki had relented on the footwear, deciding after he had awakened to go along with Yama-moto and the others' choice to wear airman's boots. He was glad he had, for he found the lighter boots easy to put on and pull off and quite comfortable. He was also satisfied with the group decision to go with a green khaki uniform for the inspection tour. Studying himself, he thought the outfit made him look "gallant." He could not conceal his surprise, however, when he first saw the admiral dressed in green, Yamamoto's signature look being a crisp white uniform. "It suited him very well," Ugaki thought, "but looked a bit strange, perhaps because we were not accustomed to seeing him in such a uniform."

They were joined at the airfield by various staff officers who were accompanying them. Two Betty bombers were idling and waiting. The Japanese communication about the trip, the one the US code breakers had intercepted, had mentioned only one Betty bomber, but the final travel plan had the two high-ranking officials flying separately for security purposes, just as they had done when they'd flown from Truk down to Rabaul in early April. Wasting no time, Yamamoto and his party boarded the Betty bomber num-bered 323, while Ugaki boarded the one numbered 326. Yamamoto was directed to the "skipper's seat," the seat directly behind the pilot. Ugaki took the same seat on his plane, removing the awk-wardly long sword from his belt and handing it to a staffer to store out of the way.

Yamamoto's Betty 323 was the first to take off from the crushed coral runway. He was right on time, which, factoring in the two-

hour time difference, was 8:00 a.m. on Guadalcanal, while for him
on Rabaul, it was 6:00 a.m. Ugaki followed in the second Betty, and
then the six escort Zeros took off in pairs. From the sky they looked
down upon the volcanoes towering over Rabaul. Yamamoto's command plane climbed to a cruising altitude of 6,500 feet as it turned
southeast. The Zeros climbed even higher, to about 8,500 feet, a
perch from which they were to maintain a close eye on their country's cherished naval leader.

The worry so many of Yamamoto's aides had expressed about the
trip had not changed things. To be sure, basic safekeeping measures
were taken: Yamamoto and Ugaki flying separately and the six-plane
escort by the prized Zero fighter planes. But a palpable casualness
defined the day trip. For one thing, the Zeros were not equipped
with radios. The pilots had removed them, as they often did. The
standard-issue radios jammed easily, created more static than audi-
ble words, and were viewed as deadweight. To lighten the aircrafts,
the pilots had tossed them out. The two Bettys carrying Yamamoto
and Ugaki, meanwhile, had barely any ammunition on board. Instead of loading extra boxes of cartridges, given the fact that the Imperial Japanese Navy's commander in chief was a passenger, only
one ammunition belt per gun had been allotted. That was because
the flight was viewed solely as a transport operation, not an attack
mission. Ammunition boxes were heavy, and adding extra boxes for
the plane's three machine guns and single 20 mm cannon was seen
as unnecessary. Besides, during their preflight briefing no one had
mentioned a word to the pilots about the nearest US bases or the
chances of encountering trouble along the way.

Yamamoto and Ugaki did not expect problems, comfortable
in the knowledge that between Rabaul and Bougainville the Japanese maintained air superiority and controlled the skies. Ugaki
was relaxed as he settled in for the short flight. To pass the time,
he pulled out an aviation map to follow the topography below. His
plane flew in tandem with Yamamoto's, so close to its left rear that
Ugaki thought his plane's wingtips might touch Yamamoto's. He
looked out the window and could see people moving about in the
first plane. "I could clearly see the profile of the commander in chief
in the skipper's seat." Yamamoto was sitting erect, his white-gloved

hands clasping a sword. Yamamoto had not put his sword aside as Ugaki had, perhaps because his held special meaning; it had been a gift from his older brother. From Ugaki's point of view, the window was a picture frame, capturing in profile the man embodying Japan's naval glory: a portrait of perfection.

CHAPTER 16

DEAD RECKONING

Pilots, Yamamoto mission
Courtesy of the National Museum of the Pacific War

MAJOR JOHN MITCHELL AND HIS BAND OF FIGHTER PILOTS COM-
pleted the first and longest leg of their outbound flight in the fifty-
five minutes he had specified. It meant that Mitch had accurately
accounted for weather, wind speed, fuel, and cruising speed to cover
the leg's 183 miles in the allotted time. He led the sixteen-plane formation on the first course change, into a more northerly direction
but still far west of any of the Japanese-held islands. Meanwhile, the
plane carrying their target, Isoroku Yamamoto, had settled into its

flight, heading south from the base at Rabaul toward the island of Bougainville.

The first leg was uneventful to a fault, as were the second and third legs, covering another 188 miles into the rising sun: mile upon mile of flying in stuffy hot cockpits with the dull sameness of a wavy, heaving sea fifty feet beneath their aircrafts. "Nothing, except waves," Mitch noted, "and one wave looks like another." For D. C. Goerke, flying so low for so long proved taxing; he was hot, nervous, and anxious. The reality that the odds were against them was ever present, but Goerke stayed motivated nonetheless: the slightest chance of success made the mission worthwhile. Besby Holmes, usually calm in flight, fought the jitters. He worried that they might encounter a flight of Zeros on the way and have to stop to fight. The eagle-eyed Doug Canning took to looking for sharks, whales, men-o'-war—anything to occupy his mind and not get disoriented. He counted forty-eight sharks. One pilot became so lulled by the unvaried sea and hypnotic drone of the plane's turbocharged engines that he drifted down to the water's surface. Mitch wouldn't call the pilot due to the radio silence he'd imposed, and he watched nervously as the fighter plane's props began throwing up spray. Fortunately, the startled pilot instantly made a flight correction. And it wasn't as if Mitch was immune—he began to doze a couple of times but was able to shake himself and clear his head. "I got a light tap on the shoulder," was how he thought about it, saying a voice had told him, "John, hold the course."

Staying the course was the challenge. Every one of his men, and anyone else who'd ever piloted a plane for that matter, knew that to get from point A to point B you usually passed something—a city, a town, a river, or a road—that gave you an idea of where you were, especially on long trips like this one. But on this trip the pilots saw nothing but the never-ending Pacific Ocean. "Not a rock in sight," Mitch said. To assist him, he had an airspeed indicator, his Elgin watch, and his navy compass—that was it. He kept looking at his watch, like a nervous tic, over and over again—a thousand times, it seemed—as he made sure to stick to the course headings exactly the way he'd charted them, finishing one leg and turning into the next, making certain the men in the fifteen other planes stayed with him in formation. The question he kept asking himself was: Would they be on time? "That was crucial," he knew. "Of utmost importance. The timing of it."

SHORTLY AFTER 9:00 A.M., HAVING FLOWN ABOUT FOUR HUNDRED miles, Mitch eased to the right in a northeasterly direction and, five minutes later, turned harder to the right for the final leg of the flight. During the trip the formation had spread out quite a bit, as Mitch had directed, but now he needed his pilots to tighten up. He turned the control yoke back and forth slightly to rock the wings on his Lightning, the signal for everybody to close in, and they did as they were told. Flight discipline and radio silence were the bedrock of the mission. Not a peep had been uttered over the radio at any point during the two-hour trip.

Mitch looked straight ahead. They should now be heading toward Bougainville, according to his calculations, about five minutes from landfall. He squinted, trying to pick up the coastline, but couldn't see it. The sun was in his eyes, and it was hazy on the "deck," meaning at sea level. He'd thought he'd at least be able to spot Bougainville's central mountain range, with its tallest peak at 8,900 feet. He couldn't understand why he wasn't picking it up, if not the mountains in the distance, then at least Empress Augusta Bay, the big X on his map, the point he'd chosen for the interception because it spanned about forty miles and was easily identifiable. He thought he should be able to see that first checkpoint, the only one that truly counted, because it would confirm the completion of the five-legged journey over the seemingly endless ocean. But he still saw nothing through the haze. He was getting nervous, feeling itchy. He worried that they weren't going to make it, that something was wrong. There was no option, though, but to stick to the flight plan and see it through, which meant climbing up and away from the ocean's surface with the other fifteen planes in tow, and it was when he did that, as the formation ascended above the haze, that he saw Bougainville: the bay, the coastline, and the mountain range beyond. They were about three miles offshore and closing in.

Mitch knew what that meant: he was right on time.

Whatever relief he felt lasted only seconds, however, overtaken by a new worry. The sun shone brightly across the sky, a clear, blue sky but a sky that was empty. Mitch scanned ahead for the distant specks he was hoping to find, the tiny dots in the sky that would be Yamamoto's bomber and his Zero escort. But he saw nothing. Every pilot was doing the same, squinting across a sprawling sky, desperately seeking their mark.

Then, for the first time in two hours, Mitch's radio crackled, as did the radios in every other fighter plane, as Doug Canning broke their collective vow of silence.

"Bogeys, 10 o'clock high," Canning said. The plainspoken Nebraskan, with eyes like binoculars, had spotted sharks, whales, and men-o'-war and now had apparently found their target.

Canning's four words were electrifying. "Hot damn, and we didn't even practice this one," thought Major Lou Kittel, a flier in the cover flight. Kittel was stunned at the notion that a plan built on so many what-ifs and assumptions regarding speed, altitude, and flight routes was right on the money. Besby Holmes, one of the four pilots in the killer group, recalled the surprise raid on Pearl Harbor, when he had been attending morning Mass as the bombs began falling all around and he'd pulled out a pistol to fire back. The thought formed: "This time Yamamoto didn't have surprise on *his* side." But Mitch did not share their excitement—quite the opposite. When he looked and locked in on what Doug Canning had spotted, his first thought was that something was terribly amiss—because instead of one Betty bomber he saw two, each a glittering silver, bright and brand-new looking. This was not the guy, Mitch decided. Yamamoto was supposed to be in one bomber, that was all.

His heart sank to his boots.

THE BETTY BOMBER CARRYING ADMIRAL YAMAMOTO WAS ABOUT ninety minutes into its flight, cruising along a southerly course and ahead of the bomber carrying his chief of staff, Matome Ugaki. Leading the way, Yamamoto's pilot had set the pace, making slight adjustments to the plane's speed to keep on schedule. Overhead the escort Zeros, split into two groups of three, flew on either side and slightly behind.

The Yamamoto trip had been uneventful, so much so that Ugaki, in his plane, had dozed off soon after departing Rabaul. The pilots were relaxed, chatting with their crews. No one was studying the sky ahead with any particular rigor since the preflight briefing had not included any mention or warning of possible enemy activity in the area. Everyone simply had to keep an eye on the lead plane carrying Yamamoto and maintain his position.

The northwestern tip of Bougainville Island was visible, and the plane carrying Yamamoto, the one with "323" painted on its rudder, had begun a slow and unremarkable descent from its cruising altitude of 6,500 feet to about 4,500 feet. It was nearly 9:30 a.m., and the island's coastline and thickly jungled lowlands were clearly in sight—mile upon mile of mangrove swamps, palm trees, and inlets. Yamamoto, right on schedule, began to make ready for landing in fifteen more minutes, as his aircraft continued its steady descent. Neither he, Ugaki, nor any of the pilots of the eight Japanese planes was aware that a few thousand feet below and off to their right, sixteen P-38 Lightnings were emerging from the haze over the ocean's surface.

HAVING COUNTED TWO BETTY BOMBERS WHEN THERE WAS SUPposed to be only one, Mitch's mind raced. What to do? Take these guys or not? Or let them go and hope that Yamamoto would come on later? He had to decide in an instant, and he chose to go after the two bombers, the Japanese Bettys being the equivalent of the proverbial bird in hand. "Take what we can get," he thought. Maybe Yamamoto was seated in one of the bombers, maybe not. But, Mitch figured, they hadn't flown up from Guadalcanal for nothing. This was what was offered to them, so they'd take it. In the next instant, he saw something else: three Zeros flying in tandem behind the Bettys to one side and another three flying on the Bettys' far side. Six Zero fighters in all, matching perfectly the intelligence about Yamamoto's trip. Mitch was reassured. He knew they had their target. His momentary angst was replaced by a cold, calculating calm. Yamamoto was in sight. All they had to do was kill him.

Mitch continued his ascent and turned slightly to the right in order to fly parallel to Yamamoto and in the same direction along Bougainville's coastline. His squadron, in tight formation now, followed closely behind. There was no time to think about anything other than the task at hand. But one thing Mitch couldn't help noticing was that they were closing in on Yamamoto so quickly that if they had been at 4,500 feet, where the admiral was, they would have collided—the very thing one of Vice Admiral Mitscher's aides had recommended as a way of underscoring the magnitude of the mission. Ram into Yamamoto's plane if necessary, the aide had said. It was unbelievable, Mitch thought, how close they were. And being in the lead, he had to resist the instinct to rush ahead—not to ram Yamamoto but to use the P-38's "cone of fire" to seek and destroy the two Betty bombers. It was what Doug Canning, flying right behind Mitch, wanted to do. "Why the hell don't we go in and get 'em?" he wondered, since they were nearest to the enemy planes. But that would have ruined the mission discipline Mitch had demanded so that his fliers wouldn't suddenly peel off in different directions and make a mess of the plan to have a killer unit target Yamamoto while a larger cover unit protected them against the anticipated force of fifty or more Zero fighters.

Instead, with radio silence now moot, Mitch pushed the microphone button in his cockpit and issued the order "Skin 'em," meaning "Shuck the external belly tanks." The pilots turned the fuel selector valves to change the fuel lines to the planes' internal tanks. Then they hit the switches to release the two auxiliary tanks that together had contained 465 gallons of extra fuel. The P-38s became faster and more maneuverable, and as they continued their ascent, Mitch signaled the squadron to split into two formations. He turned upward, showing off one of the features that had made the twin-engine Lightnings legendary: its ability to soar or, as one pilot put it, "climb like homesick angels." The eleven others in the cover unit kept up with him. They headed high up past the Japanese planes carrying Yamamoto and kept climbing to fend off the expected attackers. It was time for the four-man killer formation to veer toward the two Japanese Betty bombers.

"Tom, he's your meat," Mitch told Lanphier, the leader of the killer unit.

Right away, though, the attack hit a snag. Besby Holmes could not shed his belly tanks. He flipped the switch to jettison them, but nothing happened. He hurriedly checked the circuit breakers and tried again. Still nothing. "Dammit!" he screamed. "Drop!" But the tanks remained stuck in their shackles. Then, instead of forging ahead, Holmes broke off from the killer unit and turned back over the water, rocking his plane and hoping that would shake the tanks loose. Mitch, watching from above, was incredulous. Holmes should have gone on in anyhow. The move was all wrong, Mitch thought, a failure in mission discipline. "Here was the rabbit right down at the dog's nose, and he pulled off." Moreover, the belly tanks were nearly empty and hardly the same drag on the P-38's maneuverability that they were when full, and on a mission of such consequence, Mitch was unequivocal: Holmes should have stayed with him. To make matters worse, wingman Ray Hine dutifully followed Holmes out to sea, as he should have, to protect Holmes.

Mitch was agitated. He was also struck by a kind of planner's remorse. He had continued his climb, leading the eleven other Lightnings in his cover flight to nearly 15,000 feet, where they expected to confront Zeros attacking from every direction. Everyone was bracing for high-altitude combat and dogfights. But when they got there, the sky was empty. There was not an enemy fighter in sight, except for the six Zeros way down below. Had they known that—had he and Mitscher and the top brass involved in planning known that against all logic Japan's greatest admiral would not be greeted by a massive aerial force—Mitch would have done what Canning had been wishing to do when first sighting the Japanese bomber: get Yamamoto. He could have taken his own flight in for the kill, Mitch thought, as he had been in the front and in the perfect position to do just that. He would have certainly welcomed a do-over, but there was nothing he could do now.

He could hear Holmes struggling with the tanks. "Wait a minute, Tom," Holmes called out to Lanphier on the radio. "I'll tear my tanks off, just give me a second." But there were no seconds to spare. It was up to Tom Lanphier and Rex Barber, the killer unit cut in half—two P-38 Lightnings against a Japanese flight of two bombers and six Zero fighters.

YAMAMOTO, HIS AIDES, AND THE CREW EXPECTED IN A MATTER of minutes to be on the ground in Buin, where the base commander and his men lined both sides of the runway waiting to greet the commander in chief of the Imperial Japanese Navy. Though the pilots in the six escort Zeros had not been expecting trouble, it was simply part of flying—never mind their duty—to instinctively scan the horizon for other planes. They had looked mainly up, however, the result of hard lessons learned from encountering the new P-38. The Japanese pilots knew that their US counterparts, to fully exploit the advantages of the P-38's speed and high-altitude capability, liked to attack by dropping down swiftly from above. They'd never thought to look down below.

But Yamamoto and his flight crew were suddenly startled by the pilot in the first Zero, who had surged to the front and was dipping his wings, frantically trying to get their attention. With Yamamoto watching, the pilot kept pointing downward, directing Yamamoto's crew to look down and to the right. They did, and they saw—and, in that instant, the other escort Zeros on Yamamoto's right also saw what the lead Zero pilot had discovered, and all three Zeros turned sharply to take on the two P-38 Lightnings that were fast approaching. Simultaneously, Yamamoto's pilot did two things: he pushed the throttle to accelerate rapidly and then dived toward the jungle's treetops, where it would be far more difficult for an attacker to hit them. Most machine guns in fighter planes pointed up slightly, making the best firing position a tad beneath a target, and an attack plane trying to slip under Yamamoto flying across the treetops might easily end up in the jungle. His pilot's maneuver was a smart evasive action, if only he could get them there in time.

Ugaki's bomber did the same as the lead bomber, following Yamamoto's sharp descent, a move that startled the vice admiral and sent crew members scrambling to their combat stations. Both cabins erupted in chaos and noise, with men yelling and the wind rushing in at the gun ports, which were open. But Ugaki's pilot "overspeeded," meaning that he had accelerated so quickly, the Betty bomber's fuselage began to vibrate. To slow and stabilize the aircraft, the pilot pulled hard on the throttle, causing the plane to fall further behind Yamamoto's. Ugaki was struggling to make sense of it all when he saw bullets passing over his aircraft, including tracers, "the special bullets which make a light," as one crew member said.

The bullets missed and kept going as Ugaki yelled over the noise, "What happened to Admiral Yamamoto's plane?"

FROM HIS PERCH MITCH TRIED TO FOLLOW THE QUICKENING ACtion. With Holmes out over the water, trying to shake his belly tanks, and Hine accompanying him, Lanphier and Barber were on their own. The two had headed for the Betty bombers about five miles ahead. Lanphier took the shortest route possible, flying directly toward the bombers from a ninety-degree angle. To their surprise, the Japanese flight was continuing to hold its formation over the Bougainville coastline and in the direction of the base at Buin. It meant that they still hadn't been sighted. Barber couldn't help but think, "They were a little bit asleep." He and Lanphier were hoping to bank in on the bombers' tails and into the prime spot from which to open fire. But when they were about a mile away, they saw three Zeros, flying behind and to the right of the two Betty bombers, begin a very steep dive, throwing off their belly tanks.

Even as they pushed their P-38s as fast as possible, Lanphier and Barber could tell they were not going to reach the Betty bombers before the Zeros reached them. They'd be turning in behind the bombers as the Zeros turned onto their tails. The two flights—Lanphier and Barber from one direction and the Zeros from the other—were coming together at speeds of nearly three hundred miles per hour, and when the distance between them had closed to only about 1,000 feet, or the length of three football fields and then some, Lanphier made the split-second decision to turn left and head up straight toward the enemy fighter planes. It was a bold move providing cover of sorts for Barber, although Mitch worried that his pilots were about to get caught up in a dogfight while letting the bombers slip away. "Get the bombers," he said over the radio. "Damn it all, the bombers!"

Running interference, Lanphier fired at the Zeros, and they returned fire. Barber kept going, catching up to the Betty bombers within seconds. He rolled his Lightning to the right to move in behind but banked so sharply that his wing momentarily blocked his view. He could not see either bomber, and when he rolled back and leveled off, he saw only one. He didn't know where the other one had gone. In fact, flying at top speed, Barber had banked right over the top of the second Betty bomber and leveled off behind the lead one. The circumstances he found himself in were hardly ideal. He was traveling faster than he should for "good gunnery," meaning that his speed would affect his shooting accuracy. Then there was his position—behind the bomber. It made *him* vulnerable. He was looking up right at the tail gun. He'd have preferred a high-angle attack, shooting on the curve of pursuit. Finally, there was the plane's altitude—or lack thereof. He was getting so low at that point that he had to worry about clipping a tree. But such concerns were inconsequential in light of the fact that he was staring at the first Betty bomber, the one numbered 323, directly ahead of him at treetop level. Barber was lined up and could not delay a second longer.

He pressed the trigger on the four machine guns located in the nose of the P-38 and pushed the button to fire the 20 mm cannon, unleashing a burst of firepower from just fifty yards away that was everything the Lightning had to offer. He hit the bomber's right engine, which immediately started smoking. He fired another burst that tore through the fuselage and into the left engine, and then he fired across the bomber's left wing and back again, causing pieces of the bomber's tail to split off. Each of Barber's bursts sent bullets ripping through the thin aluminum skin of a plane that by design was lacking in heavy armor in order to increase its flying range, and as the P-38 Lightning's four machine guns blasted away and the bomber shuddered with each round, Isoroku Yamamoto turned left in his seat to look back. In that millisecond two of the hundreds of .50-caliber bullets raking the plane found him. One bullet entered his left shoulder, causing a wound from which he could have recovered, but the second one penetrated his jaw on the lower left side, tore through his brain in a slightly upward trajectory, and exited at the right temple. It was the second wound that was incompatible with life.

Barber was so close now that he could see the bomber's rear cannon and realized no one was manning it, which explained why he wasn't receiving any fire, and as he centered his fire on the fuselage one more time, the bomber suddenly seemed to stop in midair, as if stalled. The plane rolled slightly to the left, "a quarter snap." Barber nearly collided with it, barely missing the upturned wing as he flew past and over the bomber in a flash. He figured the plane's sudden stall was what happened when a pilot was killed and involuntarily yanked back on the controls. But he couldn't know for sure, and the only thing he knew as he looked over his shoulder was that the Betty bomber was spewing black smoke, a thick, heavy smoke from a fuel-stoked engine fire. He saw the bomber level off while continuing to drop, less than a hundred yards now from the tops of the trees, and that was all he saw, because he spotted a Zero approaching from behind, and in the next instant he heard the hollow noise that gunfire makes hitting a plane, a series of pings. They were unloading on him, shooting at him heavily. He hunched under the cockpit's armor plate and began his own evasive action, hugging the treetops so that the enemy would have a harder time hitting him and managing as he turned sharply back toward the coastline to shout out on the radio, "I got one of the bombers!"

Rex Barber no longer could see the Betty bomber, which headed overland in the direction of Buin but would never come close to reaching its destination. The plane, engulfed in flames, brushed the jungle canopy before plunging into it, the left wing ripped off by treetops. Upon impact it exploded into flames and broke up, the biggest chunks being the tail assembly, engines, and parts of the wings. The eleven men on board were hurled in every direction, all but one burned beyond recognition. The eleventh passenger was propelled clear of the wreck, still strapped tightly into his seat. The sight was an odd one: an officer seated upright in the brush, dressed in a green khaki uniform, medal ribbons on his chest, his hands in white gloves. The left hand clutched a sword. Admiral Isoroku Yamamoto was a picture of repose, except that he wasn't resting. He was dead.

MITCH CONTINUED CIRCLING AT 15,000 FEET, DOING HIS BEST TO watch for enemy Zeros while straining to see far below. He saw a fire, then another, in the Bougainville jungle but not much else. He did not see Lanphier's dogfight with the escort Zeros or Lanphier's next moves to elude the Zeros and go after the Betty bombers. He did not see Besby Holmes shake loose his belly tanks over the ocean and, with his wingman Ray Hine, turn toward Bougainville in time to see the first bomber crash in the jungle. It looked to Holmes as though it had gone straight in, given the size of the fire—a crash just after he'd heard Rex Barber shouting on his mic about getting one of the Bettys. The action continued nonstop beneath Mitch, with Holmes noticing that the second Betty bomber was trying to flee over water, so he and Hine performed a maneuver called a barrel roll in order to give chase. Meanwhile, Barber was dodging Zeros, zigzagging to avoid their gunfire and then getting an assist from Holmes and Hine, who scattered the Zeros as they went speeding past overhead toward the sea. That's when Barber, still flying as low as he could, saw the other Betty bomber, too, hugging the water, right along the edge of the island, so low that its props were kicking up a wake behind it. Holmes and Hine went into a steep dive toward that bomber, and Barber raced to join them, enemy Zeros buzzing all around. The three fighter pilots opened fire, and when Barber closed in, he fired a burst into the fuselage. The plane exploded before crashing into the ocean. Pieces flew and hit Barber's Lightning, one chunk striking the left wing, another cutting a gash in the canopy above Barber.

Practically nothing was going on with Mitch and the eleven other pilots, save for their mounting anxiety about fuel consumption. The longer they circled, the less fuel they'd have to return to Guadalcanal. Doug Canning, for one, almost got a taste of combat at one point when he stumbled upon a Zero. Undetected, he turned in behind the fighter plane, but then his canopy fogged up. He couldn't see. He tried wiping the canopy by hand, but that made matters worse, smudging everything. He was forced to let the Zero get away, and when he looked around he was all by himself, with no idea where Mitch and the others in the cover flight had gone. The only thing to do was to get more altitude. He climbed to about 18,000 feet, looked some more, but still could not find anyone. He had apparently drifted ten miles out to sea and was on his own. He circled alone.

The action was all down at sea level, and although Mitch did not have his eyes on his four pilots, he'd heard snippets of heated radio calls and seen puffs of gunfire here and there along the coast, as well as one, maybe two fires in the jungle. To him the fires were key evidence that planes had been shot down. He needed to make a ruling, seated as a one-man jury of sorts, and he needed to make one quickly, given the fuel concerns, the fact that the interception was nearing its ten-minute mark, and, most alarming, the fact that dust clouds had begun stirring around the Japanese base in Buin, indicating planes scrambling to take off. Lacking proof beyond a reasonable doubt, he nonetheless reached two conclusions: the radio noise he had overheard meant his killer pilots were still okay, and the fires in the jungle meant they'd accomplished the job he'd given them—taking out the enemy bombers. He shouted, "Let's get the hell out and go home."

The flight of P-38s was turning in a southeasterly direction for Guadalcanal when Mitch heard Lanphier calling for help. He hesitated, shouted to the others to continue, and studied the area far below. He spotted a P-38 off the eastern tip of Bougainville. It was just above the water, with an engine smoking and trailing smoke. He saw a Zero sitting on the P-38's tail, little puffs of smoke coming from the Zero's guns. Right away he and his wingman, Jack Jacobson, rolled down from 15,000 feet, diving as fast as they could to help. They reached 400 miles per hour and were on the verge of buffeting when the Zero suddenly peeled away from its target, its pilot apparently noticing Mitch and Jacobson fast approaching. They both fired bursts at the aircraft as it fled. It sped toward the Buin airfield, and they let it go.

Mitch and Jacobson were unable to find the P-38. They wondered if another Zero had swooped in to finish off their crippled colleague, but neither of them had seen a plane go down. They took another minute to look around but knew they had to get going if they were to have enough fuel to make the three-hundred-mile trip back. Initially Mitch had thought Lanphier was the flier in the disabled P-38, because he had been the one calling for help just as Mitch had issued the order for everyone to clear out. But Lanphier turned out to have been elsewhere, frantically eluding a Zero, when he had called. And having outraced the Zero, he was already heading back to Guadalcanal, ahead of Mitch.

In fact, the pilot who had gone down was Jacobson's friend and tent mate, Ray Hine. Besby Holmes and Rex Barber had lost sight of Hine after all three had gone after the second Betty bomber. "There's one of the finest boys I've ever known," Jacobson would write in his diary that night. "Hope he made land." But First Lieutenant Raymond K. Hine, just twenty-three and from Indiana, never did return. Neither he nor any trace of his plane was ever found.

HEADING BACK, MITCH AND ALL THE OTHER PILOTS, ESPECIALLY the three from the killer unit, were thinking that no one could have survived their attacks on the two Betty bombers—the first one exploding into a ball of fire upon impact in the jungle, the second crashing into the ocean. They were wrong about the second Betty, though. Three of the twelve men aboard were alive: Yamamoto's chief of staff, Matome Ugaki, the plane's pilot, and one of the officers. Their plane had crashed into the water at full speed, rolling over onto its left side, bursting into a blaze, but the impact had apparently catapulted Ugaki and the pilot through the canopy covering the cockpit, while the third officer had somehow managed to escape the wreckage. Ugaki was hurt badly and struggled to reach the beach two hundred yards away, grabbing hold of a large toolbox from the wreck to use as a float. Soldiers from a nearby barracks came running and, after realizing Ugaki was friend and not foe, retrieved him, the pilot, and the officer. Ugaki's condition was the worst: bruises and abrasions over his entire body; left eye swollen shut; clots of blood on his face; a broken left rib; and a right forearm compound fracture. He was carried to the barracks, where a medical orderly began treating his injuries. But the vice admiral insisted that he had urgent business to attend to. He ordered that the base commander at Buin be contacted and told that communications about the crashes "should be made by confidential telegram and be restricted as much as possible; this is from the Chief of Staff." Ugaki then turned his attention to the chance that others had survived. He made sure that rescuers returned to the beach to look for any from his aircraft, and, after asking about Yamamoto's aircraft, was told that a search plane was already flying to the crash site.

Mitch and his men, though wrong about Ugaki's bomber, were spot on with their hunch about the first bomber: there were no survivors. Even Ugaki, told about the search plane while he was being treated for his injuries, was not hopeful. He'd witnessed the aircraft staggering southward, just brushing the jungle top with reduced speed, emitting black smoke and flame. Ugaki had gasped, muttering to himself, "My God," then grabbing the shoulder of a crew member. "Look at the commander in chief's plane!" he'd said, pointing out the window. Within seconds, Yamamoto's plane was gone. Ugaki had strained his eyes but could not see it, only a pall of black smoke rising from the jungle to the sky.

Within a few hours, the aerial searchers located the crash site and reported seeing no sign of life either at the wreck or in the immediate vicinity. By midafternoon Sunday, a military cable was sent notifying the chief of the naval general staff and the navy minister in Tokyo of Yamamoto's crash and disappearance. The secret cable prompted an emergency gathering of the country's top naval officials. In Rabaul, Yamamoto's senior aide, Captain Yasuji Watanabe, who had pressed the admiral the prior evening to permit him to join him, hurried to fly from the base to join the search. The devastated Vice Admiral Shigeru Fukudome, a longtime friend, spoke for many when he lamented, "There could be only one Yamamoto, and nobody could take his place."

THE REMAINING FIFTEEN P-38 LIGHTNINGS, SCATTERED DURING the attack, flew back not as a single unit but in various combinations. Mitch and his wingman, Jack Jacobson, were together, cruising at about 10,000 feet. Doug Canning, unable to find his wingman, D. C. Goerke, saw Besby Holmes running out of gas, struggling along, and he accompanied Holmes to be sure that Holmes was okay. Having been separated, Tom Lanphier and Rex Barber flew alone.

On the way back to Guadalcanal, Lanphier couldn't contain himself. "I got the son-of-a-bitch!" he shouted triumphantly on the radio. "He won't dictate terms in the White House now." Not every flier heard him, but those that did were incredulous. Never mind that the assertion Lanphier was making—that he was the one who had shot down Yamamoto's plane—was premature, to say the least, and hardly standard protocol. "Lanphier had not cross-checked results with other shooters," noted Major Lou Kittel of the cover flight. Even if Lanphier had somehow managed to attack a bomber, instead of showing the poise of a seasoned fighter pilot, he was acting like a gold digger racing to a land office to stake out a claim before anyone else could. Worse was the security breach, putting at risk the code break-ing that the US commanders were obsessed with guarding. Mitch had stressed that very concern, another pilot, First Lieutenant Roger J. Ames, recalled as he listened to Lanphier's stunning remarks. "Just get out of the island," Mitch had instructed. "Don't say anything we weren't supposed to, to let the Japanese know, in any way, shape or form that we might have broken their code." To Ames, Lanphier on an open radio clearly referring to Yamamoto by invoking the White House line for which the enemy admiral was notorious was the epit-ome of recklessness. If Japanese monitors picked up Lanphier, thought Ames, "that would be enough . . . they could maybe put it together."

Lanphier would not let up. Landing at Fighter Two at about 11:00 a.m.—after a number of pilots from the cover flight had al-ready returned—he continued proclaiming his achievement to anyone within earshot, making no mention of Rex Barber or any other pilot for that matter but saying he'd looped around after tak-ing on the escort Zeros, blasted the famous enemy admiral's plane, and then watched as it plunged into the jungle. An officer who was one of the first to meet him on the ground heard Lanphier, still in his cockpit, claim victory over Admiral Yamamoto in no uncertain terms. The officer, Lieutenant Joseph Young, was surprised by the comments. He didn't know Lanphier well but did know his reputa-tion as very collected, mature—and a superior pilot. The histrionics left him astounded, and he considered Lanphier's dramatics to be irrational. Lanphier next climbed into a jeep for a ride down the runway toward the operations tent, a ride that was tantamount to a victory lap as he stood in the back calling out "I got him! I got him! I got that son-of-a-bitch!" Some found the display off-putting, with one crew chief later composing a poem to lampoon Lanphier: "I got him, I got him, is what Capt. Lanphier said / I alone shot Yamamoto dead / Let no other man make this claim / He'd steal my honor and soil my name."

Lanphier's posturing was a distraction to the celebratory recep-tion the returning pilots were receiving at Fighter Two, with at least one P-38 performing a roll and buzzing the field to the delight of cheering onlookers. Pilots leaped from their planes and excitedly gathered to greet the next one to land, and when their mission leader touched ground at about 11:30 a.m., a bunch of them began jumping up and down and cheering Mitch, "like it was a football game that we had won," Mitch recalled. He unstrapped his parachute and climbed down from the cockpit to a hero's welcome from all the crew chiefs.

Mitch's gas tanks were nearly empty, as were nearly everyone else's, especially the pilots in the killer unit who had burned through their fuel in combat with the Japanese. In fact, Holmes was delayed on his return, having to stop along the way; his main wing tank was empty from battle damage, and he had only four gallons left in another tank when he landed on the Russell Islands on a runway under construction. Holmes's plane wasn't the only one damaged; Lanphier's had a few bullet holes in one of the horizontal stabilizers, the small, fixed wings at the rear that keep a plane in control and flying straight. But it was Barber's that took a while to assess—104 holes from 52 bullets fired into the rear section of his P-38 by Zeros chasing him and a chunk of metal from the second Betty bomber still stuck in the skin of the plane. Barber had returned just prior to Mitch, and they both caught up to what everyone else was hearing: Lanphier's claim for the Yamamoto shootdown.

Barber was not pleased. He and Lanphier were soon going at it. "How in the hell do you know you got Yamamoto?" Barber snarled. Barber was certain he'd shot down the lead bomber, but he knew he could not claim with equal certainty that Yamamoto had been aboard that plane. He doubted that Lanphier had even hit a Betty, but even if Lanphier had, how could he know positively, as another pilot put it, "that he had shot the big man down"? Lanphier persisted with his version of events, calling Barber a liar, and tensions got even worse when Besby Holmes returned. Learning that he and the missing Ray Hine had been left out altogether in the various recappings of the interception, he erupted in anger. Holmes pressed his case that he'd shot down the second Betty bomber and also some Zeros.

No formal debriefing ensued, just competing claims conveyed during informal discussions with the two intelligence officers who'd helped the day before in plotting the mission, Navy Lieutenant Joseph E. McGuigan and Army Captain William Morrison. The developing tempest threatened to overshadow the outpouring of accolades.

Pete Mitscher, the vice admiral who'd seen the men off that morning, pulled up in a jeep at the Fighter Two operations tent with a case of I. W. Harper bourbon whiskey. "You raised hell with 'em, boys," the elated Mitscher said, going around shaking the fliers' hands. "They've been jabbering so much on the radio that we think you got him. I want you to know how good a job you did. By God, that's what I call flying." Marine Major John Condon was equally impressed, calling the mission "a marvel of performance on Mitchell's part."

An all-out donnybrook among the three pilots was averted when an apparent solution took hold—the notion that there must have been three Betty bombers. The theory would later prove inaccurate, but at the time it seemed to make sense, given each pilot's claims, and it meant that Barber, Lanphier, and Holmes could each take credit for one. With that, peace was restored and the killer unit pilots calmed down. Lanphier, who'd worked briefly as a reporter and knew how to turn a phrase, even sought to reassure Barber and Holmes, telling them he would work on the "Fighter Report" with McGuigan and Morrison to make sure they got it right. Meanwhile, Mitch, ever the leader, had done his best to cool things down, stressing the bigger picture, not anyone's individual glory. The mission was a team effort. "I couldn't have cared less who shot him down," he said. "We got him, that's what we went up there for, and we got him. Joe Blow shot him down, fine with me. We got him."

The men partied into the night, "singing like a bunch of high school kids after a ball game," wrote *Yank* correspondent Mack Morriss from inside his tent at nearby Henderson Field, unaware of why the army fighter pilots at Fighter Two were making a ruckus. "Three or four men howl like dogs," Mack Morriss continued. "They've sung everything I can think of and I think—I hope—they've about exhausted their repertoire." But they carried on all night long, fueled by another accolade—the equivalent of a hearty backslap—from Bull Halsey, whom Mitscher had notified. "Congratulations to you and Major Mitchell and his hunters," wrote Halsey in a cable from New Caledonia. "Sounds as though one of the ducks in their bag was a peacock."

For the first time in two days, Mitch could let down. He wasn't going to let anything get into the way of savoring the moment—

the mission's huge and unexpected outcome, beyond his wildest dreams, save for the loss of Ray Hine. Going in, he had figured they'd suffer significant losses in combat against the anticipated legions of Zeros that, fortuitously, had never arrived. In short order and relying largely on dead reckoning, he'd overseen one of the longest ever aerial interceptions to carry out the targeted kill of an enemy leader.

John Mitchell had told Annie Lee just a few weeks earlier that he was not ready to come home. He was responding to her lobbying for his return, and he had told her he didn't think he'd done enough yet for the cause. He was feeling different now, on this night of April 18, 1943. He'd just made history: Yamamoto had been the architect of the Pearl Harbor attack, and he was the architect of Yamamoto's demise. "We'uns is coming home!!" he announced in his next letter to his wife, continuing in a lower, less exclamatory tone, as if he were whispering in her ear. "Speaking of action," he said, "I have a wonderful story to tell you concerning a little action I led the boys on this last time."

EPILOGUE

Yamamoto funeral
Courtesy of the National Museum of the Pacific War

FOLLOWING THE SHOOTDOWN OF YAMAMOTO'S PLANE ON SUN-
day, April 18, 1943, the admiral's devoted aide, Captain Yasuji Watanabe, tried to fly to Bougainville to join the search but was delayed
in Rabaul by a severe tropical storm. When Watanabe reached Buin
the next day, he learned that an eleven-man crew of army soldiers,
working road construction eighteen miles west of the base, had been
redeployed to find the crash site. The crew cut its way through the
thick, dark jungle and, near dusk, discovered the remains of the
Betty bomber. The air smelled of a repulsive mixture of charred
flesh, smoke, and gasoline. Ten bodies were scattered amid the
wreckage, and when the searchers saw an eleventh person strapped

in a seat thrown from the plane, they were initially fooled into thinking they'd found a survivor. The man's head drooped forward, however; he was dead. Positive identification was made by virtue of the corpse's two missing fingers; it was Isoroku Yamamoto.

Yamamoto and the ten other bodies were carried to Buin the next day, Tuesday, April 20, where, following autopsies, they were cremated. Watanabe put the admiral's ashes into a small wooden box lined with papaya leaves. That same day, he transmitted official notice of the admiral's death to Japanese Imperial Navy headquarters in Tokyo, which kept the devastating news shrouded in secrecy. To maintain the chain of command, a successor was immediately named to take over as commander in chief of the Combined Fleet. His name was Mineichi Koga. Few people, if any, considered him Yamamoto's equal. Even Koga knew that. "There was only one Yamamoto," he said following his promotion. "His loss is an unsupportable blow to us."

Escorting Yamamoto's ashes, Watanabe and the wounded chief of staff, Matome Ugaki, flew back to Rabaul. Ranking officers were allowed in on the secret of Yamamoto's death and were able to pay their respects at a hastily assembled private wake. Within days, Watanabe and Ugaki left Rabaul for the fleet's main anchorage at Truk. From there Ugaki departed on May 7 for Tokyo on the flagship *Musashi*, to bring the admiral's remains home. Japan's leaders, meanwhile, continued to suppress the news. Fearful of a nationwide panic, they needed time. But with the flagship due back in Japan on May 21, navy officials in Tokyo knew they could wait no longer. On May 19, they notified Yamamoto's wife, Reiko, and the family.

Then, early the next morning, retired Vice Admiral Teikichi Hori drove to Chiyoko Kawai's modest home in the Tokyo neighborhood of Kamiyacho. When Chiyoko opened the front door, she immediately recognized her lover's old friend. She had not heard from Isoroku in nearly two months, since his departure for Rabaul, when he'd written one of his longest love letters, saying he was heading south to the war front and would not write again for a while. It had been a letter full of passion and had included a lock of Yamamoto's hair. Chiyoko studied the unexpected visitor standing in the doorway. She knew Hori well and could tell by his formal bearing and pale demeanor that something was wrong.

Hori said, "I am very sorry to tell you this sad and unexpected news."

Chiyoko Kawai felt her body stiffen.

"Yamamoto has been killed in action."

Chiyoko felt faint, free-falling in grief to a place she called "bottomless sorrow." Her only thought was "Everything is finished."

Chiyoko's sorrow became a nation's. The news of Yamamoto's death was finally released on Friday, May 21, a month after its occurrence and just as the flagship bearing his ashes dropped anchor in Tokyo Bay. When it was first reported on the radio, the newscaster's voice broke while reading the statement the government had provided to the media: "Admiral Isoroku Yamamoto, commander in chief of the Combined Fleet, gallantly met death aboard a warplane which engaged in an encounter with the enemy while directing strategic operations during April this year." The statement included news of Admiral Koga's appointment. It was all covered in less than fifty words, and when the announcer was done reading them, he cried.

Mourners lined the rail tracks two days later as a special train carrying Yamamoto's elder son, Yoshimasa, and various high-ranking officers traveled from the shipyard to Tokyo. Yasuji Watanabe held up the box containing Yamamoto's ashes in the window along the way. The widow, Reiko, other family members, and a host of military officials met the train at Tokyo station. The government announced Yamamoto's posthumous promotion to fleet admiral and his award of the Grand Order of the Chrysanthemum, First Class, the country's highest honor. The two weeks leading up to the state funeral on June 5, held in Hibiya Park in central Tokyo, saw a continuous outpouring of sadness. Thousands upon thousands of citizens turned out along the city streets the day of the funeral to pay their last respects, and in attendance at the service itself were hundreds of government and military leaders, including Prime Minister Hideki Tojo and emissaries of the emperor. The navy band played Chopin's funeral march during the procession to the park, and Chiyoko Kawai watched discreetly from a spot not far from her home. Afterward, Yamamoto's ashes were divided in half, with one container interred in a cemetery on the outskirts of Tokyo, next to the country's most

famous naval leader, Admiral Heihachiro Togo, the second transported north to Yamamoto's hometown of Nagaoka. On June 7, the
ashes were buried there in a simple plot near a Zen temple, marked
by a hand-carved stone with the words "Killed in action in the South
Pacific, April 1943."

In the weeks leading up to the funeral and during the weeks afterward, Chiyoko Kawai was visited several times by officials, including men she'd once considered friends. The point of the visits
was clear—to suppress the story of the long love affair and thus to
keep untarnished the reputation of the country's naval hero. Teikichi
Hori came one day and left taking away with him "letters sent from
my darling," as she recorded in her diary. More than once Chiyoko
was advised to commit suicide, and in her grief, there were moments
she considered doing so. But she did not. She left Tokyo and moved
to Numazu City on the coast, where she opened a restaurant, kept
a low profile, and, for more than a decade, steadfastly maintained
her silence—until a reporter came calling in the spring of 1954.
The article that resulted appeared in the *Weekly Asahi* on April 18
of that year, eleven years after Yamamoto's death. It was titled "Admiral Yamamoto's Sweetheart," and subtitled "The War God Was
a Human Being, Too." Most of the couple's letters had been seized
and destroyed, but Chiyoko had saved a batch of the most precious
ones, along with her diaries. She opened up to the interviewer and
talked emotionally about her love for Isoroku Yamamoto, recalling
how they'd first met, how they'd stayed together, and how, after his
death, she'd plummeted into a bottomless sorrow. But with the war
now over and from the perspective of Japan's defeat, she'd arrived
at a different view. She told the reporter, "It might have been rather
fortunate for Yamamoto to be killed in action." Had he lived through
to the end, she said, he might have "been hanged as a war criminal.
So, I no longer feel sorrow."

IN THE UNITED STATES, THE ANNOUNCEMENT OF YAMAMOTO'S
death was greeted triumphantly. "Public enemy number one of the
American Navy is dead!" the young Chet Huntley announced on CBS
radio to start a lengthy newscast on the death of a Japanese military
icon who, at Pearl Harbor, "in about sixty minutes inflicted more

hurt on that force than ever occurred in all our history." In newspapers across the country Yamamoto's death was front-page news, with nearly every story repeating the jingoistic boast he had supposedly made *after* the Pearl Harbor attack—that he would march victoriously into the White House—when in fact his statement, made *before* the attack, had expressed deep worry about war with the United States. The distorted version was now a permanent tagline, demonizing him to the fullest. Typical was the *Washington Post* headline: "Yamamoto Killed; Boasted He'd Dictate Terms in White House." Much of the reaction in the United States was powered by revenge. Commander Edwin T. Layton, the intelligence chief who before the war had known Yamamoto personally and who as intelligence chief to Admiral Chester W. Nimitz had hand delivered the code breakers' work on Yamamoto's fatal trip, certainly felt that way. "All of us who were there that day, and experienced the shock and violence of Japan's Pearl Harbor attack—and who lost shipmates and friends there—can be forgiven, I hope, if we were pleased in our inner hearts when Pearl Harbor was avenged, in part at least, by Yamamoto being ambushed and killed," he wrote later.

In Washington, a larger-than-usual gaggle of reporters gathered late in the morning of May 21 for a press conference in President Roosevelt's office. The president, seated at a desk decorated with a fresh bouquet of red roses, seemed to enjoy the extra numbers. He was dressed in a light gray suit with a blue necktie and casually smoked a cigarette as he bantered with the newsmen in their typical cat-and-mouse game where he dodged most of their questions and offered up only the information he wanted to. He began by noting Maritime Day was the next day, May 22, and he reminded the assembled press that a year ago he'd used the occasion to praise the nation's shipyard workers. The reporters grew restless. They wanted the president's take on the big wartime news of the day: the word from Japan about Yamamoto's death. But FDR said nothing, just looked across the desk through a haze of blue smoke. The press persisted: this was the admiral who'd said he would dictate peace terms in the White House; didn't the president have a comment?

FDR's eyebrows circled high in "calculated, Barrymorish astonishment," as a *Time* magazine correspondent wrote later that day in a memorandum to his editor. Finally, FDR spoke: "Gosh!" That was it. The president was feigning surprise, as if the report out of Japan about Yamamoto was news to him—good news, to be sure, but completely unexpected. Every reporter in the room could tell the president was role-playing. "Reporters got it," the *Time* correspondent noted, and everyone began laughing along with the president.

One reporter asked, "May we quote that, 'Gosh'?"

The president chuckled "an affirmative."

The reporters did not know the exact extent of the president's knowledge, but they did know he knew much more about the Yamamoto shootdown than he was letting on with his phony wonderment. Indeed, his coyness was part of the administration's larger playbook for not doing or saying anything that might put at risk the United States' decryption of Japan's codes as the basis of the shootdown. It was better to seem bewildered and pleased at the serendipitous turn of events, and there was no question that FDR, who'd been briefed weeks earlier on the work of Mitch's men, was very pleased. Several days later, he composed a faux letter to Yamamoto's widow—a note he shared only privately with staff—revealing a dark, even cruel gloating. It read:

Dear Widow Yamamoto:

Time is a great leveler and somehow I never expected to see the old boy at the White House anyway. Sorry I can't attend the funeral because I approve of it.

Hoping he is where we know he ain't.

Very sincerely yours,
Franklin Delano Roosevelt

IT WASN'T AS IF THE US LEADERS OR THE GENERAL PUBLIC thought that Yamamoto's death would instantly result in Japan's defeat. Instead, the death was a capstone of a critical pivot in the Pacific war, a war that had begun with a frightening Japanese dominance in the aftermath of Pearl Harbor but that more recently had turned in favor of the United States with the victory at Midway the previous year and the long and bloody battle for control of Guadalcanal in the fall of 1942. Yamamoto's death boosted Americans' confidence, while in Japan morale plunged. It was a blow that one American journalist called "comparable to what the loss of Gen. Douglas MacArthur would be to the United States." The government-controlled media in Japan would continue to try to inspire the shocked and disheartened public by invoking Yamamoto's memory. "His stirring fighting spirit lives on in the Imperial Japanese Navy" was a typical rallying cry. But the reality was that the navy's leading strategist was gone, and, as one military scholar noted, "the Japanese felt lost at sea without their beloved admiral." The Pacific war would continue, brutally, for two more years, but from that point on it tilted always in the United States' favor, with Japan's navy never again prevailing in a significant clash. Yamamoto's inspired leadership and originality had proved to be unique.

FOR MITCH AND HIS MEN IN THE SOUTH PACIFIC, THE NEWS FROM Tokyo was welcome confirmation. "We had been 99 percent sure we'd killed the SOB," Mitch said, "but it was possible he'd bailed out and we hadn't seen him." Listening to the report on the radio, Mitch noticed that "the Japs gave the date but little else." The details didn't matter at that point, however. "We knew for sure then we'd gotten him."

Mitch and his men had awakened the Monday after the April 18 mission with serious hangovers from their late-night partying. Still, they'd managed to gather on the airfield for a group photograph. The fifteen mission pilots posed in front of several P-38 Lightnings, with Mitch crouched front and center, Rex Barber to his immediate right. Missing, of course, was Ray Hine. Vice Admiral Pete Mitscher, meanwhile, had ordered another flight to head north from Fighter Two toward Bougainville. The idea was to carry on, business as usual, to make it seem as though the Yamamoto encounter had been coincidence rather than the result of advance knowledge. Following Mitch's flight with another one would encourage the idea that the first flight had not been a special operation, and Mitscher continued sending P-38s to the Bougainville area for the next several weeks.

But Mitch was not involved in the cover-up flights. Nor were any of the other pilots who had flown the Yamamoto mission. They were done—grounded that first day and the next and the one after that. It didn't have anything to do with their wild partying; it was because they knew too much. "They wouldn't let us fly any more on Guadalcanal," Doug Canning said. "We knew the code had been broken and they didn't want a leak if we were captured." They were on their way out of the war. Mitch, still at Fighter Two when the Japanese announced Yamamoto's death, was ready to depart any day. Others were already gone; Canning, Mitch's wingman, Jack Jacobson, and Besby Holmes were in New Caledonia, while Tom Lanphier and Rex Barber were in Auckland, New Zealand, enjoying a rest leave.

In the days since the shootdown, the senior command, especially Admiral Nimitz and his intelligence chief, Layton, had pressed upon Mitscher and Bull Halsey on New Caledonia the need to keep lips sealed regarding the prior intelligence of Yamamoto's trip. The last thing anyone wanted was a scare or worse, like the one following Midway when the *Chicago Tribune* had run a story disclosing that Nimitz and the US Navy had known about Yamamoto's battle plans ahead of time. That story had certainly put the code breakers' work at risk.

On Guadalcanal the problem was that the Yamamoto shootdown became an open secret the moment Tom Lanphier began claiming he'd shot down the enemy admiral. The challenge became to contain the news. Making matters more difficult, accolades kept coming for Mitch and the others, praise that did not reference Yamamoto but was hardly subtle. The top army general in the Pacific, Lieutenant General M. F. Harmon, for example, was over the moon about the performance of his army air forces fliers. In letters to Mitch and to the Army's top brass in Washington, DC, during the weeks after the shootdown, Harmon was mindful of the "Navy's insistence on secrecy" given that the "public knowledge of all the facts involved would adversely affect the intelligence sources." Yet he wrote a letter hailing Mitch "for the superior manner" in which he'd led "an *important air action* [author's italics] in the vicinity of Kahili, Bougainville Island, on April 18." Said Harmon about the mission leader, "This operation was excellently conceived and splendidly executed." And before the Japanese had even made public

Yamamoto's death, Vice Admiral Mitscher told the P-38 pilots that promotions were forthcoming and paperwork was under way for everyone to get honors. Eleven of the pilots were in line to receive a Navy Cross for heroism, while Mitch, Lanphier, Barber, Holmes, and Raymond Hine (posthumously) were up for the Congressional Medal of Honor, the highest possible award for bravery in combat, a medal dating back to the Civil War and presented by the president on behalf of Congress.

Mitch could hardly contain himself. "I'm tickled to death over being recommended for the Medal of Honor," he told Annie Lee in his May 3 letter, the one saying he was coming home. "Of course, I haven't got it yet, but enough of the brass hats got their signatures and endorsements on it, and I'm pretty sure it will go through.

"The President hangs that one on—and you will be with me."

SOON ENOUGH, HOWEVER, THE PROSPECT FOR HIGH HONORS went sour. Tom Lanphier and Rex Barber, careless in the company they kept while enjoying a breather in New Zealand, ruined things for everyone. They hit the golf course every day the week they were there, playing with one of General Harmon's aides. To fill out the foursome, they agreed to let J. Norman Lodge join them. Lodge was a seasoned war correspondent for the Associated Press. He'd heard about the shootdown and had been gathering material for a story, which wasn't too difficult; by early May, anyone who'd been on Guadalcanal knew at least the general outline of the mission. While Lanphier and Barber were focusing on their stroke and having a good time, the enterprising reporter was using the outings to nail down his information. "He would talk about the mission," Barber said, "and ask us if this or that was about right. We would agree or, if he was incorrect on details, correct him."

Lanphier and Barber should have known better. The next week the pilots arrived at the navy base at Noumea, New Caledonia, and not two days passed before they and their army officer golfing buddy were summoned to Admiral Bull Halsey's ship. The three men were informed that reporter Norman Lodge had submitted a detailed account of the mission to censors for clearance. The blockbuster story had landed on Halsey's desk, and, although it had been suppressed,

Halsey's legendary anger was not. The admiral screamed at them for talking to the reporter. He roared about security and their responsibility regarding classified information. He said they were unfit and should be court-martialed. Then he grabbed papers off a nearby table. "See these pages," he shouted. "Your citations for the Congressional Medal." He waved five pieces of paper—citations not just for Lanphier and Barber but for Mitch, Besby Holmes, and Raymond Hine. He trashed all of them, saying they would not be getting the medals. "You're not worthy," he said in disgust. But, and only because of the mission's importance, he would forward paperwork for five Navy Crosses.

The tirade finally ended. Barber stood there stunned. He and the other two saluted, but Halsey just glared and pointed to the door. "That was the end of our meeting," Barber said afterward. "We were not asked one question nor did any one of us speak."

THE FIRST WEEK OF JUNE SAW MITCH FINALLY MAKING THE LONG trip home, landing first in San Francisco and flying from there to San Antonio for a grand reunion with Annie Lee—their first time under the same roof since his departure for the South Pacific in January 1942. The couple had less than a week together, however. The Army Air Forces had plans for Mitch, and by mid-June, he was on his way to Washington, DC, where he and Tom Lanphier received a hearty welcome in military circles. It seemed that plenty of people knew about their role in bringing down Japan's greatest naval leader, and the Army's top commanders wanted somehow to showcase them. They couldn't do it in connection with the Yamamoto mission; nothing had changed about keeping its details secret. But Mitch and Tom could be presented as ace pilots who'd helped turn the tide in the Pacific piloting the new fighter plane, Lockheed's P-38 Lightning. They attended a confidential debriefing on June 15 with officials in the Office of Air Staff Intelligence and then huddled with the Army's press officers. On June 17, the Department of War issued a press release: "Two Outstanding Fighter Pilots of AAF Tell of South Pacific Air Battles." They next appeared on the popular radio show *Army Hour*, rolling out their main talking points on the prowess of "our baby, the P-38," which Lanphier called the "best all-purpose fighter plane ever built. It will go faster, higher, and as far as any fighter plane ever built, and deal out more punishment."

Then they hit the road. The two traveled to bases around the country to meet with pilots in training and pump them up with stories of aerial combat against the Japanese. And they did press interviews at nearly every stop, trumpeting the P-38 Lightning. It became the Mitch and Tom road show, or, more accurately, the Tom and Mitch show. Tom seemed to do most of the talking and get most of the press attention—favoritism fueled in part by Lanphier's father, the army colonel who worked the Washington press corps on his son's behalf. Within days of the pilots' mid-June press debut, for example, the *Washington Post* ran a story about Tom, "son of Col. Thomas G. Lanphier," with no mention of Mitch. The headline: "Lanphier Stars as 'Hot' Pilot in Pacific War." The *Los Angeles Times*' front-page story at least mentioned Mitch but quoted only Tom. The photograph that ran with the story showed Mitch deferentially holding a match to light Tom's cigarette.

The two zigzagged along the West Coast on their speaking tour, mostly together but sometimes splitting up. They spoke to cadets at March Field in Riverside, California; at the training school in Merced; at Muroc Army Airfield (later renamed Edwards Air Force Base); and at Minter Field in Bakersfield. At the Cal-Aero Academy at Chino Airport, Lanphier flew a demonstration in "the art of knocking Japs out of the sky." While at Minter Field, Mitch put on an air show to climax his talk to about eight hundred cadets. Besby Holmes showed up to watch, and the two went out that night for drinks and a spaghetti dinner.

The star pilots spent the rest of the summer in California, managing to mix in plenty of nights out on the town. They even had their Hollywood moment: staying at the Hollywood Knickerbocker, mingling with film stars and directors in the splendor of the Spanish Colonial Revival–style hotel. Tom Lanphier was taken one night to Jack Benny's house for a dinner party hosted by Jack's wife, Mary Livingstone, herself an actress and comedian. Other nights, top executives of Lockheed Aircraft, maker of the P-38 Lightning, hosted them at their Hollywood mansions, where at parties Mitch and Lanphier socialized with the film stars Ellen Drew, Wallace Beery, and Jimmy

Stewart, the actor turned air force captain. Stewart flew the B-24 Liberator, a heavy bomber, and over drinks and dinner, he couldn't stop asking Mitch and Lanphier about the speedy P-38. By night's end, the group had hatched a plan to get Stewart up in a Lightning. Mitch and Lanphier were due to report the next morning to the base in Santa Ana for a meeting with the commander of West Coast pilot training. They had Stewart tag along and, after the meeting, let the actor try flying a P-38. Mitch wrote to Annie Lee all about it: "Stewart was really tickled. He is a fine fellow and we really enjoyed having him around."

In August, Mitch sneaked off to see Annie Lee in San Antonio. One day while there, he took a plane from a nearby base and, in fun, buzzed her parents' house in El Campo. Another day, he and Annie Lee went out with his former wingman, Jack Jacobson, who happened to be passing through with his wife. But after a few days Mitch was back on the road, traveling beyond California to Oregon and Washington and then across country to a slew of bases in the Deep South, including army airfields in Greenfield and Grenada, Mississippi. Mitch spent a weekend in Enid with his father, Noah, and his stepmother, Eunice, at the family homestead. Father and son caught up; it had been several years since they'd seen each other. They played checkers, and Mitch told him about the war.

One night late in the fall, Mitch had a dream: he'd gone out to eat, and when he returned, there was a message for him to call President Roosevelt, the secretary of war, and some senators. "I couldn't figure it out, but figured finally that I was receiving the Congressional Medal of Honor." Awake, he knew that the dream was a fantasy; he'd learned months earlier about Bull Halsey's tantrum following Rex Barber and Tom Lanphier's foolishness with the newspaper reporter. In fact, he'd already received an honor that was less prized than the medal originally intended for him. It had happened during a stopover in San Francisco. The ceremony, held in a general's office, was simple, brief, and anticlimactic: he and Lanphier had been presented with Navy Crosses "on the authority of Admiral W. F. Halsey, commander of the South Pacific area." The award cited their "extraordinary heroism and distinguished service" and mentioned Mitch's leadership, his "record of eight Japanese planes shot down,

approximately 100 operational missions and 200 hours in combat."
Even so, the moment was bittersweet. "Of course, I was glad to get
it," Mitch wrote Annie Lee. "But I had been hoping for the Medal of
Honor, so it was disappointing."

By the time of the dream, Mitch had been spending a fair amount
of time wondering what his next job would be. He'd left Guadalcanal in June, and now it was November. Five months on the road,
the song and dance of mainly talking about, rather than flying, the
P-38, had grown tiresome. Then, in early December, he finally got
word: he was assigned to take command of the 412th Fighter Group
at Muroc Army Airfield, a newly created unit for training pilots and
testing fighter planes. Mitch wasn't so sure he wanted to do that. He
was itching for more combat, realizing after months of inactivity
that a life of all talk, no action wasn't really for him.

THE US MILITARY'S TARGETED KILL OF YAMAMOTO ON APRIL 18,
1943, proved historic, but there have been many more "targeted-kill
operations" since, particularly with the dawn of the twenty-first
century. The most famous was on May 2, 2011. That was the day
US Navy Seal Team 6 raided a compound in Abbottabad, Pakistan,
and killed Osama bin Laden, the founder and leader of the terrorist
group Al-Qaeda. Whereas Yamamoto had been behind the devastating Pearl Harbor raid, Osama had been behind the September 11,
2001, terrorist attacks that had killed nearly three thousand people
and injured another six thousand as two hijacked planes flew into
the World Trade Center's twin towers in New York City, another
plane crashed into the Pentagon in Washington, DC, and a fourth
plane crashed in a Pennsylvania field. In its wake, Congress granted
the president broad powers "to use all necessary and appropriate
force" to hunt down Al-Qaeda terrorists in legislation titled "Authorization for Use of Military Force" (AUMF). Practically overnight,
targeting leaders of terrorist groups—"Decapitation of high-value
individuals"—became vital to counterterrorism programs overseen by presidents George W. Bush and Barack Obama, especially
through the use of cruise missile strikes and armed, pilotless drones.

The expanding practice drew criticism. Human rights monitors, international lawyers, and the American Civil Liberties Union questioned

the lawfulness of targeted-kill operations and condemned their collateral damage—the killing of innocent civilians. In response, Obama's attorney general, Eric H. Holder, Jr., made clear the administration's view. "It is entirely lawful—under both United States law and applicable law of war principles—to target specific senior operational leaders of al Qaeda and associated forces," he told an audience at Northwestern University School of Law on March 5, 2012. "This is not a novel concept. In fact, during World War II, the United States tracked the plane flying Admiral Isoroku Yamamoto—the commander of Japanese forces in the attack on Pearl Harbor and the Battle of Midway—and shot it down specifically because he was on board."

The Yamamoto shootdown nearly always comes up in today's discussion of targeted-kill operations, not just as a historical marker but as a model of success. Critics of targeted kills often argue that they are not as effective as their proponents claim. "Decapitation does not increase the likelihood of organizational collapse," concluded the scholar Jenna Jordan in "When Heads Roll: Assessing the Effectiveness of Leadership Decapitation," an analysis of nearly three hundred incidents occurring over a fifty-year period.

In contrast, the Yamamoto mission, hastily planned and overseen by Major John Mitchell, is cited as an exemplar of a targeted kill. "It was highly effective," wrote US Air Force Major Adonis C. Arvanitakis in a more recent study examining the Yamamoto mission. The reason? Yamamoto was no ordinary naval leader but larger than life. "What they understood in 1943 and what we need to understand today is that the targeted killing needs to be more than eliminating the next person in the chain of command," Arvanitakis wrote. The architect of the Pearl Harbor attack "was truly unique, irreplaceable. His elimination had long lasting effects in terms of Japanese morale and will." US military leaders were hardly naive, however. They didn't expect Japan's immediate collapse, noted Arvanitakis; rather, they saw killing Yamamoto as "necessary to helping a larger cumulative effect."

MITCH CONTINUED TO CONTRIBUTE TO THAT CUMULATIVE EF-fect, although it took a while for him to see combat again. He took over the new 412th Fighter Group at the start of 1944, and for the next eighteen months he trained cadets and oversaw aircraft testing. The fact that the 412th was the Army Air Forces' first jet unit made the job tolerable. He and his pilots focused primarily on testing two turbojet-powered fighters in development, the Lockheed P-80 Shooting Star and the Bell P-59 Airacomet. Mitch had little good to say about the P-59, calling it "a colossal flop," while he thought the P-80 showed promise; eventually, in fact, the P-80 became the first jet fighter used by the Army Air Forces in combat operations. In terms of pilots, he was proud of his work training them to fly the new jets. For him, a highlight was being the first army air forces pilot to fly a jet across the continental United States, departing from Buffalo, New York, and landing in Palmdale, California, just north of Los Angeles. During that time, he also won a promotion to lieutenant colonel, and on January 24, 1945, he and Annie Lee welcomed their firstborn, Theresa Ann Mitchell, whom they nicknamed Terri.

Even so, Mitch jumped at the chance to return to the South Pacific a few months after Terri was born, when his friend and former commander, Vic Viccellio, asked if he wanted to help lead the 15th Fighter Group. "Snapped up Col. Viccellio's offer," Mitch wrote in his diary. He packed up and drove Annie Lee and baby Terri to Texas so they could stay with her parents in El Campo, and on April 15, 1945, he departed from Fairfield, California. It was two years to the day from when he had been on Guadalcanal, expecting to be shipped out, and had been summoned to take on the Yamamoto mission.

Five days later, after stopovers in Hawaii, Johnston Atoll, Kwajalein Island, and Saipan, Mitch arrived at his destination: Iwo Jima, the volcanic island the US Marines had just captured from the Japanese. "On the way over my feelings and thoughts were varied," he wrote. "Sometimes I felt elated at returning to combat and then I felt depressed at leaving my dear wife and little baby so far away and in such a lonesome spot there in El Campo." He found he liked Iwo Jima better than Guadalcanal; the men slept in tents on cots with rubber mattresses, and, as he noted in his diary, "there are *no mosquitoes.*" Perhaps the best thing was that, unlike Guadalcanal, Iwo Jima wasn't getting bombed all the time. "So when we crawl in the sack at night we know we are going to sleep well."

Over the next several months, Mitch flew a North American Aviation P-51 Mustang, a long-range fighter, escorting B-29 Super-fortresses on raids over Japan. He flew some twenty-odd missions, each covering about six hundred miles and lasting more than six hours, much longer than what he'd been used to during his service on Guadalcanal. Once again he displayed his combat skills, shooting down four enemy aircraft. "I got one Zeke and should have got another," he wrote after one of the kills. "I saw the bastard trying to get out, but he never made it."

Mitch was on Iwo Jima when VE Day, or Victory in Europe Day, was celebrated on May 8, 1945, marking the formal surrender of Nazi Germany. "A great day for us all," he wrote excitedly in his diary, while also noting "No celebration here. The question here is will the Japs elect to fight it out or will they give up." They elected to fight it out on their own until, three months later, the United States dropped atomic bombs on the Japanese cities of Hiroshima and Nagasaki. "It's almost unbelievable," Mitch wrote, "and yet it was announced by President Truman, so it must be so." Following Japan's surrender, Mitch returned to Texas, where he reunited with Annie Lee and baby Terri.

Whereas other members of his squadron on Guadalcanal had left the service in the months after the Yamamoto shootdown, Mitch had chosen to stay on, not just through the end of the war but well beyond. Promoted to colonel in 1945, he spent the next several years relocating his family from one army base to another, in Mississippi, Alabama, Kentucky, Texas, even Alaska. While they were in Mississippi, a second daughter, named Joan, was born on November 4, 1946. Then, upon the outbreak of war in Korea in June 1950, Mitch volunteered for duty. He flew more than 110 combat missions as commander of the 51st Fighter Wing, during which he shot down four MiG-15s, the Russian-built jet fighters used by the North Koreans. He returned home after an armistice was signed in late July 1953. His final posting was at Fort Custer in Battle Creek, Michigan, as commander, Detroit Air Defense Sector. His son, John William Mitchell, Jr., nicknamed Billy, was born there on October 14, 1958. He retired soon afterward, having served more than two decades. Mitch and Annie Lee moved to California's Marin County and bought a home in San Anselmo, just north of San Francisco. They were only ten miles from the base at Hamilton Field where in 1940, fresh out of cadet school, Mitch had discovered that he was born to fly. He became a cofounder of an oil and gas business, dabbled in commodities, and stayed in touch with a number of his friends from the war, namely Rex Barber, Doug Canning, and his former wingman Jack Jacobson.

IT WAS IMMEDIATELY AFTER JAPAN SURRENDERED THAT THE question of who actually had shot down Isoroku Yamamoto escalated from an argument at Fighter Two, when Tom Lanphier had staked his claim, into open hostility. With the war over, the military and the pilots no longer had to be coy about the shootdown mission—either about the source of the admiral's itinerary being the code breakers' decryption or about the interception itself. The Army Air Forces was ready: on September 11, 1945, it issued a publicity release providing putative answers to the aerial drama and identifying Tom Lanphier as the single slayer of Yamamoto. The news release was based on the so-called Fighter Interception Report written by two intelligence officers at Fighter Two after Mitchell and his men had returned from the mission. None of the pilots had ever seen a copy of the report; many later recalled Lanphier saying that he'd assist in its preparation and take care of things. Which he apparently did—in his favor. The report described Lanphier battling the escorting Zeros and then shooting down the lead Betty bomber, the one with Yamamoto aboard. Lanphier, the report said, had "nosed over, and went down to the tree tops after his escaping objective [the Betty]. He came into it broadside—fired his burst—a wing flew off and the plane went flaming to earth." The debriefing report became the official record of the historic mission.

Lanphier was ready, too. For weeks the onetime newspaperman had been writing a dramatic retelling of the mission, and on the same day as the Army Air Forces' press release, the North American Newspaper Alliance distributed the first installment of his six-part series. Newspapers all around the country began publishing his account. In its coverage that first day, the *New York Times* accompanied the first-person story "describing how he shot down Admiral Yamamoto" with a separate news article reporting Lanphier's feat. The front-page headline: "Yamamoto's Killer Identified by Army." The story began, "Lieut. Col. Thomas G. Lanphier, 29, Army Air Forces pilot, was identified by the War Department today as the man who shot down the plane carrying Admiral Isoroku Yamamoto, commander in chief of the Japanese Navy, in April 1943." The story mentioned Mitch but no other pilots. Lanphier's place in history seemed secured.

When they read Lanphier's series, Mitch, Rex Barber, and other pilots were apoplectic. "He's just got a big mouth," Mitch wrote Annie Lee from Iwo Jima on September 16, having heard the news of Lanphier's heroics on the radio. Mitch's father, Noah, sent him Lanphier's series. Mitch was disgusted, calling Lanphier a publicity hound. "He really hogs the show. Of course, I didn't shoot Yamamoto down, but he doesn't know if he or Barber did." The feud was on, and it lasted for decades, as Lanphier was officially credited with the Yamamoto kill, a status he exploited in the 1960s as a top executive at Convair, working his connections in the military to win lucrative contracts for the aircraft maker. Those protesting the Lanphier narrative were frustrated—the equivalent of fighting city hall. Making matters worse was the eventual realization that there had not been three Betty bombers in Yamamoto's inspection flight—as the pilots had concluded back at the base and as had been recorded in the interception report. Based on information released during the 1950s by Japanese officials, there had been just the two bombers—the one carrying Ugaki, which was shot down over water, and the other carrying Yamamoto, which was shot down over jungle. When previously it had been thought there were three, Barber and Lanphier had been credited as having shot down two of them over the jungle. The dispute was about Lanphier claiming that Yamamoto had been in the bomber he'd shot down, when there was no way to determine which of the two bombers he had been in. With the correction, it was now clear that only one Betty had been shot down over jungle—and that it had held Yamamoto.

No one had ever disputed Barber's account that while Lanphier had courageously veered off to deflect the three escort Zeros, Barber had attacked the Betty bomber over the jungle. The new information thus gave credence to the view that Barber had shot down Ya-

mamoto and that Lanphier hadn't gotten Yamamoto at all or, if he
had somehow looped back, he'd fired a few shots at a wounded plane
that was already going down. The official mission report came under
increasing attack. Barber, Mitch, and other veterans who supported
Barber argued that Lanphier's account was a concoction. Finally, in
the 1970s, the Yamamoto kill was reviewed by two historians for the
Office of Air Force History. With only one Betty bomber destroyed
over the jungle, it was decided in 1973 that the two pilots had shot
at the same plane. The official credit for the kill was therefore split
between Barber and Lanphier.

The revision did not end the hostility among the mission pilots,
however. Lanphier never budged, insisting he'd been wronged and
had been solely responsible for killing Yamamoto. He wrote an article for *Reader's Digest* titled "I Shot Down Yamamoto" and then,
in the mid-1980s, wrote his life story in a manuscript called "At All
Costs Reach and Destroy." He was never able to find a publisher for
the 352-page manuscript. Lanphier died of cancer on November 26,
1987, in La Jolla, California, one day shy of his seventy-second birthday. Meanwhile, Rex Barber and his supporters, including Mitch,
lobbied for years to convince air force officials to award Barber sole
credit for the kill. They spoke out at veterans' gatherings and panels;
filed appeals with various military review boards, such as the Air
Force Board for Correction of Military Records; and even went to
US District Court. Each step of the way Barber presented new evidence, some of which had been gathered from trips to the crash site
on Bougainville, but the ruling splitting the credit has never been
changed. Rex Barber died on July 26, 2001.

IN THE SPRING OF 1943, JOHN MITCHELL AND HIS PILOTS USUALLY
awakened to the sun rising east of Fighter Two, a first light that
skimmed across the ocean's surface before reaching the tents set
against the small hill overlooking the airfield. Seventy-six years later,
on May 9, 2019, the sun rose at 6:24 a.m., its morning rays moving
across the ocean's surface to brighten the area where Mitch, Rex Barber, Tom Lanphier, and the others had once been bivouacked. Except
instead of tents, there are now tiny homes—shacks, really. And instead of a runway below the hill, there are the green fairways of the

Honiara Golf Course. The far end of the airfield, where the planes once set down, is home to a thriving beer brewery.

Whether at the former Fighter Two airfield or about two miles away at the former Henderson Field, it's nearly impossible today to find any sign of the US presence during the war years of 1942 and 1943. To be sure, there is the Guadalcanal American Memorial, sitting atop Skyline Ridge overlooking the Matanikau River, a monument honoring the US and Allied soldiers and sailors who fought during the Guadalcanal campaign. And there is the Vilu War Museum on the island's northeast coast, beyond the point where the Japanese were concentrated once the Americans invaded in August 1942. It's a private outdoor museum where, over the years, a family has collected a variety of Japanese and US guns and relics, including a rusted, broken-down P-38 Lightning. But actual battlefields and encampments are not memorialized as war sites elsewhere have been, whether in Europe to commemorate action there or in the United States to remember the Civil War. The absence of monuments on Guadalcanal is no surprise, really. Neither Japan nor the United States had had any prior history on or control of the island; rather, during the Pacific war the troops of the two countries came, fought, and left. Today, the island's downtown is a bustling port, its main street congested with traffic, its friendly residents filling a central market featuring fresh fruits, vegetables, fish, and handmade crafts. The port of call, Honiara, is the capital of all of the Solomon Islands, with its more than 900 islands, of which 147 are inhabited.

It takes knowledge and imagination to picture Mitch hustling down from his tent, climbing into the cockpit of a P-38, and heading after Washing Machine Charlie to end the nightly, noisy raid that disrupted everyone's sleep. Or to picture the Yamamoto mission pilots returning from Bougainville, landing one after another to congratulations all around—before, that is, Lanphier and Barber had words over who actually got the admiral.

The bitter, prolonged controversy has tended to dominate the treatment of the historic mission, overshadowing, as some historians have noted, the true heroes: the code breakers at Station Hypo, the planners in Admiral Pete Mitscher's tent, and, most of all, mission commander John W. Mitchell. Although Mitch's key role rarely goes unacknowledged, discussion is often consumed by the contentious question: Who got Yamamoto? It's unfortunate. "None of the accounts of the Yamamoto Mission give full credit to the superb leadership and navigational ability of Major John Mitchell," Lieutenant Colonel Besby Holmes wrote in an article for *Popular Aviation* in the late 1960s. Pilot Doug Canning, a panelist at a Yamamoto Mission Retrospective in 1988, said, "I'd also like to say one more time, the greatest fighter pilot that ever lived on this earth, is John Mitchell."

Mitch was due the Congressional Medal of Honor for the Yamamoto mission. It was something he had dreamed about but was denied through no fault of his own. Starting in the late 1980s, some veterans sought to correct that error. They called themselves the Second Yamamoto Mission Association (SYMA). The group rounded up letters of support and then persuaded Representative Michael Bilirakis (R-FL) of the House Armed Services Committee to sponsor legislation in 1993 to authorize the president to award Mitch the medal. The goal was to make it happen before Mitch died. Mitch, turning eighty, was in treatment for pancreatic cancer. House Resolution 3017 eventually attracted sixty-eight cosponsors, divided evenly along party lines, but the measure never gained momentum and stayed stuck in committee. Colonel John W. Mitchell died in San Anselmo, California, on November 15, 1995. He was eighty-one.

Tom Lanphier never showed remorse for his conduct on the golf links in New Zealand that had proved so damaging, costing not only himself but innocents, such as Mitch, the Congressional Medal of Honor as well—men who had had nothing to do with leaking information to an enterprising journalist. Lanphier waved off the indiscretion, saying that everybody on Guadalcanal was talking about the shootdown; his chatting up a reporter was much ado about nothing. Rex Barber, however, was different. He was racked with guilt. "I have felt badly over this ever since—not for myself or Lanphier, as we should not have talked at all," he wrote to a historian in 1989. For Barber, who passed away in 2001, it was all about Mitch and the damage done. Mitch, he said, was so deserving, "and should have received the Congressional Medal. I wish *to this day* I could do something about this injustice."

ACKNOWLEDGMENTS

WRITING A NONFICTION NARRATIVE BUILT AROUND WORLD WAR II army fighter pilot John W. Mitchell, the architect of the mission to intercept Japanese Admiral Isoroku Yamamoto, would not have been possible without the cooperation of the Mitchell family: Theresa "Terri" Mitchell Cleff and her husband, Paul, and John W. "Billy" Mitchell, Jr., and his wife, Stacy. They opened up their homes and their voluminous primary resource materials—hundreds of letters their father had written to their mother, starting before Pearl Harbor and continuing through the end of World War II; letters their mother had written to her aunt during the war; and a handful of letters written by their grandfather Noah Mitchell; a diary and flying notebook their father had kept after he had left California for the South Pacific in January 1942. Both Terri and Billy hosted me at their homes in Illinois and California, respectively, so that I could painstakingly review and copy these and other materials—such as newspaper articles, military papers, and photographs—documenting the history of their parents' families in Mississippi and Texas. The Mitchell family materials have never been used before in prior coverage of the historic Yamamoto mission.

I thank Rebecca Williams and Tim Hodgdon at the Louis Round Wilson Special Collections Library at the University of North Carolina at Chapel Hill, who assisted with access to the Burke Davis Papers. Davis, a historian at UNC, maintained extensive, detailed records for his book, published more than fifty years ago, on the Yamamoto mission. He interviewed John Mitchell and others as part of his research, asking the same questions I would have asked were the pilots alive today. In Davis's unedited interview transcripts I found details about events and people that never made it into his fine book and that became material for a new and updated narrative. In the Davis collection was a copy of a diary Mitchell had kept for a few months, which was not part of the personal material in the Mitchell family's possession.

Similarly, I thank Patricia Nava of the Special Collections and Archives Division, History of Aviation Collection, University of Texas Library, which holds the C. V. Glines Papers. The Glines papers contained interview notes and correspondence with Mitchell and others that proved helpful in rounding out the mission and life at Fighter Two.

Over the years, the Nimitz Education and Research Center at the National Museum of the Pacific War, Fredericksburg, Texas, has recorded and collected numerous oral histories with veterans. Four interviews proved especially helpful: those with pilots Rex Barber, Doug Canning, and Jack Jacobson, and Japanese pilot Kenji Yanagiya. The Museum hosted a weekend symposium in 1988 called the Yamamoto Mission Retrospective, at which Mitchell, Barber, and others appeared, and I want to thank archivists Chris McDougal and Reagan Grau at the museum for sharing video taken at the time of the speakers participating in panel discussions, as well as photographs from the archives.

I thank Rich Remsberg, a researcher specializing in photo, film, artifact, and vintage music, who was a font of research tips and provided photographs of Yamamoto and audio of radio newscasters reporting the admiral's death in May 1943. I also want to thank Susan Coop Howell, archivist/librarian's assistant, the Webb School, Bell Buckle, Tennessee, for digging up photographs of John Mitchell's father, Noah, and for information about his attendance at the school; Fred Allison, PhD, Chief, Oral History Section, USMC History Division, Marine Corps University, for providing a transcript of Alva B. "Red" Lasswell's oral history; Attorney Clifford "Joe" Cole of Piggott, Arkansas, and the Lasswell family for sharing biographical materials about Alva Lasswell; Chequita Wood, assistant managing editor, *Air Force Magazine*, for providing materials about First Lieutenant Wallace Dinn, including an article Dinn wrote that appeared in the March 1943 issue of the magazine.

For their research assistance, I thank William Ashley Vaughan, who scoured old newspapers in Mississippi and Tennessee for stories about Noah Mitchell and the Mitchell family in Enid, Mississippi, in the early 1900s; George Cully, premier Air Force historian, who gathered information about various fighter squadrons from the Air Force archive at Maxwell Air Force Base in Montgomery, Alabama; and David Morriss, son of the journalist Mack Morriss, who was reporting from Guadalcanal for several months in 1943. On Guadalcanal Island, I thank Moses Kenny for his good company while driving me around the island during my stay there, showing me the World War II sights and especially Fighter Two and the Henderson Field area, where John Mitchell and his pilots were located.

In my research I consulted many books and articles; along the way I found the work of a handful of historians especially helpful. I want to acknowledge them: Burke Davis, John Wible, Carroll V. Glines, R. Cargill Hall, Daniel Haulman, Ian Toll, and Hiroyuki Agawa. Pacific Wrecks, a website founded and maintained by Justin Taylan, is an excellent resource.

I want to especially thank my friend and colleague Mitch Zuckoff. We've written a book together, talked endlessly about writing and journalism, and for years have cheered each other on. Mitch was the one who first told me the story of the Yamamoto shootdown. He'd written the best-selling *Lost in Shangri-La: A True Story of Survival, Adventure, and the Most Incredible Rescue Mission of World War II* and, in the course of his work, had run across all kinds of stories from the Pacific War, some told, some not. Without his input and wisdom and even a pile of his research materials, I would never have gotten off the ground.

I thank Tom Fiedler, former Dean of the College of Communication at Boston University, where I teach journalism. Tom retired in 2019 after a decade as dean, and I'll miss him. He's been a champion of storytelling, not just mine but across the college. I thank, too, Bill McKeen and Susan Walker, cochairs of the Journalism Department, Jennifer Underhill, Sarah Kess, Maggie Mulvihill, John Hall, and Tobe Berkovitz for their support and friendship. Two BU graduate students in journalism provided key assistance along the way: Mariya Manzhos transcribed videotape of John Mitchell, Rex Barber, and others; Geoffrey Line read the manuscript with a keen eye. My thanks to Jacob

Boucher, Steve Theer, Joey Campos, Jake Kassen, and Tristan Olly in the college's technology office for their help in keeping my laptop and technology humming along.

I also want to thank a longtime friend, Bill Cole, who drove me around southern California in his camper so that I could visit with Billy Mitchell and his wife and spend time at the Planes of Fame Air Museum in Chino. Bill made the research trip more fun than if I'd done it alone, and from the start he has been a fan of the story. He often sent me research tips and at one point even tried to arrange a flight for me in a working P-38 Lightning.

Bill was also one of my readers, a group I want to acknowledge and that included another longtime friend, David Holahan, my sons, Christian and Nick, my brother, John, and Donald Altschiller, research librarian, Mugar Memorial Library, Boston University. Each brought a different perspective, and their careful reading made for a better book.

I'm lucky to have literary agent Richard Abate in my corner. He's the best. He and Rachel Kim of 3 Arts Entertainment were always available whenever needed, from start to finish. My heartfelt appreciation to Jonathan Jao, my editor at HarperCollins, Sarah Haugen and the entire team there, as well as Roger Labrie for his careful reading and helpful comments.

Most of all, I want to thank my family. My sons, Nick and Christian, offered plenty of helpful insights along the way; my daughters, Holly and Dana, made sure I kept one foot in the day-to-day and wasn't completely lost to time; and my wife, Karin, held it all together.

I have dedicated this book to my father, John F. Lehr, a staff sergeant in the US Marine Corps. He was seventeen when he enlisted in April 1945. Several months later, he was sent to Guam, where he was stationed for the next two years. Arriving after the war ended, he never saw combat and was part of an occupying force. When I was growing up, he never said much about his service, and, frankly, I'm not sure I would have listened if he had. I was more interested in playing ice hockey or swimming and simply running around than hearing a grown-up's talk about something that had happened long ago and far away. I wish that hadn't been the case, that I'd been curious and asked questions, just as I wish today that I could talk to him about the research for and writing of this book. There is, however, one thing I do recall: while on Guam, my father befriended a Japanese prisoner of war. I learned that because every once in a while a letter arrived from Japan at our home in Connecticut. It was from the former POW. He and my father had apparently become occasional pen pals. I don't recall much about the correspondence. It's not as though my father read the letters to us. But he did explain the reason for them: the former POW was grateful for the way my father had treated him. In addition to news about his career, family, and life in Japan, the letters were expressions of appreciation. I remember my father being a bit embarrassed, feeling that the thanks were over the top; he didn't think he'd done anything special, offering a cigarette at one time or another or, overcoming the language barrier, engaging in conversation. Simple things like that. But for the prisoner the small gestures were apparently a big deal, and I think the reason must have had something to do with the fact that my father had treated him as a person, not some caged, vile savage. I don't know for sure; as I said, my father never said much about Guam. But that one memory—of their letter writing—has stayed with me, not just because of the simple gestures that carried so much meaning but because of the fact that two enemy soldiers, their countries such fierce adversaries, had become friends.

NOTES

Prologue: The Day

1   thousands of men and women assigned to Pearl Harbor were off duty: Susan Wels, *Pearl Harbor: America's Darkest Day* (New York: Time-Life Books, 2001), 138.

2   a bomb plummeting through the roof exploded: Ibid., 115.

2   THIS IS NOT DRILL: Bill McWilliams, *Sunday in Hell: Pearl Harbor Minute by Minute* (New York: E-rights/E-Reads, 2011), 280.

2   three-blast alarm signaling an air raid: Ibid., 324.

2   slender black masses in the waters below: Ibid., 286.

2   The *Oklahoma*'s hull was struck: Wels, *Pearl Harbor*, 118.

2   a high-flying Japanese bomber hit a bull's-eye: McWilliams, *Sunday in Hell*, 326–28.

3   the catastrophic first wave of attack planes was followed by a second: Ibid., 431.

3   "a day which will live in infamy": Wels, *Pearl Harbor*, 166–67.

4   "One of our greatest enemies is malaria": "Knox Twice Bombed on Trip," *Boston Globe*, February 1, 1943.

4   nearly two thousand men: John Miller, Jr., *Guadalcanal: The First Offensive* (Washington, DC: Center of Military History, United States Army, 1949), 140.

4   intestinal infections and diarrhea: John Hersey, *Into the Valley: A Skirmish of the Marines* (New York: Schocken Books, 1942), xvi–xvii.

4   a letter to "My Darling": John Mitchell, letter to Annie Lee, April 16, 1943, MFP.

6   "Yamamoto represented": Burke Davis, interview with Edwin T. Layton, April 8, 1968, DA.

6   temporary command at Rabaul: Burke Davis, *Get Yamamoto* (New York: Random House, 1969), 100.

6   she seemed "strong" and "sturdy": Isoroku Yamamoto, letter to his older brother, circa 1917, quoted in Hiroyuki Agawa, *The Reluctant Admiral: Yamamoto and the Imperial Navy* (Tokyo: Kodansha International, 1969), 67.

6   only later that his identity changed: Ibid., 18; Carroll V. Glines, *Attack on Yamamoto* (Atglen, PA: Shiffer Military History, 1993), 44.

7   Yamamoto was the architect: Glines, *Attack on Yamamoto*, 52; Agawa, *The Reluctant Admiral*, 264; Wels, *Pearl Harbor*, 133–34.

7   he wanted to thank his men: John Toland, *The Rising Sun: The Decline and Fall of the Japanese Empire, 1936–1945* (New York: Random House, 1970), 553.

7   "I have to go": Agawa, *The Reluctant Admiral*, 347.

8   "decapitation operation": Lieutenant Commander Victor D. Hyder, US Navy,
      School of Advanced Military Studies, United States Army Command and Gen-
      eral Staff College, Fort Leavenworth, KS, "Decapitation Operations: Criteria
      for Targeting Enemy Leadership." In his monograph, Hyder examined five de-
      capitation operations the United States undertook during the twentieth cen-
      tury, two of which occurred prior to the Yamamoto mission. There were the
      1901 manhunt to capture General Emilio Aguinaldo, the Philippines, and the
      1916 operation against Francisco "Pancho" Villa, Mexico. Neither occurred
      during a formally declared war, however, making the Yamamoto mission
      the first US wartime decapitation operation. The Aguinaldo operation was
      undertaken after the United States had defeated Spain in the Philippines in
      the Spanish-American War in 1898; Aguinaldo continued a guerrilla fight for
      independence against the Americans, a conflict, or "war," that lasted for three
      years, until 1902. Aguinaldo was captured in early 1901 and offered clem-
      ency. "Pancho" Villa was targeted as an outlaw and revolutionary seen by US
      leaders as a threat to US-Mexican relations. Following Villa's raid across the
      border into New Mexico, President Woodrow Wilson authorized a "punitive
      expedition" to hunt for Villa in Mexico. The yearlong mission failed.

8   the fear of a "yellow peril": Timelines found at rohwer.astate.edu and encyclo
      pedia.densho.org.

9   "Chinese Reporter—NOT Japanese—Please": "How to Tell Your Friends
      From the Japs," *Time* magazine, December 22, 1941, 33.

9   "ferocious killing with no holds barred": E. B. Sledge, *With the Old Breed: At
      Peleliu and Okinawa* (New York: Ballantine, 2010), 33.

9   "existential struggle for annihilation": Victor Davis Hanson, Sledge, introduc-
      tion to Sledge, *With the Old Breed*, xix.

9   "That we could defeat the enemy": Isoroku Yamamoto, letter to Yoshiki
      Takamura, December 22, 1941, in Adam Leong Kok Wey, *Killing the Enemy:
      Assassination Operations During World War II* (London: I. B. Taurus, 2015), 129.

Chapter 1: Johnnie Bill and the Moon

13   "Rainbow in My Heart": The exact date the poem was written is unclear, but
      it was either late 1942 or early 1943; it was contained in one of Mitchell's letters
      to his wife from that time period (MFP). The entire poem is as follows:

> There is a rainbow in my heart,
> But no pot of gold lies at the end.
> The pot contains my cherished hopes
> Of my dear one's return again.
> For the rainbow is God's covenant,
> A promise from Heaven above
> That hope shall remain eternal
> For those we dearly love.
> So each day I view my rainbow,
> With its tints of gold and blue.
> Each day I renew my fondest hopes
> Of life's sweet dream with you.

Shine on, dear Arc of Hope, shine on,
Let not thy light grow dim,
But let me cherish within my heart
Most loving thoughts of him.
And when this grim war is over
And he comes home at last,
Let us keep that hope eternal
That ALL such wars have passed.

14 a survey crew started a settlement: James F. Brieger, *Hometown Mississippi* (Jackson, MS: Town Square Books, 1997), 673.

14 William Carson Mitchell, arrived on October 1, 1845: US Census; W. C. Mitchell's gravestone in Enid Oakhill Cemetery.

14 enlisted in the Confederate Army: Burke Davis, interview with John Mitchell, September 27–28, 1967, DA.

15 state's slave population exceeded its white population: John F. Marszalek and Clay Williams, "Mississippi Soldiers in the Civil War," *Mississippi History Now*, 2009, http://mshistorynow.mdah.state.ms.us/articles/175/mississippi-soldiers-in-the-civil-war.

15 Confederate general Nathan Bedford Forrest: Burke Davis, interview with John Mitchell, September 27–28, 1967, DA.

15 "a thriving place": From the writings in 1965 of Enid resident Irene Kuykendall Gaines, contained on page 3 of materials about Enid assembled in 1993 by Mavis H. Newton from interviews and other sources and provided to the author by the Mitchell family.

16 Noah was on the small side: Information about the Webb School, its founders, and Noah Mitchell's time at the school was provided by school archivist Susan Howell in January 2018.

16 John Mitchell would be born into the modest, unpretentious life: Lehr, interview with Terri Mitchell Cleff, MFP.

17 "Just a wide place in the road": Noah Mitchell, letter to Annie Lee Miller, May 1941, MFP.

17 The village's population: Mississippi History Timeline, online.

18 "The fruit is excellent and abundant": "Enid," *Mississippi Sun*, September 18, 1924.

18 "Birds seem to know those who love them": Noah Mitchell, "Around Home," MFP.

18 the sights and sounds of Johnnie Bill's home: "A Glimpse into the Enid Emporium History," compiled by Mavis H. Newton in 1993 from interviews and previously written materials, MFP.

19 "the spirit of ruthless brutality": Mitchell Yockelson, "America Enters the Great War," *Prologue Magazine* 49, no. 1 (Spring 2017), https://www.archives.gov/publications/prologue/2017/spring/wwi-america-enters.

19 Mitchell went to his local draft office: Noah Mitchell, World War I draft registration card, MFR. Draft statistics from Mississippi from Richard V. Damms, "World War I: Loyalty and Dissent in Mississippi During the Great War," Mississippi History Now, http://mshistorynow.mdah.state.ms.us/articles/237/World-War-I-the-great-war-1917–1918-loyalty-and-dissent-in-mississippi.

19 the influenza pandemic swept around the planet: Louie Matrisciano, "Spanish Influenza in Mississippi (1918)," Historical Text Archive, http://www.hancock countyhistoricalsociety.com/vignettes/the-influenza-epidemic-of-1918/.

19 Johnnie Bill's young life was not without tragedy: The cause of death of Elizabeth Mitchell was not discoverable in available official and Mitchell family records. Lillian Mitchell's certificate of death from the State of Tennessee Board of Health, Bureau of Vital Statistics, File No. 472, February 17, 1922. See also "Enid," *Mississippi Sun*, February 23, 1922.

20 "Ancient Greece had her Homer, modern America has her David W. Griffith": Ned McIntosh, *Atlanta Constitution*, December 7, 1915.

21 "If it is necessary every Negro in the state will be lynched": Quoted in Ray Hill, "The White Chief: James K. Vardaman of Mississippi," *Knoxville Focus*, March 10, 2013.

21 "Negro Mammy": Newspaper clipping of the poem in the family Bible, MFP.

21 "Everyone declared it to be a *real party*": "Miss Edith Mitchell Hostess Dinner-Dance," *Mississippi Sun*, December 8, 1927.

21 "Mr. N. B. Mitchell, our handsome townsman," was married: "Mitchell-Massey," *Mississippi Sun*, July 19, 1928.

22 his days were filled with sports, schoolwork, and the outdoors: Burke Davis, interview with John Mitchell, September 27–28, 1967, DA; Herbert Hoover, "Principles and Ideals of the United States Government," October 22, 1928, https://millercenter.org/the-presidency/presidential-speeches/october-22 -1928-principles-and-ideals-united-states-government.

23 "I always liked airplanes": George T. Chandler, interview with John Mitchell, May 18, 1989, MPF. The unpublished videotaped interview was provided to the author.

23 a prized $500 scholarship: "Award Scholarship," *Mississippi Sun*, July 30, 1931, 5; "John W. Mitchell of Enid Wins Columbia University Award," *Mississippi Sun*, October 1, 1931, 1.

24 "In order that no foreign soldier put foot upon our shores": Major General Andrew Hero, Jr., "New Year's Greeting," January 1930, https://sill-www.army .mil/ada-online/coast-artillery-journal/_docs/1930/1/Jan%201930.pdf.

25 "world of greenness": John Mitchell, letter to Annie Lee Miller, April 16, 1945, MFP.

25 he decided to re-up for another hitch: Burke Davis, interview with John Mitchell, September 27–28, 1967, DA.

25 "clean, intelligent and courteous": John Mitchell, letter to Annie Lee Mitchell, October 28, 1941, MFP.

26 Annie Lee was born on November 2: Various family history records and, in particular, an unpublished memoir by Joe H. Miller titled "My Story," MFP.

28 "real paradise": Miller, "My Story," MFP.

28 "God's Gift to Teachers": "Comments on The 1933 Seniors," *The Rice Bird*, El Campo High School, May 25, 1933, MFP.

29 "Not another flyboy": Letters; Lehr interviews with Terri Mitchell and Billy Mitchell, 2017, MFP.

30 "They would think we were fools": Annie Lee, letter to Aunt Ludma, April 9, 1940, MFP.

31 mostly there were flowers and fun and a romance that steadily gained traction: Annie Lee, letters to Aunt Ludma, April 8, June 20, and June 29, 1940, MFP.

31 John William Mitchell received his pilot wings: John Mitchell, letter to Annie Lee, December 11, 1940, MFP. Note: The United States Army Air Corps became the United States Army Air Forces (USAAF) in June 1941.

31 "You pinned my first pair of wings on me": John Mitchell, letter to Annie Lee, July 26, 1945, MFP.

31 "talked of little things": John Mitchell, letter to Annie Lee, October 21, 1940, MFP.

31 "singing some of our old time songs": John Mitchell, letter to Annie Lee, August 26, 1940, MFP.

31 telling his father about her: Ibid.

32 "telling my position from the sun, moon and stars": Burke Davis, interview with John Mitchell, September 27–28, 1967, DA.

32 "Yours truly put in about an hour and ten minutes soaring": John Mitchell, letter to Annie Lee, September 4, 1940, MFP.

33 "This is one of the most beautiful nights": John Mitchell, letter to Annie Lee, September 13, 1940, MFP.

Chapter 2: Isoroku, Aeroplanes, and a Geisha Girl

36 The rising star of the Imperial Navy: Willard Price, "America's Enemy No. 2: Yamamoto," *Harper's Magazine*, April 1942, 449–58; Alistair Horne, *Hubris: The Tragedy of War in the Twentieth Century* (New York: HarperCollins, 2015), 23.

36 "This is an American editor": Price, "America's Enemy No. 2," 450.

36 "the crushing of White superiority": Ibid.

37 The age range of the Takano offspring spanned decades: Hiroyuki Agawa, *The Reluctant Admiral: Yamamoto and the Imperial Navy* (Tokyo: Kodansha International, 1969), 122–23.

37 he was taught briefly by an American missionary: Michael D. Hull, "Profiles: Japan's Naval War Leader," *WWII History*, Vol. 11, No. 7 (Fall 2012): 28; Carroll V. Glines, *Attack on Yamamoto* (Atglen, PA: Schiffer Military History, 1993), 42; Agawa, *The Reluctant Admiral*, 75. Note: Price, "America's Enemy No. 2," described Yamamoto was being taught as a boy to despise Americans, encouraged by a father, who told his son that Americans were "hairy barbarians, creatures with an animal odor owing to their habit of eating flesh." To contrary, the best available evidence shows Yamamoto admiring the United States and, before the war, enjoying relationships with a number of Americans. Most historians note that even if his father did describe Americans that way, Yamamoto rejected it.

38 admission to the Naval Cadet School: Glines, *Attack on Yamamoto*, 42.

38 "This upstart Asian David": Horne, *Hubris*, 22–32.

39 Japan launched a surprise torpedo attack: Ibid.; Glines, *Attack on Yamamoto*, 42.

39 "I found I was not afraid": Glines, *Attack on Yamamoto*, 43.

39 an unusual wartime moment between two navy commanders: Horne, *Hubris*, 105–107.

40 He became involved with one girl: Agawa, *The Reluctant Admiral*, 8–9, 204–7.

40   his adoption by the Yamamoto family: Glines, *Attack on Yamamoto*, 44.

41   "leather-faced, bullet-headed": Price, "America's Enemy No. 2," 458.

41   His first tour began in May 1919: Agawa, *The Reluctant Admiral*, 65–71.

42   Yamamoto discovered Abraham Lincoln: Ibid., 85.

43   the 5:5:3 agreement: Glines, *Attack on Yamamoto*, 46.

43   Yamamoto's luck at roulette: Agawa, *The Reluctant Admiral*, 75.

43   director of studies at a new air training school: Burke Davis, *Get Yamamoto* (New York: Random House, 1969), 25; Glines, *Attack on Yamamoto*, 45; Agawa, *The Reluctant Admiral*, 72, 76–82.

44   "Poker was his favorite": Ellis M. Zacharias, *Secret Missions: The Story of an Intelligence Officer* (New York: G. P. Putnam's Sons, 1946), 93.

44   "exceptionally able, forceful, a man of quick thinking": Ibid., 92.

45   *The Great Pacific War:* William H. Honan, *Visions of Infamy: The Untold Story of How Journalist Hector C. Bywater Devised the Plans That Led to Pearl Harbor* (New York: St. Martin's Press, 1991), xiii–xv.

45   the novel became the talk of the town: Ibid., 168, 172.

45   Yamamoto followed the heated commentary: Ibid., 177; Glines, *Attack on Yamamoto*, 50.

45   "The aircraft carrier, the combination of sea and air power, was an obsession": Zacharias, *Secret Missions*, 94.

46   the first solo flight across the Atlantic Ocean: Agawa, *The Reluctant Admiral*, 86–87.

46   "he was interested in war": Zacharias, *Secret Missions*, 91.

46   "Japan will lose if she adopts the traditional defensive strategy": Honan, *Visions of Infamy*, 185.

47   a massive attack of Pearl Harbor: Ibid., 184–86.

47   Japanese defense industry racing to build aircraft carriers and destroyers: Davis, *Get Yamamoto*, 27; Glines, *Attack on Yamamoto*, 46.

47   the London Naval Conference: Glines, *Attack on Yamamoto*, 46; Agawa, *The Reluctant Admiral*, 88.

48   "he did not wear a false face": Burke Davis, interview with Edwin T. Layton, April 10, 1968, 1–5; Layton, interview, Oral History Program of the United States Naval Institute, May 1970, from edited transcript available in 1975, 58–59.

48   Lieutenant Joseph Rochefort: Hervie Haufler, *Codebreakers' Victory: How the Allied Cryptographers Won World War II* (New York: New American Library, 2003), 118–19.

49   Yamamoto's caution: Agawa, *The Reluctant Admiral*, 12ff.

49   "Father seemed indifferent and undemonstrative": Yoshimasa Yamamoto, quoted in ibid., 69.

49   "You're lucky to be in love with your wife": Ibid.

50   Chiyoko Kawai: Donald M. Goldstein, and Katherine V. Dillon, *The Pearl Harbor Papers: Inside the Japanese Plans* (New York: Brassey's, 1993), 126–27.

51   "I scarcely remember such an event or you either": Ibid., 127.

51   He invited Chiyoko to meet him: Ibid., 127–28.

51   "I was fired with high spirits": Isoroku Yamamoto, letter to Chiyoko Kawai, May 1935, quoted in ibid., 128.

Chapter 3: The Flyboy from Enid

54 "My cold bothered me so much": John Mitchell, letter to Annie Lee, September 13, 1940, MFP.

54 "something to write home about": John Mitchell, letter to Annie Lee, September 11, 1940, MFP.

55 "our war situation grows more serious": John Mitchell, letter to Annie Lee, October 2, 1940, MFP.

55 "we are heading for a war": John Mitchell, letter to Annie Lee, October 10, 1940, MFP.

55 "second motor failure": John Mitchell, letter to Annie Lee, October 2, 1940, MFP.

55 the plane stood up on its nose: John Mitchell, letter to Annie Lee, October 22, 1940, MFP.

55 the pilot who came in too low: John Mitchell, letter to Annie Lee, December 17, 1940, MFP.

55 His engine suddenly cut out: John Mitchell, letter to Annie Lee, January 20, 1941, MFP.

56 P-40 stalled at 12,000 feet: John Mitchell, letter to Annie Lee, September 25, 1940, MFP.

56 aerial search party for two fliers: John Mitchell, letters to Annie Lee, October 20 and October 21, 1940, MFP.

56 "You may have seen something about it": John Mitchell, letter to Annie Lee, March 12, 1941, MFP.

56 "Nine-tenths of all these accidents are due to actions of the pilot": John Mitchell, letter to Annie Lee, September 25, 1940, MFP.

56 "They are really not fit to fly": John Mitchell, letter to Annie Lee, October 10, 1940, MFP.

57 "loops, slow rolls, Immelmanns and chandeliers": John Mitchell, letters to Annie Lee, October 9 and October 10, 1940, MFP.

57 a game of follow the leader: John Mitchell, letter to Annie Lee, January 8, 1941, MFP.

57 "really had no effect on me": John Mitchell, letter to Annie Lee, October 9, 1940, MFP.

57 "I fully enjoyed every bit of it": John Mitchell, letter to Annie Lee, November 5, 1940, MFP.

57 a mock dogfight: John Mitchell, letter to Annie Lee, October 2, 1940, MFP.

58 "He got me the first time": John Mitchell, letter to Annie Lee, October 20, 1940, MFP.

58 "Uncle Sam's prized possession": Ibid.

58 "Roosevelt will lead us in to war": John Mitchell, letter to Annie Lee, November 5, 1940, MFP.

59 "I have never had so much fun flying": John Mitchell, letter to Annie Lee, November 18, 1940, MFP.

59 learning on the job: Burke Davis, interview with John Mitchell, September 27–28, 1967, DA.

59 "I'm bragging": John Mitchell, letter to Annie Lee, November 18, 1940, MFP.

59 the flyboy from Enid had the right stuff: Burke Davis, interview with Henry Viccellio, February 12, 1968, DA.

59   "These new boys": John Mitchell, letters to Annie Lee, December 3 and December 12, 1940, MFP.

60   twelve-plane formation: John Mitchell, letter to Annie Lee, January 16, 1941, MFP.

60   each man had to do his part: John Mitchell, letter to Annie Lee, January 27, 1941, MFP.

60   "kept on dogfighting": John Mitchell, letter to Annie Lee, January 16, 1941, MFP.

61   stuck on the ground: John Mitchell, letter to Annie Lee, January 24, 1941, MFP.

61   wins, losses, and his teams' run: John Mitchell, letter to Annie Lee, March 26, 1941, MFP.

61   joyride in the secondhand convertible: John Mitchell, letter to Annie Lee, April 13, 1941, MFP.

61   "I availed myself of fried chicken": John Mitchell, letters to Annie Lee, December 3 and December 15, 1940, MFP.

62   "my last drinking": John Mitchell, letter to Annie Lee, January 6, 1941, MFP.

62   "The old moon": John Mitchell, letter to Annie Lee, December 6, 1940, MFP.

62   disappearing moon: John Mitchell, letter to Annie Lee, December 15, 1940, MFP.

62   new moon: John Mitchell, letter to Annie Lee, February 3, 1941, MFP.

62   full moon: John Mitchell, letter to Annie Lee, October 11, 1940, MFP.

62   "Hurry with that picture": John Mitchell, letters to Annie Lee, September 4, 6, 13, 19, 20, and 23, 1940, MFP.

62   "I got your sweet letters": John Mitchell, letters to Annie Lee, September 27 and 29, 1940, MFP.

63   "you are what you are": John Mitchell, letter to Annie Lee, January 13, 1941, MFP.

63   long-distance love: John Mitchell, letter to Annie Lee, December 17 and 18, 1940, MFP.

64   "married life is the life for me": John Mitchell, letter to Annie Lee, August 20, 1940, MFP.

64   her birthday present: Annie Lee, letter to Aunt Ludma, November 3, 1940, MFP.

64   "stepping out on you": John Mitchell, letter to Annie Lee, October 21, 1940, MFP.

64   a way to cope with the loneliness: John Mitchell, letter to Annie Lee, September 29, 1940, MFP.

65   "cross-country flying time": John Mitchell, letters to Annie Lee, November 20 and November 22, 1940, MFP.

65   he could not go to Texas: John Mitchell, letters to Annie Lee, December 13 and December 17, 1940, MFP.

65   he rushed to mail them: John Mitchell, letter to Annie Lee, January 4, 1941, MFP.

65   the wound of not being together on Christmas Day: John Mitchell, letter to Annie Lee, December 20, 1940, MFP.

65   "Let's get married!": Ibid.

66 "I know for sure that it's Johnnie I want": Annie Lee, letter to Aunt Ludma, March 3, 1941, MFP.

66 "there is always tomorrow": John Mitchell, letter to Annie Lee, December 17, 1940, MFP.

66 this was what he'd trained for: John Mitchell, letter to Annie Lee, April 4, 1941, MFP.

Chapter 4: No Ordinary Strategy

67 Yamamoto set sail for the United States: Hiroyuki Agawa, *The Reluctant Admiral: Yamamoto and the Imperial Navy* (Tokyo: Kodansha International, 1969), 24.

68 "all nations are entitled equally to enjoy a sense of national security": "Japan Will Urge Big Cut in Navies," *New York Times*, October 9, 1934.

69 "one of Japan's most brilliant naval officers": "Yamamoto Declines to Reveal Proposal," *New York Times*, October 8, 1934.

69 "I have never looked upon the United States as a potential enemy": "Japan Will Urge Big Cut in Navies."

70 "Japan objects to the ratio system": Agawa, *The Reluctant Admiral*, 25–26; "Deadlock Feared in Talks on Navies," *New York Times*, October 17, 1934.

70 "outspoken seadog": Associated Press, "Naval Building Race Threatens," *Boston Globe*, October 23, 1934, 11.

70 all were working to gain an edge: William H. Honan, *Visions of Infamy: The Untold Story of How Journalist Hector C. Bywater Devised the Plans That Led to Pearl Harbor* (New York: St. Martin's Press, 1991), 225.

70 relied on an interpreter: Agawa, *The Reluctant Admiral*, 36.

70 "We sailors get on admirably together": Honan, *Visions of Infamy*, 224.

71 always ready to initiate a poker or bridge game: Agawa, *The Reluctant Admiral*, 40.

71 arms control and world peace: Ibid., 38–44; Honan, *Visions of Infamy*, 222.

71 a pawn for the militarists: Isoroku Yamamoto, letter to Chiyoko Kawai, September 1935, quoted in Donald M. Goldstein and Katherine V. Dillon, *The Pearl Harbor Papers: Inside the Japanese Plans* (New York: Brassey's, 1993), 128; letter of May 1, 1935, Agawa, *The Reluctant Admiral*, 58.

72 a hypothetical Pacific war: Honan, *Visions of Infamy*, 223–26.

73 Japan's immovable rejection of the ratio system: Ibid., 223–25; Hector C. Bywater, "U.S. Policy in Naval Talks: Firm Line Urged Upon Britain," *Daily Telegraph*, December 4, 1934.

73 "Abandonment now of the principles": Charles A. Selden, "Davis Says Japan Upsets Security of All in Pacific; Fears Costly Naval Race," *New York Times*, December 7, 1934.

73 "Japan and the United States are far apart": Special to the *New York Times*, "Japan and Britain to Halt Navy Talk," *New York Times*, December 29, 1934; Charles A. Selden, "Davis Leaves London," *New York Times*, December 30, 1934.

73 When Yamamoto left London: Agawa, *The Reluctant Admiral*, 49–50; 124; Edwin Hoyt, *Yamamoto: The Man Who Planned the Attack on Pearl Harbor* (Guilford, CT: Lyons Press), 84.

74 "impossible to reach any agreement": Agawa, *The Reluctant Admiral*, 51.

74   popular image as strongman: Ibid., 52.

74   two competing groups: Ibid., 53; Hoyt, *Yamamoto*, 84–85.

75   "beguiling and beautiful": Quoted in Agawa, *The Reluctant Admiral*, 56–58.

75   "I dreamed that we were driving": Quoted in ibid., 64.

75   his appointment as chief of the navy's Aeronautics Department: Hoyt, *Yamamoto*, 85; Agawa, *The Reluctant Admiral*, 90, 95.

75   Yamamoto sought to build a fleet: Carroll V. Glines, *Attack on Yamamoto* (Atglen, PA: Schiffer Military History, 1993), 47; Hoyt, *Yamamoto*, 85–87.

76   "suddenly shifted to political duties": Agawa, *The Reluctant Admiral*, 119.

77   he was not the only important man in Chiyoko's life: Ibid., 65, 204–207.

77   preparing to do battle to rule the Pacific region: Louis Morton, "The Japanese Decision for War," *United States Naval Institute Proceedings* 80, no. 12 (December 1954): 1325–35.

77   the army more than doubled in size: Ibid., 1326.

77   radio programs delivered cheerleading daily updates: Samuel Hideo Yamashita, *Daily Life in Wartime Japan, 1940–1945* (Lawrence: University Press of Kansas, 2015), 11.

77   He continued to argue forcefully: Captain Roger Pineau, "Admiral Isoroku Yamamoto," in *The War Lords: Military Commanders of the Twentieth Century*, ed. Sir Michael Carver (Boston: Little, Brown, 1976), 390–403, at 396.

78   outspoken in opposing the war in China: Morton, "The Japanese Decision for War," 1326–27.

78   a tripartite pact: Agawa, *The Reluctant Admiral*, 186; Glines, *Attack on Yamamoto*, 49.

78   "the crisis threatening the nation could not be graver": Agawa, *The Reluctant Admiral*, 125.

79   Yamamoto had a target on his back: Pineau, "Admiral Isoroku Yamamoto," 396.

79   "Yamamoto, the Stern, Silent Admiral": Agawa, *The Reluctant Admiral*, 5–6, 169–71.

79   Yamamoto's departure from Tokyo: Ibid., 9–11; Hoyt, *Yamamoto*, 2–3, 104.

79   Yamamoto was essentially sea based: Agawa, *The Reluctant Admiral*, 174, 179.

80   the pinnacle of his powers: Burke Davis, *Get Yamamoto* (New York: Random House, 1969), 32.

80   his high expectations for fliers: Glines, *Attack on Yamamoto*, 49–50; Pineau, "Admiral Isoroku Yamamoto," 396.

80   "air attack on Hawaii may be possible now": Glines, *Attack on Yamamoto*, 50; Davis, *Get Yamamoto*, 32 (quoting Yamamoto slightly differently).

81   "It is a mistake to regard Americans as luxury loving and weak": Pineau, "Admiral Isoroku Yamamoto," 397.

81   "I do only what is best for my country": Ibid., 396–97.

82   "This moment is the critical time": Isoroku Yamamoto, letter to Mitsuari Takamura, November 4, 1940, quoted in Goldstein and Dillon, *The Pearl Harbor Papers*, 114.

83   "deciding the fate of the war on its first day": Isoroku Yamamoto, letter to Navy Minister Koshiro Oikawa, January 7, 1941, quoted in Goldstein and Dillon, *The Pearl Harbor Papers*, 115; Hoyt, *Yamamoto*, 108; Honan, *Visions of Infamy*, 252–53.

83 Operation Z: Glines, *Attack on Yamamoto*, 51; Honan, *Visions of Infamy*, 254; Hoyt, *Yamamoto*, 109.

83 flew in the face of conventional wisdom: Hoyt, *Yamamoto*, 109.

83 "the great all-out battle": Goldstein and Dillon, *The Pearl Harbor Papers*, 1–2.

83 the warmongers' ignorant miscalculation: James A. Field, Jr., "Admiral Yamamoto," *United States Naval Institute Proceedings* 75, no. 10 (October 1949): 1006–07; Agawa, *The Reluctant Admiral*, 291. Note: The translation of Yamamoto's letter of January 24, 1941, is slightly different in Agawa: "If there should be a war between Japan and America, then our aim, of course, ought not to be Guam or the Philippines, nor Hawaii or *Hong Kong, but a capitulation at the White House, in Washington itself.*" (Italics mine.)

84 his objective was to land a preemptive blow: Pineau, "Admiral Isoroku Yamamoto," 396–97.

84 a decision diametrically opposed to my personal opinion: Isoroku Yamamoto, letter to Rear Admiral Teikichi Hori, November 11, 1941, quoted in James A. Field, Jr., "Admiral Yamamoto," *United States Naval Institute Proceedings* 75, no. 10 (October 1949): 1112; Yamamoto, letter to Hori, although this source dates the letter as October 11, 1941, quoted in Goldstein and Dillon, *The Pearl Harbor Papers*, 124. The translation varies slightly: "Because I have been assigned the mission, entirely against my private opinion, and also I am expected to do my best. Alas, maybe, that is my fate."

85 "tabletop maneuvers": Honan, *Visions of Infamy*, 256–57; Morton, "The Japanese Decision for War," 5–6.

86 deep worry about his country: Isoroku Yamamoto, letter to Vice Admiral Shigetaro Shimada, October 24, 1941, quoted in Goldstein and Dillon, *The Pearl Harbor Papers*, 118.

86 "Combined Fleet Top Secret Operational Order Number 1": Pineau, "Admiral Isoroku Yamamoto," 398; Glines, *Attack on Yamamoto*, 51–52; Honan, *Visions of Infamy*, 253; Goldstein and Dillon, *The Pearl Harbor Papers*, 148.

87 "Climb Mt. Niitaka": Hoyt, *Yamamoto*, 128–29.

87 "the picture of hatchet-faced solemnity": Ian W. Toll, "A Reluctant Enemy," *New York Times*, December 6, 2011.

87 "Watch for the time when the petals of these flowers fall": Isoroku Yamamoto, letters to Chiyoko Kawai, December 1941, quoted in Goldstein and Dillon, *The Pearl Harbor Papers*, 128–29; Agawa, *The Reluctant Admiral*, 242–47; Vice Admiral Ryunosuke Kusaka, memoir, quoted in Goldstein and Dillon, *The Pearl Harbor Papers*, 154. Note: For the sake of consistency and clarity, times and dates are in Pacific Standard Time (PST). For example, Yamamoto sent the telegram "Climb Mt. Niitaka" on December 2 in Japan; in Hawaii, the date was December 1.

Chapter 5: A Taste of War

90 "we didn't draw a sober breath from the night we left": John Mitchell, letter to Annie Lee, April 20, 1941, MFP.

90 going to England to fly with Royal Air Force: Ibid.

91 the news unsettled Annie Lee: Annie Lee Miller, letter to Aunt Ludma, April 8, 1941, MFP.

91   "It makes my heart stand still": Annie Lee Miller, letter to Aunt Ludma, May 27, 1941, MFP.

91   John William Mitchell will come home: Noah Mitchell, letter to Annie Lee, May 1941, MFP.

92   Willkie was the pilots' kind of guy: John Mitchell, letter to Annie Lee, April 22, 1941, MFP.

93   the P-38 Lightning: John Mitchell, letter to Annie Lee, April 23, 1941, MFP. See https://www.pacificwrecks.com/aircraft/p-38/tech.html.

93   the importance of the mission: John Mitchell, letter to Annie Lee, April 25, 1941, MFP.

94   one of his full-moon moments: John Mitchell, letter to Annie Lee, May 11, 1941, MFP.

94   The men grew antsy: John Mitchell, letter to Annie Lee, May 25, 1945, MFP.

94   Lieutenant Commander Ian Fleming: Copy of Fleming's hotel registration in William Thomas, "Birthplace of James Bond," *St. Catharine Standard*, March 29, 2013, https://www.thejamesbonddossier.com/content/james-bond-where -it-all-began.htm.

95   an audience with the queen of England: John Mitchell, letter to Annie Lee, May 25, 1941, MFP.

96   Brits rubbed him the wrong way: John Mitchell, letter to Annie Lee, June 9, 1941, MFP.

97   "she's just the gal you need": Noah Mitchell, letter to John Mitchell, May 25, 1941, MFP.

97   found London a tough place for ordinary people: John Mitchell, letter to Annie Lee, June 9, 1941, MFP.

97   his eye on a Harris tweed suit: John Mitchell, letters to Annie Lee, June 9 and July 7, 1941, MFP.

97   "the old spirit was lacking": John Mitchell, letters to Annie Lee, June 9 and June 17, 1941, MFP.

98   sucker punched: John Mitchell, letter to Annie Lee, June 23, 1941, MFP.

98   "I say let's get married": Ibid.

99   he'd wrecked the RAF Spitfire: Ibid.

100   spotted by enemy ground troops: John Mitchell, letter to Annie Lee, July 20, 1941, MFP; Burke Davis, interview with John Mitchell, September 1967, DA. Mitchell's flight over the Channel became exaggerated over time; some ac- counts include the dramatic but inaccurate report that Mitchell shot down a German fighter plane before returning to England.

100   "We'uns coming home!": John Mitchell, letter to Annie Lee, July 4, 1941, MFP.

100   he'd learned plenty as an observer: Burke Davis, interview with John Mitchell, September 1967, DA.

100   he also waxed poetic: John Mitchell, letter to Annie Lee, June 23, 1941, MFP.

100   "I'm anxious to see him and yet I'm a little afraid": Annie Lee Miller, letter to Aunt Ludma, July 22, 1941, MFP.

102   they began making plans: John Mitchell, letter to Annie Lee, September 25, 1941, MFP.

102   "9,836,431 mosquitoes": John Mitchell, letter to Annie Lee, October 6, 1941, MFP.

102 "fighting skeeters with one hand and slinging ink with the other": John Mitchell, letter to Annie Lee, October 20, 1941, MFP.

103 Charlotte, North Carolina, was a five-star hotel: John Mitchell, letter to Annie Lee, October 28, 1941, MFP.

103 the plane descended at full speed: "Lightning-Fast Plane Plunges to Earth Near Here, Killing Promising Young Army Pilot," *Desert Sun of Palm Springs, California*, November 14, 1941.

104 another P-38 test pilot had crashed: "Pilot Killed as P-38 Crashes into House: Experimental Plane Loses Tail Assembly at 400 M.P.H." *Los Angeles Daily Mirror*, November 5, 1941.

104 twin tragedies: John Mitchell, letter to Annie Lee, November 14, 1941, MFP.

105 Operation Flipper: Adam Leong Kok Wey, "Case Study of Operation Flipper," in *Killing the Enemy: Assassination Operations During World War II* (London: I. B. Taurus & Co., 2015), 131–58.

106 slated to return to Hamilton Field: John Mitchell, letter to Annie Lee, November 11, 1924, MFP.

106 their window of opportunity: John Mitchell, letter to Annie Lee, October 28, 1941, MFP.

106 where to marry: John Mitchell, letter to Annie Lee, October 25, 1941, MFP.

106 meet up in Reno: John Mitchell, letter to Annie Lee, November 3, 1941, MFP.

106 He plotted their honeymoon: John Mitchell, letter to Annie Lee, November 26, 1941, MFP.

106 "I've always thought a wedding was the woman's show": John Mitchell, letter to Annie Lee, November 24, 1941, MFP.

106 Mitch would get to see one of his sisters: John Mitchell, letter to Annie Lee, November 30, 1941, MFP.

107 a feature story in the local paper: John Mitchell, letters to Annie Lee, December 1 and December 4, 1941, MFP.

107 a telegram that would change the course of history: Burke Davis, *Get Yamamoto* (New York: Random House, 1969), 40.

Chapter 6: When the Rose Petals Fell

109 the operations room of his flagship: Hiroyuki Agawa, *The Reluctant Admiral: Yamamoto and the Imperial Navy* (Tokyo: Kodansha International, 1969), 257–58.

110 Yamamoto was waiting for word: Carroll V. Glines, *Attack on Yamamoto* (Atglen, PA: Schiffer Military History, 1993), 51; Agawa, 243.

110 a metaphor for the imminent surprise attack: Isoroku Yamamoto, letter to Chiyoko Kawai, December 5, 1941 (Tokyo time), quoted in Donald M. Goldstein and Katherine V. Dillon, *The Pearl Harbor Papers: Inside the Japanese Plans* (New York: Brassey's, 1993), 129.

110 waiting and worrying: Agawa, *The Reluctant Admiral*, 250–51.

111 proper diplomatic practice would be followed: Louis Morton, "The Japanese Decision for War," *United States Naval Institute Proceedings* 80, no. 12 (December 1954): 8–9; Agawa, *The Reluctant Admiral*, 272–73.

111 aware of what was afoot: Agawa, *The Reluctant Admiral*, 278–79; Morton, "The Japanese Decision for War," 9–10; Bill McWilliams, *Sunday in Hell: Pearl Harbor Minute by Minute* (New York: E-rights/E-Reads, 2011), 177–83.

112   his final war message: Edwin Hoyt, *Yamamoto: The Man Who Planned the Attack on Pearl Harbor* (Guilford, CT: Lyons Press), 131; Vice Admiral Ryunosuke Kusaka, memoir, quoted in Donald M. Goldstein and Katherine V. Dillon, *The Pearl Harbor Papers: Inside the Japanese Plans* (New York: Brassey's, 1993), 155.

112   a signal flag was hoisted: Agawa, *The Reluctant Admiral*, 253; Kusaka, memoir, quoted in Goldstein and Dillon, *The Pearl Harbor Papers*, 155.

112   final intelligence report from Tokyo: Kusaka, memoir, quoted in Goldstein and Dillon, *The Pearl Harbor Papers*, 156; Agawa, *The Reluctant Admiral*, 254.

113   the petals fell from her roses: Letters between Isoruko Yamamoto and Chiyoko Kawai, quoted in Goldstein and Dillon, *The Pearl Harbor Papers*, 129.

113   "the apprehension that had made us worry so long disappear suddenly": Kusaka, memoir, quoted in Goldstein and Dillon, *The Pearl Harbor Papers*, 154.

113   "all hands were fired up": Ibid., 155.

114   "First bomb in the war on America": Agawa, *The Reluctant Admiral*, 254.

114   the attack flight would include 183 planes: McWilliams, *Sunday in Hell*, 196.

114   daybreak on December 7: Kusaka, memoir, quoted in Goldstein and Dillon, *The Pearl Harbor Papers*, 159.

114   a flag signaling "Take off": Ibid.

114   the entire first attack wave of 183 aircraft was airborne: McWilliams, *Sunday in Hell*, 6, 198; Kusaka, memoir, quoted in Goldstein and Dillon, *The Pearl Harbor Papers*, 159.

115   350 weapons-loaded planes into the sky: McWilliams, *Sunday in Hell*, 199, 209.

115   The day had come: Burke Davis, *Get Yamamoto* (New York: Random House, 1969), 40; Agawa, *The Reluctant Admiral*, 256–57; McWilliams, *Sunday in Hell*, 279.

115   Second Lieutenant Besby Holmes: Interview with Besby Holmes, in Eric Hammel, *Aces Against Japan II: The American Aces Speak*, vol. 3 (Pacifica, CA: Pacifica Press, 1996), 5.

116   enamored with flying: Burke Davis, interview with Besby Holmes, February 14, 1968, DA; Davis, *Get Yamamoto*, 47; interview with Besby Holmes, in Hammel, *Aces Against Japan II*, 3.

116   March 1941 to enlist as an air cadet: Interview with Besby Holmes, in Hammel, *Aces Against Japan II*, 4–5.

117   the Japanese were attacking Pearl Harbor: Ibid.; Burke Davis, interview with Besby Holmes, February 14, 1968, DA; Davis, *Get Yamamoto*, 46.

118   Lieutenant Rex T. Barber: Burke Davis, interview with Rex T. Barber, October 1, 1967, DA.

119   a fighter pilot on paper who would learn on the job: Ibid.; Davis, *Get Yamamoto*, 51–52.

120   Second Lieutenant Thomas G. Lanphier, Jr.: Burke Davis, interview with Thomas G. Lanphier, Jr., via correspondence, January 25, 1968, DA.

120   Lanphier was a military brat: Davis, *Get Yamamoto*, 55; Burke Davis, interview with Thomas G. Lanphier, Jr., January 25, 1983, DA; Lanphier, unpublished autobiography, 65, 75–76, 87–88, DA.

121   The first time he flew was as a teenager: Lanphier, unpublished autobiography, 46, DA.

121 his father was the one who would insist that he enlist: Lanphier, unpublished autobiography, 91–92, DA.

122 Lanphier certainly had a lot to live up to: Davis, *Get Yamamoto*, 57; Lanphier, unpublished autobiography, 95, DA.

122 Captain Henry "Vic" Viccellio: Burke Davis, interview with Henry Viccellio, February 12, 1968, DA; Lanphier, unpublished autobiography, 99, DA.

123 bad weather had kept him from flying south: John Mitchell, letter to Annie Lee, December 4, 1941, MFP.

124 more than 3,400 Americans dead or wounded: Susan Wels, *Pearl Harbor: America's Darkest Day* (New York: Time-Life Books, 2001), 133–34.

125 War was now at the forefront: Davis, *Get Yamamoto*, 49; Burke Davis, interview with John Mitchell, September 27–28, 1967, DA.

125 "Surprise attack successful": Agawa, *The Reluctant Admiral*, 258–59; Hoyt, *Yamamoto*, 133.

125 the successful bombing and invasion of other targets: Hoyt, *Yamamoto*, 134–35; Glines, *Attack on Yamamoto*, 52–53.

125 the private letter Yamamoto had penned: J. A. Field, Jr., "Admiral Yamamoto," *United States Naval Institute Proceedings* 75, no. 10 (October 1949): 1105–06.

126 he was indisputably the hero: Noda Mitsuharu, oral history on *Nagato*, quoted in Haruko Taya Cook and Theodore F. Cook, *Japan at War: An Oral History* (New York: New Press, 1992), 81.

126 "I am longing only for letters from you night and day": Isoroku Yamamoto, letter to Chiyoko Kawai, December 28, 1941, quoted in Goldstein and Dillon, *The Pearl Harbor Papers*, 129.

126 "We were blessed by the War God": Isoroku Yamamoto, letter to Sankichi Takahashi, December 19, 1941, quoted in Goldstein and Dillon, *The Pearl Harbor Papers*, 120.

127 "dreamed of going to America": Noda Mitsuharu, oral history on *Nagato*, quoted in Cook and Cook, *Japan at War*, 81.

Chapter 7: Wedding Bells and Pacific Blues

132 a pit-stop wedding: John Mitchell, diary, January 22, 1942, MFP; John Mitchell–Annie Lee Miller marriage certificate, San Antonio, TX, December 13, 1941.

133 war and separation weren't going to be easy: John Mitchell, letter to Annie Lee, December 17, 1941, MFP.

133 "The nicest Xmas I EVER HAD!": John Mitchell, letter to Annie Lee, December 19, 1941, MFP.

134 "Johnny, she hardly knew ye": John Mitchell, diary, January 22, 1942, MFP.

134 the first wartime convoy to the Pacific: Burke Davis, interview with Henry Viccellio, February 12, 1968, DA; John Mitchell, diary, January 22, 1942, MFP.

135 "How long it was to be before we again would see that bridge": John Mitchell, diary, January 22, 1942, MFP.

135 They played records, drank, dined: Ibid.

135 the *Monroe's* maiden voyage: Burke Davis, interviews with Rex Barber, September 29 and October 1, 1967, DA; John Mitchell, diary, January 29, 1942, MFP; Doug Canning, oral history interview at Nimitz National Museum of the Pacific War, October 4, 2001.

136   "a trail of hootch bottles": Burke Davis, interview with Doug Canning, September 24, 1967, DA; Doug Canning, oral history interview at Nimitz National Museum of the Pacific War, October 4, 2001.

136   Mudville: John Mitchell, diary, February 12, 1942, MFP; John Mitchell, letter to Annie Lee, March 29, 1942, MFP; Burke Davis, interview with Rex Barber, September 29 and October 1, 1967, DA; Burke Davis, *Get Yamamoto* (New York: Random House, 1969), 73.

136   crates containing P-39 Airacobras followed: John Mitchell, diary, February 12, 1942, MFP.

137   working against the clock: John Mitchell, diary, January 27, 1942, MFP.

137   plans for destroying documents and armaments: John Mitchell, diary, February 12, 1942, MFP.

137   all joined in to assemble the puzzle pieces: Burke Davis, interview with Henry Viccellio, February 12, 1968, DA.

137   he carelessly flipped a jeep in the mud: John Mitchell, diary, February 12, 1942, MFP.

138   "we are tearing 'em up faster than we are putting them together": John Mitchell, diary, March 1, 1942, MFP.

138   They found a movie theater in Nausori: Jeanne T. Heidler, *Daily Lives of Civilians in Wartime Modern America* (Westport, CT: Greenwood, 2007), 89; Mark Harris, *Five Came Back: A Story of Hollywood and the Second World War* (New York: Penguin, 2014), 1–5.

139   two months since his wedding: John Mitchell, diary, February 14, 1942, MFP.

139   couldn't help but notice the additional warships: John Mitchell, diary, February 12 and 14, 1942, MFP.

139   the men completed a move to Nadi: John Mitchell, diary, February 12, 1942, MFP.

140   the squadron began flying drills: John Mitchell, diary, May 26, 1942, MFP; Burke Davis, interview with Henry Viccellio, February 12, 1968, DA; Davis, interview with Doug Canning, September 24, 1967, DA.

140   He drilled the men hard: Burke Davis, interview with Doug Canning, September 24, 1967, DA.

140   the promotions became official: John Mitchell, diary, March 17, 1942, MFP; John Mitchell, letter to Annie Lee, March 20, 1942, MFP.

141   Mitch had a thing for Texans: DA: Lanphier, unpublished autobiography, 110.

141   drove into the middle of the village: John Mitchell, diary, March 8, 1942, MFP; John Mitchell, letter to Annie Lee, March 29, 1942, MFP.

142   "I found three letters from my honey": John Mitchell, letters to Annie Lee, March 28 and May 12, 1942, MFP.

143   the incompatibility of marriage and war: John Mitchell, letter to Annie Lee, March 29, 1942, MFP.

143   he found comfort in her reports: John Mitchell, letter to Annie Lee, June 13, 1942, MFP.

143   The Pacific blues: John Mitchell, letters to Annie Lee, April 27, May 13, and May 26, 1942, MFP.

143   rumor had it: Wilmott Ragsdale, "New Caledonia," *Time*, March 26, 1942, 2.

144 enemy planes: John Mitchell, diary, April 7 and April 27, 1942, MFP.

144 "It's a waiting game": John Mitchell, letter to Annie Lee, May 13, 1942, MFP.

144 "We receive news flashes": Ibid.

144 You attack Pearl Harbor, we bomb Tokyo: Colonel C. V. Glines, "Some Little Known Facts About the Doolittle Raid," *Popular Aviation*, March–April 1967, 16–21; Dan Sewall, "75 Years After the Doolittle Raid, the Last Survivor Remembers How the US Struck Back at Japan in WWII," Associated Press, April 18, 2017. Of the eighty men in the raid—sixteen five-person crews consisting of pilot, copilot, bombardier, navigator, and mechanic gunner—three died trying to reach China and eight were captured by Japanese soldiers. Of those, three were executed and a fourth died in captivity. The others made it back.

144 "inspired by the imagination, guts and skill": Lanphier, "Fighter Pilot," 121.

145 The Doolittle Raid: Glines, "Some Little Known Facts About the Doolittle Raid."

145 killed thousands upon thousands of Chinese: Alistair Horne, *Hubris: The Tragedy of War in the Twentieth Century* (New York: HarperCollins, 2015), 245.

145 "my screaming for action of some kind": John Mitchell, letter to Annie Lee, June 13, 1942, MFP.

Chapter 8: Unfinished Business

148 the Moby Dick of subs: "Giant Submarine Cruiser May Have Attacked Oil Field," *Los Angeles Times*, February 24, 1942; "Coast Alert for New Raids: Army Vigilant for Sub Return," *Los Angeles Times*, February 25, 1942.

148 "Shells were exploding": "Faulty Shells Blast Homes," "Chilly Throng Watches Shells Bursting in Sky," "Five Deaths Laid to Blackouts," "Plane Signaling Suspects Seized," *Los Angeles Times*, February 26, 1942; Evan Andrews, "History Stories: World War II's Bizarre 'Battle of Los Angeles,'" March 7, 2019, https://www.history.com/news/world-war-iis-bizarre-battle-of-los-angeles.

149 manhunt for a Japanese pilot: Bill McWilliams, *Sunday in Hell: Pearl Harbor Minute by Minute* (New York: E-rights/E-Reads, 2011), 519. The Japanese airman was eventually captured and killed, and a local Japanese American, or nisei, who had assisted the pilot killed himself.

149 "Japanese hunting licenses": Michiko Kakutani, "When History Repeats," *New York Times Sunday Review*, July 15, 2018, 1.

149 ten tips for patriots to differentiate Japanese from Chinese immigrants: "How to Tell Your Friends from the Japs," *Time*, December 22, 1942, 33.

150 "We must remove the Japanese": "Immediate Evacuation of Japanese Demanded: Southern Californians Call for Summary Action by Army After Submarine Attack," *Los Angeles Times*, February 25, 1942.

150 "looking forward to dictating peace in the United States": "'I'll Capture White House,' Jap Admiral Bragged Year Ago," *Washington Post*, December 17, 1941.

150 demonic sketch of Yamamoto: "Japan's Aggressor: Admiral Yamamoto," *Time*, December 22, 1941.

151 a "hard chunk of man": Willard Price, "America's Enemy No. 2: Yamamoto," *Harper's Magazine*, April 1942, 449–58.

151 "he was a peculiarly personal foe": James A. Field, Jr., "Admiral Yamamoto," *United States Naval Institute Proceedings* 75, no. 10 (October 1949): 1105–13.

151   The raw numbers told the story: Susan Wels, *Pearl Harbor: America's Darkest Day* (New York: Time-Life Books, 2001), 133–34.

151   Japanese losses had turned out to be minimal: Carroll V. Glines, *Attack on Yamamoto* (Atglen, PA: Schiffer Military History, 1993), 52; Hiroyuki Agawa, *The Reluctant Admiral: Yamamoto and the Imperial Navy* (Tokyo: Kodansha International, 1969), 264.

152   an end to the despised 5:5:3 ratio: Edwin Hoyt, *Yamamoto: The Man Who Planned the Attack on Pearl Harbor* (Guilford, CT: Lyons Press), 139.

152   The flyboys on Fiji were shaken: Burke Davis, interview with Henry Viccellio, February 12, 1968, DA.

152   radio stations began playing music: Agawa, *The Reluctant Admiral*, 288.

153   the Imperial Declaration of War was broadcast: Samuel Hideo Yamashita, *Daily Life in Wartime Japan, 1940–1945* (Lawrence: University Press of Kansas, 2015), 18.

153   "I wrote *kwu kwu*": Isoroku Yamamoto, letter to Chiyoko Kawai, January 8, 1942, quoted in Donald M. Goldstein and Katherine V. Dillon, *The Pearl Harbor Papers: Inside the Japanese Plans* (New York: Brassey's, 1993), 129.

153   "victory disease": Roger Pineau, "Admiral Isoroku Yamamoto," in *The War Lords: Military Commanders of the Twentieth Century*, ed. Sir Michael Carver (Boston: Little, Brown, 1976), 390–403, at 398.

153   "It is easy to open hostilities": Ibid.

154   "we've just got to get a peace agreement": Agawa, *The Reluctant Admiral*, 292.

154   the aircraft carriers—remained at large: Bill McWilliams, *Sunday in Hell: Pearl Harbor Minute by Minute* (New York: E-rights/E-Reads, 2011), 283.

154   the US Pacific Fleet's recovery could happen faster: Hoyt, *Yamamoto*, 137; Agawa, *The Reluctant Admiral*, 262–64; H. P. Willmott, *Pearl Harbor* (London, UK: Cassell, 2001), 161.

154   "we could lose it": Glines, *Attack on Yamamoto*, 53.

155   "Official reports should stick to the absolute truth": Agawa, *The Reluctant Admiral*, 288.

155   the national spirit has shown itself ready and able: Arthur Krock, "Six Months After Pearl Harbor," *New York Times*, June 7, 1942.

156   "BUY WAR BONDS": *Washington Post*, October 10, 1942.

156   "a right nice nest egg": John Mitchell, letter to Annie Lee, April 22, 1942, MFP.

156   "Now comes the adults' hour": Isoroku Yamamoto, letter to Niwa Michi, exact date unknown, but written in days following Doolittle attack, April 18, 1942, quoted in Agawa, *The Reluctant Admiral*, 299.

156   Midway Island: William H. Honan, *Visions of Infamy: The Untold Story of How Journalist Hector C. Bywater Devised the Plans That Led to Pearl Harbor* (New York: St. Martin's Press, 1991), 262.

157   "press the nation's political leaders to initiate overtures for peace": Ibid.

157   "Midway will draw out the enemy's carriers and destroy them": Pineau, "Admiral Isoroku Yamamoto," 398–99.

158   unnerved by the news of the raid: Ibid., 399.

158   the Battle of the Coral Sea: Hervie Haufler, *Codebreakers' Victory: How the Allied Cryptographers Won World War II* (New York: New American Library, 2003), 150.

Chapter 9: Midway: Yamamoto's Lament

162 "MO" meant Port Moresby: Eliot Carlson, *Joe Rochefort's War: The Odyssey of the Codebreaker Who Outwitted Yamamoto at Midway* (Annapolis, MD: Naval Institute Press, 2011), 279.

162 Nimitz's forces fought Yamamoto's to a standstill: Hervie Haufler, *Codebreakers' Victory: How the Allied Cryptographers Won World War II* (New York: New American Library, 2003), 149–50; Carlson, *Joe Rochefort's War*, 276–81.

162 Red Lasswell: Alva B. "Red" Lasswell, oral history, Historical Division, US Marine Corps, Marine Corps Historical Center, 2.

163 Rochefort saw Finnegan as complementary to Lasswell: Carlson, *Joe Rochefort's War*, 100–03.

163 "The Dungeon": Carlson, *Joe Rochefort's War*, diagram of Dungeon in photograph insert, 103; John Prados, *Combined Fleet Decoded: The Secret History of American Intelligence and the Japanese Navy in World War II* (New York: Random House, 1995), 315.

163 "unless you do a good job of translating the whole [effort] is lost": Joe Rochefort, oral history, US Naval Intelligence, 102, quoted in Carlson, *Joe Rochefort's War*, 104.

164 intense rivalry flared: Lasswell, oral history, 16; Haufler, *Codebreakers' Victory*, 119.

164 the navy program would prove itself far superior: Haufler, *Codebreakers' Victory*, 118–19.

164 "my good right arm": Lasswell, oral history, 17.

164 Lasswell followed Finnegan: Lasswell, oral history, 23. Lasswell later called his Midway decryption his greatest wartime achievement.

165 "an intelligence officer has one task": Haufler, *Codebreakers' Victory*, 128.

165 "I was just as surprised": Lasswell, oral history, 30.

165 superenciphement: Haufler, *Codebreakers' Victory*, 119–20; Carlson, *Joe Rochefort's War*, 359.

166 "From a monastery the place turned into a pressure cooker": Carlson, *Joe Rochefort's War*, 302.

166 two-watch system: Lasswell, memoir, 42–44, courtesy of Joe Cole and the Lasswell family.

166 the men began popping amphetamines: Carlson, *Joe Rochefort's War*, 302.

166 the distribution of new codebooks was no easy task: Michael Smith, *The Emperor's Code: Breaking Japan's Secret Ciphers* (New York: Arcade, 2000), 125; John Prados, *Combined Fleet Decoded: The Secret History of American Intelligence and the Japanese Navy in World War II* (New York: Random House, 1995), 316; Carlson, *Joe Rochefort's War*, 359.

166 looking "over Yamamoto's shoulder": Carlson, *Joe Rochefort's War*, 341.

167 The analysts recognized AF as a geographic symbol: Carlson, *Joe Rochefort's War*, 307, 317; Smith, *The Emperor's Code*, 138.

167 from A to WAG: Carlson, *Joe Rochefort's War*, 325.

168 a mistake could be catastrophic: Lasswell, memoir, 42; Haufler, *Codebreakers' Victory*, 151.

168 he detected what looked to be an important message: Lasswell, oral history, 37.

168   Washington did not buy Hypo's conclusion that AF was Midway: Lasswell, oral history, 36.

169   "The Japanese took the bait": Wilber Jasper Holmes, *Double-Edged Secrets: U.S. Naval Intelligence Operations in the Pacific During World War II* (Annapolis, MD: Naval Institute Press, 1979), 91.

169   the Japanese message used AF: Carlson, *Joe Rochefort's War*, 99, 333–36; Prados, *Combined Fleet Decoded*, 317; Haufler, *Codebreakers' Victory*, 150–52.

169   the Japanese Imperial Navy finally altered its code: Lasswell, oral history, 37.

170   Yamamoto's second chance to force an early peace: Hiroyuki Agawa, *The Reluctant Admiral: Yamamoto and the Imperial Navy* (Tokyo: Kodansha International, 1969), 311–12; Carlson, *Joe Rochefort's War*, 342, 378.

170   Chiyoko had fallen ill: Donald M. Goldstein and Katherine V. Dillon, *The Pearl Harbor Papers: Inside the Japanese Plans* (New York: Brassey's, 1993), 129; Agawa, *The Reluctant Admiral*, 309.

171   "My darling awaiting me": Goldstein and Dillon, *The Pearl Harbor Papers*, 130; Agawa, *The Reluctant Admiral*, 309.

171   "Your spiritual strength in overcoming illness": Isoroku Yamamoto, letter to Chiyoko, May 27, 1942, quoted in Goldstein and Dillon, *The Pearl Harbor Papers*, 130.

171   "I hated to loose our firmly held hands": Goldstein and Dillon, *The Pearl Harbor Papers*, 130.

171   "I wish, if I could, to desert everything": Isoroku Yamamoto, letter to Chiyoko, May 27, 1942, quoted in Goldstein and Dillon, *The Pearl Harbor Papers*, 130.

172   he was struggling with his own health issues: Agawa, *The Reluctant Admiral*, 312, 321; Prados, *Combined Fleet Decoded*, 329–30.

172   every part of the Midway plan was in place: Prados, *Combined Fleet Decoded*, 329–30; Haufler, *Codebreakers' Victory*, 150–51; Alistair Horne, *Hubris: The Tragedy of War in the Twentieth Century* (New York: HarperCollins, 2015), 258–59.

172   Nimitz, knew what was coming: Haufler, *Codebreakers' Victory*, 152.

173   "The scene was horrible to behold": Prados, *Combined Fleet Decoded*, 326.

173   "*Kaga, Soryu* and *Akagi* ablaze": Agawa, *The Reluctant Admiral*, 316.

174   looks of "indescribable emptiness" in their eyes: Yeoman Noda from *Yamato*, quoted in Prados, *Combined Fleet Decoded*, 330–31.

174   torpedoing their own ship: Agawa, *The Reluctant Admiral*, 320.

175   "a victory of intelligence": Haufler, *Codebreakers' Victory*, 156; Prados, *Combined Fleet Decoded*, 330–35; Agawa, *The Reluctant Admiral*, 319–22; Horne, *Hubris*, 274–75.

175   the grim truth suppressed: Agawa, *The Reluctant Admiral*, 321–22; Prados, *Combined Fleet Decoded*, 340; Horne, *Hubris*, 274.

175   "Pearl Harbor has now been partially avenged": "Jap Fleet Blasted in Midway Battle," *Boston Globe*, June 5, 1942; "Jap Fleet Smashed by U.S.," *Chicago Sunday Tribune*, June 7, 1942, 1.

176   Yamamoto's move on the Aleutian Islands "was a feint": Stanley Johnston, "Navy Had Word of Jap Plan to Strike at Sea," *Chicago Sunday Tribune*, June 7, 1942.

176   "secrecy of the operations and movements seems to have been leaked out": Isoroku Yamamoto, letter to Chiyoko Kawai, June 21, 1942, quoted in Goldstein and Dillon, *The Pearl Harbor Papers*, 131.

177   the crisis passed: Prados, *Combined Fleet Decoded*, 341–43; Smith, *The Emperor's Code*, 142–43.

177   "We waxed the hell out of them": John Mitchell, letter to Annie Lee, June 13, 1942, MFP.

Chapter 10: Mitchell on the Move

179   a special celebration of America's Independence Day: John Mitchell, letter to Annie Lee, July 7, 1942, MFP; Burke Davis, interviews with Rex Barber, September 29 and October 1, 1967, DA; Davis, interview with Doug Canning, September 24, 1967, DA; questionnaire submitted by A. J. Buck, January 1, 1968, DA.

180   "I'm going to buy a *big* refrigerator and keep nothing but beer in it": John Mitchell, letter to Annie Lee, August 31, 1942, MFP.

180   "scattered yellow meat": John Mitchell, diary entry, July 22, 1942, MFP.

180   "Jap Sea Power Blunted, But Still Mighty": Associated Press, Washington, DC, June 7, 1942, in the *Chicago Daily Tribune* and elsewhere, June 8, 1942.

181   "plunged Allied crypto teams into almost total eclipse": Hervie Haufler, *Codebreakers' Victory: How the Allied Cryptographers Won World War II* (New York: New American Library, 2003), 206.

181   "a chance to learn combat and weaknesses": Burke Davis, interview with Rex Barber, September 29 and October 1, 1967, DA.

181   "we have had absolutely no excitement here": John Mitchell, letters to Annie Lee, June 18 and July 13, 1942, MFP; Mitchell, diary entry, July 22, 1942, MFP.

182   comfortable stopover for a variety of bigwigs: Burke Davis, interview with Henry "Vic" Viccellio, February 12, 1968, DA; Burke Davis, *Get Yamamoto* (New York: Random House, 1969), 78–79.

182   pull together under a single command the various fighter squadrons: Questionnaire submitted by D. C. "Doc" Strother, December 7, 1966, DA; John Mitchell, diary, July 22, 1942, MFP.

183   "I feel as though I am being wasted": John Mitchell, diary, August 15, 1942, MFP.

183   a future in politics: Rex Barber, letter to Carroll V. Glines, January 19, 1989, GA; Burke Davis, correspondence with Rex Barber and Tom Lanphier, DA. In denying Barber's recollection in a July 12, 1969, letter, Lanphier said that achieving political office had never been his ambition. "My ambitions in those days in the Pacific were: to do my fighting job well; to get home alive; to get married promptly thereafter; and, as soon as the war was over, to get out and into a good newspaper job." John Mitchell, diary, August 26 and August 31, MFP, in which Mitchell envied Lanphier's account of having shot down a Zero, writing "What a lucky dog!" However, Lanphier was never given credit for the "kill."

184   "Nothing but a bunch of politicians": John Mitchell, diary, August 10, 1942, MFP.

184   "very shapely pair of pens": John Mitchell, letter to Annie Lee, July 2, 1942, MFP.

185   "The poem with the Varga [*sic*] drawing was excellent": Phil Stack, "Victory for a Soldier," *Esquire*, June 1942; John Mitchell, letter to Annie Lee, July 13,

1942, MFP. Jeanne became known as "The Varga Girl," her image the most reproduced of Varga's pin-up portraits. Some GIs even copied her image onto their aircraft for good luck. Years after the war, in 1975, the *Chicago Tribune* polled World War II veterans about what they most remembered, and a common reply was Glenn Miller's band and the Varga Girl.

185   spent the night at the craps table: John Mitchell, diary, June 23, 1942, MFP.

185   Slaughtering the cows: John Mitchell, letter to Annie Lee, July 7, 1942, MFP; Burke Davis, interviews with Rex Barber, September 29 and October 1, 1967, DA; Davis, interview with Doug Canning, September 24, 1967, DA; questionnaire submitted by A. J. Buck, January 1, 1968, DA.

185   "We carried on all night": Burke Davis, interviews with Rex Barber, September 29 and October 1, 1967, DA; Davis, interview with Doug Canning, September 24, 1967, DA; questionnaire submitted by A. J. Buck, January 1, 1968, DA; John Mitchell, letters to Annie Lee, July 2 and July 7, 1942, MFP.   ·

186   information about the airstrip on Guadalcanal: John Miller, Jr., *Guadalcanal: The First Offensive* (Washington, DC: US Army Center of Military History, 1949), 6–8.

188   the ghoulish massacre: E. B. Sledge, *With the Old Breed: At Peleliu and Okinawa* (New York: Ballantine, 2010), 33–34; John Prados, *Islands of Destiny: The Solomons Campaign and the Eclipse of the Rising Sun* (New York: NAL Caliber, 2012), 90–91; John Wukovits, "The Ill-Fated Goettge Patrol," Warfare History Network, October 16, 2016, https://warfarehistorynetwork.com/daily/the-ill-fated-goettge-patrol.

188   "the first offensive move we have made": John Mitchell, diary, August 10 and August 14, 1942, MFP.

189   the army's scramble to provide aerial support: Questionnaire submitted by D. C. "Doc" Strother, December 7, 1966, DA.

189   the two sides went after each other: Hiroyuki Agawa, *The Reluctant Admiral: Yamamoto and the Imperial Navy* (Tokyo: Kodansha International, 1969), 326.

189   Henderson Field had only a single airstrip: Questionnaire submitted by D.C. "Doc" Strother, December 7, 1966, DA.

189   "Fourteen of us are to go": John Mitchell, diary, August 31, 1942, MFP.

189   "Flight B was the best": Julius "Jack" Jacobson, oral history, National Museum of the Pacific War, Fredericksburg, TX, May 4, 1994.

190   too many pilots for too few planes: Mack Morriss, *South Pacific Diary, 1942–1943* (Lexington: University Press of Kentucky, 1996), 31, 40–43.

190   "The Zeros have it over us": John Mitchell, diary, August 15, 1942, MFP.

190   the P-38: John Mitchell, diary, September 19, 1942, MFP.

190   the Battle of Bloody Ridge: John Hersey, *Into the Valley: A Skirmish of the Marines* (New York: Schocken Books, 1942), xvii.

191   Guadalcanal's plane inventory: Miller, *Guadalcanal*, 148.

191   malaria had erupted: General Millard F. Harmon, letter to Admiral Robert Ghormley, October 6, 1942, quoted in Miller, *Guadalcanal*, 140.

191   "The Japs had beat them": Burke Davis, interview with Doug Canning, September 24, 1967, DA.

192   "super planes and supermen": John Mitchell, letter to Annie Lee, August 19, 1942, MFP.

192   He had led the bomber pilot back: John Mitchell, diary, August 14, 1942, MFP.

192   "penetrating eyes": Questionnaire submitted by D. C. "Doc" Strother, December 7, 1966, DA.

192   "there was no pretense about him": Burke Davis, interview with Jack Jacobson, October 3, 1967, DA.

192   "'I like to be responsible'": George T. Chandler, videotaped interview with John Mitchell, May 18, 1989, MFP.

192   "good fighter pilots never think defensively": John Mitchell, diary entry, September 19, 1942, MFP; Burke Davis, interviews with John Mitchell, September 27 and 28, 1967, DA.

193   "the Jap is capable of retaking Cactus-Ringbolt": General Millard F. Harmon, letter to Admiral Robert Ghormley, October 6, 1942, quoted in Miller, *Guadalcanal*, 140.

193   "every conversation swings around to home": John Hersey, notes from October 18, 1942, wired to his editor at *Time*, Dispatches from *Time* magazine correspondents, first series, 1942–1955, Harvard College Library, Special Collections, Houghton Library.

194   the enemy was all around: Doug Canning, oral history interview, October 4, 2001, National Museum of the Pacific War; Burke Davis, interview with Henry "Vic" Viccellio, February 12, 1968, DA; Robert Cromie, "Picture of War on Guadalcanal by Tribune Man," *Chicago Tribune*, October 28, 1942; Hersey, *Into the Valley*, xvi.

Chapter 11: The First Kill

198   Japanese Americans began arriving "dusty, hot and tired" to the Rohwer War Relocation Center: Digitized autobiographies of teenage internees at Rohwer: Mary Kobayashi and Alyce Okamura, Rising Above, https://risingabove .cast.uark.edu/archive/item/50; Mary Kobayashi, "My Autobiography," December 10, 1942, Rohwer Reconstructed, https://risingabove.cast.uark.edu /archive/item/53.

198   "Over one thousand bushels for sale at low prices!": N. B. Mitchell, "Don't Disgrace the Church,"*Mississippi Sun*, July 30, 1931.

198   "this camp was surrounded by barbed wire": Edward Y. Inouye (a Rohwer detainee), letter to Rosalie S. Gould, May 3, 1991, Rosalie Santine Gould– Mabel Jamison Vogel Collection, Butler Center for Arkansas Studies, Little Rock, AR.

198   "the world seemed to crumble from under my feet": Mary Kobayashi, "My Autobiography," December 10, 1942, Rohwer Reconstructed, https://rising above.cast.uark.edu/archive/item/53.

199   "I have never felt so helpless": Nobuko Hamzawa, "Autobiography," December 1, 1942, Rohwer Reconstructed, https://risingabove.cast.uark.edu /archive/item/58.

199   "I had to forget all that was taught to me about our democratic country": Takeo Shibata, "Autobiography," December 15, 1942, Rohwer Reconstructed, https://risingabove.cast.uark.edu/archive/item/51. No information was available in the archive about the teenagers' lives and whereabouts after their internment at Rohwer.

199   an operational conference: Hiroyuki Agawa, *The Reluctant Admiral: Yamamoto and the Imperial Navy* (Tokyo: Kodansha International, 1969), 331.

200   miscalculation of US strength: John Miller, Jr., *Guadalcanal: The First Offensive* (Washington, DC: US Army Center of Military History, 1949), 135–46.

200   "I fear that I have perhaps one hundred days left": Quoted in Burke Davis, *Get Yamamoto* (New York: Random House, 1969), 71.

200   gallows humor: Agawa, *The Reluctant Admiral*, 331.

201   Flies and mosquitoes: Julius "Jack" Jacobson, oral history, May 4, 1994, National Museum of the Pacific War, Fredericksburg, TX.

201   battle scarring: Burke Davis, interview with John Mitchell, September 27–28, 1967, DA; Davis, interview with Doug Canning, September 24, 1967, DA; Mack Morriss, *South Pacific Diary, 1942–1943* (Lexington: University Press of Kentucky, 1996), 52; *Wings at War Series*, no. 3, *Pacific Counterblow: The 11th Bombardment Group and the 67th Fighter Squadron in the Battle for Guadalcanal, an Interim Report* (Washington, DC: Center for Air Force History, 1992), 28.

201   "the landing force had had to fight on": John Hersey, *Into the Valley: A Skirmish of the Marines* (New York: Schocken Books, 1942), xvi–xvii.

202   "Americans are not invincible": John Hersey, "The Marines on Guadalcanal," *Life*, November 1942, 56; Hersey, *Into the Valley*, xxii.

202   the degree to which the marines despised their Asian enemy: Hersey, *Into the Valley*, 47.

204   his first kill: George T. Chandler, videotaped interview with John Mitchell, May 18, 1989, MFP; Captain John Mitchell, 339th Fighter Squadron, Combat Report, December 6, 1942, MFP.

204   did their best to get settled in: Burke Davis, interview with Henry Viccellio, February 12, 1968, DA; Davis, interview with Rex Barber, September 29 and October 1, 1967, DA; Doug Canning, oral history interview, October 4, 2001, National Museum of the Pacific War.

204   the necessity of the foxholes: Burke Davis, interview with John Mitchell, September 27–28, 1967, DA; Doug Canning, oral history interview, October 4, 2001, National Museum of the Pacific War; report that "Pistol Pete" began on October 12, 1942, in *Wings at War Series*, no. 3, *Pacific Counterblow*, 37.

204   "Washing Machine Charlie": George T. Chandler, videotaped interview with John Mitchell, May 18, 1989, MFP.

205   "you'd get up and run for a foxhole": Ibid.

205   exposing them to the malaria-carrying insects: Burke Davis, interview with John Mitchell, September 27–28, 1967, DA.

205   the noisy guns never succeeded: George T. Chandler, videotaped interview with John Mitchell, May 18, 1989, MFP; Burke Davis, interviews with Rex Barber, September 29 and October 1, 1967, DA.

205   "I just moved my cot down and slept in the fox hole": Doug Canning, oral history interview, October 4, 2001, National Museum of the Pacific War.

206   "He's a flying machine": George T. Chandler, videotaped interview with John Mitchell, May 18, 1989, MFP.

206   "the Guadalcanal, or 1,000-yard stare": Mack Morriss, diary, December 2, 1942, in *South Pacific Diary: 1942–1943*, 52.

206 The disposal of the dead: Robert Cromie, "Tribune Writer Tells of 6 Day Solomon Battle," *Chicago Tribune*, November 10, 1942.

206 the Japanese bodies stacked up: Burke Davis, interview with John Mitchell, September 27–28, 1967, DA.

207 It was kill or be killed: George T. Chandler, videotaped interviews with John Mitchell and Rex Barber, May 18, 1989, MFP.

Chapter 12: Nights to Remember

210 Three Nights to Remember: John Miller, Jr., *Guadalcanal: The First Offensive* (Washington, DC: US Army Center of Military History, 1949), 141–56; *Wings at War Series*, no. 3, *Pacific Counterblow: The 11th Bombardment Group and the 67th Fighter Squadron in the Battle for Guadalcanal, an Interim Report* (Washington, DC: Center for Air Force History, 1992), 36–40.

211 threw themselves into foxholes: Burke Davis, interviews with John Mitchell, September 27–28, 1967, 13, DA.

211 the men emerged to find shells from the battleship guns: Ibid.

211 "We don't know whether we'll be able to hold the field or not": *Wings at War Series*, no. 3, *Pacific Counterblow*, 34.

211 continued bombardment: Miller, *Guadalcanal*, 151; *Wings at War Series*, no. 3, *Pacific Counterblow*, 37.

212 aerial assets were used as best as possible: Miller, *Guadalcanal*, 150.

212 "The Seabees were marvelous": Doug Canning, oral history interview, October 4, 2001, National Museum of the Pacific War; Burke Davis, interview with Canning, DA.

212 pitching in to get the fighters into the air: Miller, *Guadalcanal*, 150; *Wings at War Series*, no. 3, *Pacific Counterblow*, 40, 67.

212 army fliers dispatched to disrupt the enemy's unloading: Mitchell, Combat Report, December 6, 1942; Burke Davis, interviews with John Mitchell, September 27–28, 1967, DA; Miller, *Guadalcanal*, 151.

213 their effort was a nuisance at best: Miller, *Guadalcanal*, 151–55.

213 three-pronged assault: Ibid.

214 Admiral William F. "Bull" Halsey, Jr.: John Wukovits, "'Dear Admiral Halsey,'" *Naval History Magazine* 30, no. 2 (April 2016).

214 morale jump on Guadalcanal: Samuel Eliot Morison, *History of United States Naval Operations in World War II*, vol. 5: *The Struggle for Guadalcanal, August 1942–February 1943* (Boston: Little, Brown, 1958), 183.

214 "to keep down some of the stink": John Mitchell, letter to Annie Lee, October 29, 1942, MFP.

215 "going to town with the Japs": John Mitchell, letter to Annie Lee, October 22, 1942, MFP.

215 "he hopes our boys will win": "Land and Naval Forces in Lead Role at Solomons: Arrival of More Men and Artillery for Foe Tends to Change Situation on Guadalcanal," *Los Angeles Times*, October 17, 1942.

215 "Johnny is in the Solomons": Annie Lee, letter to Aunt Ludma, October 18, 1922, MFP.

215 "We're going to win": Wukovits, "'Dear Admiral Halsey.'"

216 the enemy death toll exceeded 1,500: Burke Davis, *Get Yamamoto* (New York:

Random House, 1969), 86; Miller, *Guadalcanal*, 161–66; *Wings at War Series*, no. 3, *Pacific Counterblow*, 42, 67.

216 The fliers pitched in from the sky: Mitchell, Combat Report, December 6, 1942, MFP.

217 "The Marines finished the job": Ibid.

217 irked when the soldiers wanted to peddle the combat mementos for money: Burke Davis, interviews with John Mitchell, September 27–28, 1967, DA.

217 "Helmet headbands were checked for flags": E. B. Sledge, *With the Old Breed: At Peleliu and Okinawa* (New York: Ballantine, 2010), 118.

217 infantrymen took to wearing necklaces made of the teeth of Japanese soldiers: Ian Toll, *The Conquering Tide: War in the Pacific, 1941–1942* (New York: W. W. Norton, 2015), 196.

218 "There are all sorts of things to be learned": John Mitchell, letter to Annie Lee, November 2, 1942, MFP.

218 "Very good for strafing": Mitchell, Combat Report, December 6, 1942, MFP.

218 "you make your own luck": George T. Chandler, videotaped interviews with John Mitchell and Rex Barber, May 18, 1989, MFP.

218 "I really like it out here": John Mitchell, letter to Annie Lee, November 2, 1942, MFP.

218 "Dugout Sunday": Robert Cromie, "Guadalcanal's Worst Day: Rain of Bomb and Shells," *Chicago Daily Tribune*, November 11, 1942.

219 "Once you commit to a dive": Julius "Jack" Jacobson, oral history, May 4, 1994, National Museum of the Pacific War, Fredericksburg, TX.

219 Jacobson landed a bomb directly on the bow: Mitchell, Combat Report, December 6, 1942, MFP.

219 they attacked the Zeros: Ibid.

219 The Japanese losses: Miller, *Guadalcanal*, 164.

220 "Japs in all the curious postures of death lay awaiting burial": Robert Cromie, "Tribune Writer Tells of 6 Day Solomon Battle," *Chicago Daily Tribune*, November 10, 1942.

220 Japanese commanders decided to call off further attacks: Miller, *Guadalcanal*, 168–69; *Wings at War Series*, no. 3, *Pacific Counterblow*, 44–45.

221 Dinn had to bail out: Mitchell, Combat Report, December 6, 1942, MFP.

221 Dinn had made it back: Lieutenant Wallace S. Dinn, "My Solomons Canoe Trip," *Air Force*, March 1943, 23–24; John Mitchell, pilot diary, October 28, 1942, MFP; Mitchell, Combat Report, December 6, 1942, MFP.

221 "They say we are due for some rest": John Mitchell, letter to Annie Lee, November 2, 1942, MFP.

222 Mitch ended up being gone a month: John Mitchell, Combat Report, December 6, 1942, MFP; Burke Davis, interviews with Mitchell, September 27–28, 1967, DA; Mack Morriss, *South Pacific Diary, 1942–1943* (Lexington: University Press of Kentucky, 1996), 31 ff.

222 "Their uniforms, little more than rags, hung from emaciated limbs": Hervie Haufler, *Codebreakers' Victory: How the Allied Cryptographers Won World War II* (New York: New American Library, 2003), 208–09.

222 "Starvation Island": Toll, *The Conquering Tide*, 174.

223 "If they couldn't say anything better just keep quiet": Annie Lee, letter to Aunt Ludma, December 6, 1942, MFP.

223 WEEK LEAVE EVERYTHING FINE: John Mitchell, cable to Annie Lee, November 24, 1942, MFP.

223 "I felt much better": Annie Lee, letter to Aunt Ludma, December 6, 1942, MFP.

223 "itching to get going again": John Mitchell, letter to Annie Lee, December 21, 1942, MFP.

223 the "drum method": Toll, *The Conquering Tide*, 178.

223 "I have been going from 3:30 a.m. until about 10 p.m.": John Mitchell, letter to Annie Lee, December 10, 1942, MFP.

224 Yamamoto, for one, was clear-eyed: Hiroyuki Agawa, *The Reluctant Admiral: Yamamoto and the Imperial Navy* (Tokyo: Kodansha International, 1969), 334–35.

224 "definitely passing to the offensive": "Halsey Sees 'Absolute Defeat' for Axis Forces This Year: 'We're Just Starting,'" *Los Angeles Times*, January 3, 1943.

224 "Buy War Bonds": *Washington Post*, December 7, 1942.

225 "It's going to be fun to see ole Vic": John Mitchell, letter to Annie Lee, December 21, 1942, MFP.

225 Reunited were pilots: Eric Hammel, *Aces Against Japan II: The American Aces Speak*, vol. 3 (Pacifica, CA: Pacifica Press, 1996), 106.

225 "eat, sleep, fly": Questionnaire submitted by A. J. Buck, January 21, 1968, DA; "Narrative History of the 339th Fighter Squadron Two Engine from Activation Until December 31, 1943," archives of the Army Air Forces Historical Office, declassified September 1958, 5.

225 "I was always eager to get into combat": George T. Chandler, videotaped interviews with John Mitchell and Rex Barber, May 18, 1989, MFP.

225 Barber got his chance: Burke Davis, interviews with Rex Barber, September 29 and October 1, 1967, DA; William Hess, *Ace Profile* (American Fighter Pilot Series), vol. 1, no. 2: *Col. Rex T. Barber* (Tucson, AZ: Mustang International Publishers, 1993).

226 He had gotten his first kill: Thomas Lanphier, Jr., "At All Costs Reach and Destroy," unpublished manuscript, circa 1984–85, 143–48.

226 "One thing mattered to Mitch": Burke Davis, interviews with Rex Barber, September 29 and October 1, 1967, DA.

226 "Just like on a football field making a tackle": George T. Chandler, videotaped interview with Mitchell and Rex Barber, May 18, 1989, MFP.

227 "Selfish motives did begin to appear": Questionnaire submitted by A. J. Buck, January 21, 1968, DA.

227 Christmas 1942: Robert Cromie, "Tribune Writer Flies over Japs' Solomon Bases," *Chicago Tribune*, January 10, 1943; John G. Norris, "Yanks on Guadalcanal to Feast at Christmas Dinner Today," *Washington Post*, December 24, 1942.

227 arrival of the much-hyped Lockheed P-38 Lightning: "Narrative History of the 339th Fighter Squadron Two Engine from Activation until December 31, 1943," 4.

228 "The transition to the P-38 was a chore": John P. Condon, US Air Forces Oral History Project interview, March 6, 1989.

228 concern about the P-38s was soon "in a total turnaround": Ibid.

228 Right off he loved the P-38: Burke Davis, interviews with John Mitchell, September 27–28, 1967, DA; Burke Davis, *Get Yamamoto*, 88.

229 "Let me go up in a P-38": George T. Chandler, videotaped interview with John Mitchell and Rex Barber, May 18, 1989, MFP; Burke Davis, interviews with Mitchell, September 27–28, 1967, DA.

229 "We'd wait for them": Burke Davis, interviews with John Mitchell, September 27–28, 1967, DA.

229 they learned to box in a Zero: Ibid.

229 The P-38s were a lot tougher: Ibid.

230 Dinn was blown out of the sky: Ibid.

230 "I lost the very best friend I have": John Mitchell, letter to Annie Lee, January 7, 1943, MFP.

231 "I burned him. And he crashed": George T. Chandler, videotaped interview with John Mitchell and Rex Barber, May 18, 1989, MFP; Burke Davis, interviews with Mitchell, September 27–28, 1967, DA; Morriss, *South Pacific Diary*, 76. Note: Washing Machine Charlie was Mitchell's seventh "kill." He had made his fifth and sixth kills two days earlier, January 27, when he had led a flight of six P-38s in combat over Guadalcanal. Though outnumbered by an estimated thirty Zeros, Mitchell accounted for downing two of the enemy planes. In his later interview with Burke Davis, Mitchell said that after he had shot down "Charlie," another Japanese pilot had taken over the "irritating duty," appearing less frequently but still an annoyance. That "Charlie" was shot down by another pilot, as was yet another replacement, and by winter's end, said Mitchell, "that was the end of the Charlies."

Chapter 13: Moon over Guadalcanal

233 "The coming year will be filled with violent conflicts": Franklin D. Roosevelt, State of the Union address, January 7, 1943, https://millercenter.org/the-presidency/presidential-speeches/january-7-1943-state-union-address; Robert De Vore, "Roosevelt Sees Allies Nearing Victory in '43," *Washington Post*, January 8, 1943.

234 "I am distressed": Isoroku Yamamoto, letter to Koga Mineichi, January 6, 1943, and letter to Niwa Michi, February 1943, quoted in Hiroyuki Agawa, *The Reluctant Admiral: Yamamoto and the Imperial Navy* (Tokyo: Kodansha International, 1969), 335 and 336, respectively.

234 Guadalcanal was a lost cause: Agawa, *The Reluctant Admiral*, 342; Ian Toll, *The Conquering Tide: War in the Pacific, 1941–1942* (New York: W. W. Norton, 2015), 186–87; John Miller, Jr., *Guadalcanal: The First Offensive* (Washington, DC: US Army Center of Military History, 1949), 350.

235 Operation KE: Toll, *The Conquering Tide*, 182–86; Agawa, *The Reluctant Admiral*, 338–39; Hervie Haufler, *Codebreakers' Victory: How the Allied Cryptographers Won World War II* (New York: New American Library, 2003), 209; Miller, *Guadalcanal*, 337.

235 Japan's otherwise humiliating failure at Guadalcanal: Burke Davis, *Get Yamamoto* (New York: Random House, 1969), 72; Agawa, *The Reluctant Admiral*, 338.

236 "Guadalcanal is now ours": "Guadalcanal," *Washington Post*, February 11, 1943.

236 "People are going around with grins on their faces": Mack Morriss, *South Pacific Diary, 1942–1943* (Lexington: University Press of Kentucky, 1996), 91.

236 Yamamoto officially moved: Toll, *The Conquering Tide*, 201; Agawa, *The Reluctant Admiral*, 338.

237 His physical well-being was becoming an added concern: Agawa, *The Reluctant Admiral*, 338; Toll, *The Conquering Tide*, 201–02.

237 The commander in chief seemed depressed: Agawa, *The Reluctant Admiral*, 336–37.

238 how best to go after New Guinea: Edwin Hoyt, *Yamamoto: The Man Who Planned the Attack on Pearl Harbor* (Guilford, CT: Lyons Press), 240–41; Haufler, *Codebreakers' Victory*, 214; Toll, *The Conquering Tide*, 202–03; Agawa, *The Reluctant Admiral*, 339.

239 "I will go in high spirits since I have heard about you": Isoroku Yamamoto, letter to Chiyoko Kawai, April 2, 1943, quoted in Donald M. Goldstein and Katherine V. Dillon, *The Pearl Harbor Papers: Inside the Japanese Plans* (New York: Brassey's, 1993), 131; Agawa, *The Reluctant Admiral*, 340–42.

239 operational structure reorganized: Victor Dykes, "Air Command Solomon Islands," diaries, memoirs, articles, declassified materials from the Army Air Forces Historical Archives, Maxwell Air Force Base, Alabama; Burke Davis, interviews with Rex Barber, September 29 and October 1, 1967, during which Barber drew a map of Fighter Two, DA.

241 Marc "Pete" Mitscher: Davis, *Get Yamamoto*, 109–12; Admiral Marc A. Mitscher, USNR, "Biographies in Naval History," Naval History & Heritage Command.

241 "Boy, am I gonna have a sore arm from patting myself on the back!": John Mitchell, letter to Annie Lee, February 12, 1943, MFP.

241 Distinguished Service Cross: General Orders Number 56, Headquarters US-AFISPAC, APO 502, March 9, 1943, MFP.

241 "you get a general idea of how it looks": John Mitchell, letter to Annie Lee, March 31, 1943, MFP.

242 "Guess it's going to seem funny to you seeing me as a major": John Mitchell, letter to Annie Lee, February 12, 1943, MFP.

242 an awful case of jungle rot: Burke Davis, interview with Besby Holmes, February 14, 1968, DA.

242 pleased with their handling of the quirks: Tom Lanphier, unpublished autobiography, 147, DA; Burke Davis, interview with Lanphier, January 25, 1968, DA.

242 "as outstanding as any combat airmen I ever saw anywhere": Burke Davis, interview with Henry Viccellio, February 12, 1968, DA.

242 "There's a woman aboard!": United Press International, "Surprised Our Men on Guadalcanal," *New York Times*, March 3, 1943.

243 "I hope to someday fix them up with water works": John Mitchell, letter to Annie Lee, January 22, 1943, MFP.

244 "I see Veronica has what it takes": John Mitchell, letter to Annie Lee, February 18, 1943, MFP.

244 the bright full moons over Guadalcanal: Ibid.

244 the impossible tug of love and war: John Mitchell, letter to Annie Lee, November 2, 1942, MFP.

244   "You keep asking about my return": John Mitchell, letters to Annie Lee, September 19, October 22, and October 29, 1942, MFP.

244   "he should be home in January!": Annie Lee, letter to Aunt Ludma, December 6, 1942, MFP.

245   "You seem to be living with only one thought in mind": John Mitchell, letters to Annie Lee, January 7, January 22, and February 12, 1943, MFP.

246   more balanced mix of flying and frolicking: John Mitchell, letter to Annie Lee, March 12, 1943, MFP; "Narrative History of the 339th Fighter Squadron Two Engine from Activation until December 31, 1943," archives of the Army Air Forces Historical Office, declassified September 1958, 7.

246   "He Wears a Pair of Silver Wings," by Kay Kyser: The rest of the song's lyrics are as follows:

And though it's pretty tough, the job he does above
I wouldn't have him change it for a king
An ordinary fellow in a uniform I love
He wears a pair of silver wings
Why, I'm so full of pride when we go walking
Every time he's home on leave
He with those wings on his tunic
And me with my heart on my sleeve
But when I'm left alone and we are far apart
I sometimes wonder what tomorrow brings
For I adore that crazy guy who taught my happy heart
To wear a pair of silver wings
For I adore that crazy guy who taught my happy heart
To wear a pair of silver wings.

246   "We played it till it wore out": Burke Davis, interviews with Rex Barber, September 29 and October 1, 1967, DA.

246   they adapted the old folksy tune "Put on Your Old Grey Bonnet": Morriss, *South Pacific Diary*, 148.

247   "I wish every person back home could walk between the crosses": Robert Cromie, "Marines Obtain Peace in a New Flanders Field," *Chicago Tribune*, January 15, 1943; Morriss, *South Pacific Diary*, 146. The cemetery was also called Lunga Point Cemetery and the American Cemetery Guadalcanal. When World War II ended, the graves were exhumed and the remains were transported either back to the United States or to the Manila American Cemetery, National Memorial Cemetery of the Pacific, according to https://www.pacificwrecks .com/cemetery/solomons-lunga-cemetery.html. The Guadalcanal cemetery was then abandoned.

248   the men "sampled the entire menu of delights available": Tom Lanphier, unpublished autobiography, 164, DA.

248   "We got better as they got poorer": Burke Davis, interviews with John Mitchell, September 27–28, 1967, DA.

248   "the fork-tailed devil": Adonis C. Arvanitakis, "Killing a Peacock: A Case Study of the Targeted Killing of Admiral Isoroku Yamamoto," master's thesis, School of Advanced Military Studies, United States Army Command and General Staff College, Fort Leavenworth, KS, 2015, 24.

248 tasked with attacking thirty enemy flying boats and Zeros: George T. Chandler, videotaped interviews with John Mitchell and Rex Barber, May 18, 1989, MFP; Tom Lanphier, unpublished autobiography, 166–74, DA; account by Major General Robert L. Petit, USAF, https://www.pacificwrecks.com/airfields/solomons/shortland/mission-3-29-43.html.

249 The raid was cast as practically Lanphier's alone: "Capt. Lanphier Cited in Fight on Jap Vessel," *Washington Post*, May 13, 1943; Jim R. Shepley, "Lanphier," *Time* correspondent memorandum, May 21, 1943, Dispatches from *Time* magazine correspondents: first series, 1942–1955, Harvard College Library, Special Collections, Houghton Library, 1955.

250 Operation I-Go: Hoyt, *Yamamoto*, 242.

250 "the personification of the Navy": Masatake Okumiya, Jiro Horikoshi, and Martin Caidin, *Zero!* (New York: E. P. Dutton, 1956), 242–44.

250 "Condition is red": Morriss, *South Pacific Diary*, 131.

251 seventy-six fighters from the three services: Davis, *Get Yamamoto*, 114–15.

251 the United States' losses were far less than Japan's: Toll, *The Conquering Tide*, 202–03; Hoyt, *Yamamoto*, 242–45; Agawa, *The Reluctant Admiral*, 344–45. The army air force units flying P-38s lost two of their own during combat; on April 6: First Lieutenant George G. Topoll, a pilot who'd flown with Lanphier and Barber on the memorable March 27 mission; on April 7: Major Waldon Williams.

251 Yamamoto decided to head south: John Toland, *The Rising Sun: The Decline and Fall of the Japanese Empire, 1936–1945* (New York: Random House, 1970), 553; Toll, *The Conquering Tide*, 203; Agawa, *The Reluctant Admiral*, 345–46; Hoyt, *Yamamoto*, 243–45.

Chapter 14: Five Days and Counting

255 The interception: Ian Toll, *The Conquering Tide: War in the Pacific, 1941–1942* (New York: W. W. Norton, 2015), 203; Burke Davis, *Get Yamamoto* (New York: Random House, 1969), 9–12; Edwin Hoyt, *Yamamoto: The Man Who Planned the Attack on Pearl Harbor* (Guilford, CT: Lyons Press), 248; John Prados, *Combined Fleet Decoded: The Secret History of American Intelligence and the Japanese Navy in World War II* (New York: Random House, 1995), 459.

256 "a waste of talent": Red Lasswell, oral history, 32–33, Oral History Project, Historical Division, Headquarters, U.S. Marine Corps, Washington, D.C., 1968.

256 John Paul Stevens: In later interviews with biographer Bill Barnhart (2010), the *New Yorker*'s Jeffrey Toobin (2010), and author and journalist Mitchell Zuckoff (2012), Stevens discussed the Yamamoto killing. Stevens was apparently not on duty in the Dungeon the night Red Lasswell and others worked on the message outlining Yamamoto's April 18, 1943, itinerary, but he was working as a traffic analyst on April 18, two days before he turned twenty-three, when a US message was received reporting the killing of the Japanese admiral. "I happened to be on duty when the message came in advising that our pilots had bagged a peacock [Yamamoto] and two sparrows," Stevens told Barnhart. The news gave the future Supreme Court justice pause—then and afterward. As did Lasswell, Stevens wondered about the ethics of killing an enemy leader

in wartime. In the end, he decided it was justified, but that did not mean dis-
regarding the inherent moral dilemma. "I had mixed feelings at the time," he
continued to Barnhart, "because, on the one hand, it was an important and
successful operation but, on the other hand, it was a deliberate elimination of
a specific individual rather than a nameless enemy."

257   the United States continued its clear advantage in gathering intelligence: Her-
      vie Haufler, *Codebreakers' Victory: How the Allied Cryptographers Won World War
      II* (New York: New American Library, 2003), 206–07.

257   "Lasswell approached cryptanalysis like a chess player": Wilber Jasper
      Holmes, *Double-Edged Secrets: U.S. Naval Intelligence Operations in the Pacific
      During World War II* (Annapolis, MD: Naval Institute Press, 1979), 64.

258   "Southeast Area Fleet/Top Secret": Davis, *Get Yamamoto*, 6–7; David Kahn,
      *The Codebreakers: The Comprehensive History of Secret Communications from An-
      cient Times to the Internet* (New York: Scribner, 1996), 598; John T. Wible, *The
      Yamamoto Mission: Sunday, April 18, 1943* (Fredericksburg, TX: Admiral Nimitz
      Foundation, 1988), 7–8. The message also included a description of the uni-
      form of the day to be worn during Yamamoto's visit and said that if there was
      bad weather the tour would be postponed one day. For his book, Kahn was
      unable to interview Red Lasswell; he'd written to Lasswell, but Lasswell had
      never replied. It meant that more years went by before Lasswell's key role in
      both the Midway and the Yamamoto messages emerged. Lasswell said he'd
      sought clearance on occasion but never got it. Moreover, he was always re-
      luctant to discuss intelligence matters. In his oral history, Lasswell strongly
      criticized Kahn's version of the Yamamoto translation for its failure to note his
      role. "I personally did the whole thing overnight," he said.

258   "We've hit the jackpot!": Prados, *Combined Fleet Decoded*, 459; Smith, 183; Toll,
      203; Davis, *Get Yamamoto*, 8–9.

259   "I got greater satisfaction out of that than any other thing": Red Lasswell, oral
      history, 38–39, Oral History Project, Historical Division, Headquarters, U.S.
      Marine Corps, Washington, D.C., 1968.

259   "leadership decapitation": Adam Leong Kok Wey, "Case Study of Operation
      Flipper," in *Killing the Enemy: Assassination Operations During World War II*
      (London: I. B. Taurus, 2015), 131–58.

259   "glad to have it out of our hands": Holmes, *Double-Edged Secrets*, 136; Edwin T.
      Layton, *And I Was There—Pearl Harbor and Midway—Breaking the Secrets* (New
      York: William Morrow, 1985), 473–75.

259   full confidence in Nimitz: Red Lasswell, oral history, 34–35, Oral History
      Project, Historical Division, Headquarters, U.S. Marine Corps, Washington,
      D.C., 1968.

260   "I will have to find a nice quiet spot": John Mitchell, letter to Annie Lee, April
      3, 1943, MFP.

261   VERY NEAT USE OF HEAT: Copy of April 11, 1943 "secret" dispatch from Mitscher
      to various squadron commanders, including Mitchell, passing along Halsey's
      rhymed congratulatory comment, MFP.

261   the "top crust of fighting airmen": Davis, *Get Yamamoto*, 113; Rex Barber was
      also cited for his use of belly tanks as bombs in various unit histories; see,
      e.g., 70th Fighter Squadron, "Historical Record of Organization," July 1, 1943,

covering the period January 1, 1943, to June 30, 1943, 5, n.: "Capt.am Lanphier and Lt. Barber originated the use of the belly tank as a bomb." See also Jim Shepley, "Lanphier," memorandum of May 21, 1943, Dispatches from *Time* magazine correspondents: first series, 1942–1944, Harvard College Library, Special Collections, Houghton Library; "Fliers Resourceful in Firing Enemy Ship," *New York Times*, July 23, 1943.

262 "you backing me makes it 100%": John Mitchell, letter to Annie Lee, April 3, 1943, MFP.

263 "Do we try to get him?": Burke Davis, interview with Edward T. Layton, April 1968, DA; Davis, *Get Yamamoto*, 3–7; Toll, *The Conquering Tide*, 203–04.

263 plotting to kill a specific enemy leader was highly unusual: Adonis C. Arvanitakis, "Killing a Peacock: A Case Study of the Targeted Killing of Admiral Isoroku Yamamoto," master's thesis, School of Advanced Military Studies, United States Army Command and General Staff College, Fort Leavenworth, KS, 2015, 4.

264 discussion about rules of war: Daniel Haulman, *Killing Yamamoto: The American Raid That Avenged Pearl Harbor* (Montgomery, AL: New South Books, 2015), 9.

264 operational authority: Toll, *The Conquering Tide*, 202–04; Prados, *Combined Fleet Decoded*, 460–61; Davis, *Get Yamamoto*, 7–9; Wible, *The Yamamoto Mission*, 9–10. See also Joseph E. Persico, *Roosevelt's Secret War: FDR and World War II Espionage* (New York: Random House, 2001), 240: "It has been speculated that Nimitz sought FDR's permission before ordering the lethal strike. No documents exist, however, establishing that Roosevelt's approval was sought, though the decision did go up to Navy Secretary Frank Knox. Whether FDR blessed the mission or learned of it after the fact . . ."

264 a range of what-ifs: Davis, *Get Yamamoto*, 7–9.

264 "direct annihilation" of enemy forces was paramount: Edwin Layton, letter to Burke Davis, April 4, 1968, DA.

265 Keeping the source of the intelligence secret was essential: Davis, *Get Yamamoto*, 7–9; Kahn, *The Codebreakers*, 599.

265 considered the mission logistics: Davis, *Get Yamamoto*, 7; Haulman, *Killing Yamamoto*, 9; Toll, *The Conquering Tide*, 203.

266 "We'll try it": Wey, *Killing the Enemy*, 29–30.

266 "Let's leave the details to Halsey": Davis, *Get Yamamoto*, 9.

266 "the information comes from Australian Coastwatchers": Layton, *And I Was There*, 475; Davis, *Get Yamamoto*, 13.

266 Yamamoto crashed a party: Hiroyuki Agawa, *The Reluctant Admiral: Yamamoto and the Imperial Navy* (Tokyo: Kodansha International, 1969), 344–45; Davis, *Get Yamamoto*, 103; Donald Davis, *Lightning Strike: The Secret Mission to Kill Admiral Yamamoto and Avenge Pearl Harbor* (New York: St. Martin's Press, 2005), 221.

267 back at it, monitoring planes: Davis, *Get Yamamoto*, 104.

267 He held a series of meetings: Masatake Okumiya, Jiro Horikoshi, and Martin Caidin, *Zero!* (New York: E. P. Dutton, 1956), 245.

267 Yamamoto had to know: Donald Davis, *Lightning Strike*, 231; Burke Davis, *Get Yamamoto*, 104–6.

268   distressed upon learning that the itinerary had been sent by radio: Agawa, *The Reluctant Admiral*, 346–47; Matome Ugaki, *Fading Victory: The Diary of Admiral Matome Ugaki, 1941–1945* (Pittsburgh: University of Pittsburgh Press, 1991), 328–60; Burke Davis, *Get Yamamoto*, 105; Donald Davis, *Lightning Strike*, 222, 234.

268   gave a heads-up to the freshly installed commander: Interview with USN Vice Admiral William A. Read, *The Pacific War Remembered*, ed. John T. Mason, Jr. (Annapolis, MD: Naval Institute Press, 1986), 160–67.

269   "I hate those yellow bastards": Quoted in Burke Davis, *Get Yamamoto*, 112.

269   "the Peacock will be on time": Quoted in ibid., 20.

269   P-38 Lightnings were the only fighter planes suitable: Ibid., 118; Haulman, *Killing Yamamoto*, 10; Prados, *Combined Fleet Decoded*, 460; Donald Davis, *Lightning Strike*, 232; E. B. Potter, *Bull Halsey* (Annapolis, MD: Naval Institute Press, 1985), 214.

270   Mitch was the one: Burke Davis, interview with Henry Viccellio, February 12, 1968, DA; Burke Davis, *Get Yamamoto*, 160.

Chapter 15: One Day More

272   "I shouldn't even mention this": John Mitchell, letter to Annie Lee, April 16, 1943, MFP.

272   "Some mission or other": Burke Davis, *Get Yamamoto* (New York: Random House, 1969), 121; Burke Davis, interview with Henry Viccellio, February 12, 1968, DA.

273   COMAIRSOLS: Burke Davis, correspondence with Mitscher's widow, Frances Mitscher, 1968, DA, along with Davis's research notes about Mitscher; *Life*, October 23, 1944, profile photograph of Mitscher, 27.

273   the "present plight": Matome Ugaki, *Fading Victory: The Diary of Admiral Matome Ugaki, 1941–1945* (Pittsburgh: University of Pittsburgh Press, 1991), 328.

274   "policies and principles": Ibid., 329; Burke Davis, *Get Yamamoto*, 105; Donald Davis, *Lightning Strike: The Secret Mission to Kill Admiral Yamamoto and Avenge Pearl Harbor* (New York: St. Martin's Press, 2005), 233.

274   what to wear on the base inspection: Ugaki, *Fading Victory*, 352–53.

275   The revised itinerary: Jay E. Hines, interview with Betty pilot Horishi Hayashi, June 22, 1990, in Japan, in R. Cargill Hall, ed., *Lightning over Bougainville: The Yamamoto Mission Reconsidered* (Washington, DC: Smithsonian Institution Press, 1991), 148; Hall, interview with Zero pilot Kenji Yanagiya, April 15, 1988, in Fredericksburg, TX, in *Lightning over Bougainville*, 110–11, during which Yanagiya said there had been no question in his mind that the first stop was to be Buin. Note: Some have said that the code breakers must have made a mistake in the decryption by saying the first stop was Ballale, when it emerged later from pilot interviews the first stop was Buin. The code breakers knew the destination symbols, and it seems unlikely they would have mistaken the first stop. The more likely explanation, as implied in the Hayashi interview, is that the itinerary was changed slightly between the time of the decryption and the trip.

275   Mitch had been asking to lead a fighter sweep: John Mitchell, letter to Carroll V. Glines, April 16, 1989, GA.

276   a quick, one-day trip: Ugaki, *Fading Victory*, 331; John T. Wible, *The Yamamoto Mission: Sunday, April 18, 1943* (Fredericksburg, TX: Admiral Nimitz Foundation, 1988), 15; Hiroyuki Agawa, *The Reluctant Admiral: Yamamoto and the Imperial Navy* (Tokyo: Kodansha International, 1969), 347; Burke Davis, *Get Yamamoto*, 104–05, including footnote, 105; Donald Davis, *Lightning Strike*, 234–35, 239–40.

276   gaggle of high-ranking officers: George T. Chandler, videotaped interview with John Mitchell, May 18, 1989, MFP; Mitchell, videotaped comments as panelist at the Yamamoto Mission Retrospective, Admiral Nimitz Museum, Fredericksburg, TX, April 1988; Burke Davis, interview with John Mitchell, September 1967, DA; Burke Davis, question/answer with William A. Read, April 29, 1968, DA; Read, oral history, 1964, in John T. Mason, Jr., ed., *The Pacific War Remembered: An Oral History Collection* (Annapolis, MD: Naval Institute Press, 1986); John P. Condon, US Air Forces Oral History Project interview, March 6, 1969; Condon, letter to Tom Lanphier, December 5, 1984, GA; Daniel Haulman, *Killing Yamamoto: The American Raid That Avenged Pearl Harbor* (Montgomery, AL: New South Books, 2015), 10; Burke Davis, *Get Yamamoto*, 115–18.

276   the whereabouts of Admiral Yamamoto had been intercepted: George T. Chandler, videotaped interview with John Mitchell, May 18, 1989, MFP; Mitchell, letter to John Wible, May 12, 1961, GA; Mitchell, letter to Carroll V. Glines, November 17, 1988, GA.

277   were to make every effort to take out Japan's naval genius: Theodore Taylor, *The Magnificent Mitscher* (Annapolis, MD: Naval Institute Press, 1954), 151; Burke Davis, question/answer with William A. Read, April 29, 1968, DA, in which Read said that Mitscher "made it clear he wanted results, even if they had to ram Admiral Yamamoto's plane." There has also been considerable disagreement about whether Navy Secretary Frank Knox and President Roosevelt authorized the mission. Lanphier and Condon took such a position, saying they saw paperwork with Knox's name on it. But no document has ever surfaced confirming authorization beyond Nimitz's. For his part, Mitchell told Burke Davis that he had never seen FDR's name on any paperwork but thought he had seen Knox's on the order when he was first handed the message at Mitscher's dugout. But he also said he was not certain about this. There is also considerable disagreement about whether during the plenary sessions Mitchell and his pilots were given clear orders for keeping secret how the United States had learned about Yamamoto's trip. Mitchell certainly came away knowing to keep secret that Yamamoto was their target, but he did not recall being briefed prior to the mission about the cover story Nimitz and Layton had devised to protect their code-breaking prowess—that coastal watchers had spotted Yamamoto's flight and that was what had led to an interception.

277   "there's always a bunch of big mouths": John Mitchell, videotaped comments as panelist at Yamamoto Mission Retrospective, Admiral Nimitz Museum, Fredericksburg, TX, April 1988; George T. Chandler, videotaped interview with Mitchell, May 18, 1989, MFP; Burke Davis, interview with Mitchell, September 1967, DA; John Mitchell, letter to John Wible, May 12, 1961, GA;

John T. Wible, *The Yamamoto Mission: Sunday, April 18, 1943* (Fredericksburg, TX: Admiral Nimitz Foundation, 1988), 11–12.

278 how impressed he'd been with the recent flying feats of Tom Lanphier's flight: Note: Years later, among the many aspects of the mission that Mitchell disagreed with Lanphier about, Lanphier wrote that at the initial briefing Mitscher had ordered Mitchell to make Lanphier one of the shooters, in the so-called killer unit. Mitchell was outraged and told Lanphier so in a July 9, 1984, letter to Lanphier. "While I recall Mitscher stating he thought your flight should be on the mission," Mitchell wrote, "I do not believe he would be so presumptuous as to tell me who should make the hit, nor was he. This was my determination after giving the matter much thought." Later Mitchell continued, "You know as well as I that no naval commander would take the position of telling the squadron commander of an AAF unit how he should set up his flights and the exact tactics he should use to accomplish his mission, no matter that that squadron would be under his tactical control." Still further along in the letter, Mitchell continued, "You know me well enough that I would brook no interference in what I, as the squadron commander, would plan to do. When it was my JOB, my RESPONSIBILITY, and my life, I called the shots." Buttressing Mitchell's position is the fact that just moments before Mitscher, despite the views of his own staff, deferred to Mitchell on an aerial attack.

278 the problem: the compass in the P-38 cockpit: John Mitchell, videotaped comments as panelist at Yamamoto Mission Retrospective, Admiral Nimitz Museum, Fredericksburg, TX, April 1988; George T. Chandler, videotaped interview with Mitchell, May 18, 1989, MFP.

279 "A compass—one of those big Navy jobs": Adonis C. Arvanitakis, "Killing a Peacock: A Case Study of the Targeted Killing of Admiral Isoroku Yamamoto," master's thesis, School of Advanced Military Studies, United States Army Command and General Staff College, Fort Leavenworth, KS, 2015, 26; Burke Davis, *Get Yamamoto*, 124.

279 "The longest planned intercept ever": John Mitchell, letter and résumé in Mitchell album, MFP.

279 "By guess and by God": Burke Davis, interview with John Mitchell, September 27–28, 1967, DA.

280 "the Japanese would come up with possibly as many as 50 of these 75 fighters": John Mitchell, videotaped comments as panelist at Yamamoto Mission Retrospective, Admiral Nimitz Museum, Fredericksburg, TX, April 1988.

280 "We expected a hell of a battle": Burke Davis, interview with Hentry Viccellio, February 12, 1968, DA.

280 The tension among his base commanders during the day had nagged at him: Burke Davis, *Get Yamamoto*, 105–07, for which he corresponded with Watanabe.

280 selfless, not selfish, decisions: Wible, *The Yamamoto Mission*, 12.

281 "I wanted to get after those Zeros": Carroll V. Glines, *Attack on Yamamoto* (Atglen, PA: Schiffer Military History, 1993), 38.

281 there was no way Condon could put him into the right place at the right time: John Mitchell, letter to Tom Lanphier, July 9, 1984, GA.

281 "It was my squadron, my neck and my responsibility": John Mitchell, letters to Carroll V. Glines, November 27, 1988 and April 16, 1989, GA.

281 built upon assumptions: Burke Davis, interview with John Mitchell, September 27–28, 1967, DA.

281 Studying the map, he plotted: Wible, *The Yamamoto Mission*, 12, with map on 17; John Mitchell, videotaped comments as panelist at Yamamoto Mission Retrospective, Admiral Nimitz Museum, Fredericksburg, TX, April 1988; Glines, *Attack on Yamamoto*, 35–37; Burke Davis, interview with Mitchell, September 1967, DA.

282 Everyone knew about Yamamoto: Burke Davis, interview with Doug Canning, September 24, 1967, DA. Note: Canning was reciting the phrase attributed to Yamamoto to further demonize him but that was a distortion of what Yamamoto had actually said in a letter months before Pearl Harbor.

282 everybody wanted to go: Glines, *Attack on Yamamoto*, 36.

282 "about a thousand to one": John Mitchell, letter to Carroll V. Glines, November 27, 1988, GA; Glines, *Attack on Yamamoto*, 33.

283 Close to midnight, the men got the word to gather: Note: Some accounts put the briefing as early Sunday morning, before dawn, but Mitchell wrote, and said, that he had briefed the squadron late Saturday night. See John Mitchell, letter to Carroll V. Glines, November 27, 1988, GA: "I briefed the squadron after dark on the evening of April 17. The briefing was in the open on the hillside where our tents were located." Pilot D. C. Goerke, in his interview with Burke Davis in April 1968, also recalled that Mitchell's briefing was held late on the night of April 17, 1943.

283 the dogleg route: John Mitchell, videotaped comments as panelist at Yamamoto Mission Retrospective, Admiral Nimitz Museum, Fredericksburg, TX, April 1988.

283 "No one was to touch that mic button": Quoted in Glines, *Attack on Yamamoto*, 38.

284 follow him, and fly low but not too low: Burke Davis, *Get Yamamoto*, 135.

284 stick to their assigned roles: George T. Chandler, videotaped interview with John Mitchell, May 18, 1989, MFP; Wible, *The Yamamoto Mission*, 12.

284 the mission was a crapshoot: Wible, *The Yamamoto Mission*, 161.

284 The two intelligence officers were reassuring: Burke Davis, *Get Yamamoto*, 135.

284 "get the admiral at all cost": Rex Barber, letter to Carroll V. Glines, January 5, 1989, GA.

285 forced into crash-landing: Mack Morriss, *South Pacific Diary, 1942–1943* (Lexington: University Press of Kentucky, 1996), 153.

285 Glenn Miller's "Serenade in Blue": Burke Davis, *Get Yamamoto*, 137.

285 awake before dawn: Wible, *The Yamamoto Mission*, 14.

287 Everyone was ready to go. Except, suddenly, Joe Moore: John Mitchell, videotaped comments as panelist at the Yamamoto Mission Retrospective, Admiral Nimitz Museum, Fredericksburg, TX, April 1988.

288 95 degrees in the cockpit: John Mitchell, letter to Burke Davis, January 28, 1968, DA; Davis, interview with Mitchell, September 1967, DA; Davis, interview with Rex Barber, September 29–October 1, 1967, DA; Wible, *The*

Yamamoto Mission, 16–17; Glines, *Attack on Yamamoto*, 57–60; Burke Davis, *Get Yamamoto*, 141–43; Donald Davis, *Lightning Strike*, 251.

288   the admiral dressed in green: Ugaki, *Fading Victory*, 352.

289   8:00 a.m. on Guadalcanal: Note: For clarity and consistency, Guadalcanal time is used, which due to the time zone difference was two hours later than Rabaul's Tokyo time. In other words, Yamamoto took off at 8:00 a.m. Guadalcanal time, which was 6:00 a.m. Rabaul time.

289   casualness defined the day trip: R. Cargill Hall, interview with Zero pilot Kenji Yanagiya at the Yamamoto Mission Retrospective, Admiral Nimitz Museum, Fredericksburg, TX, April 1988.

289   barely any ammunition on board: Jay E. Hines, interview with Betty pilot Horishi Hayashi, June 22, 1990, in Japan, in Hall, *Lightning over Bougainville*, 142.

289   Yamamoto and Ugaki did not expect problems: Ugaki, *Fading Victory*, 353.

290   a portrait of perfection: Ugaki, *Fading Victory*, 353; Agawa, *The Reluctant Admiral*, 358–60; Arvanitakis, "Killing a Peacock," 28.

Chapter 16: Dead Reckoning

292   "Nothing, except waves": George T. Chandler, videotaped interview with John Mitchell, May 18, 1989, MFP.

292   the slightest chance of success made the mission worthwhile: D. C. Goerke, letter to Carroll V. Glines, December 15, 1988, GA.

292   Besby Holmes, usually calm in flight, fought the jitters: Eric Hammel, *Aces Against Japan: The American Aces Speak*, vol. 1, "The Hawk, Besby Holmes," 3–10.

292   Canning took to looking for sharks: Burke Davis, interview with Doug Canning, September 24, 1967, DA.

292   One pilot became so lulled: John Mitchell, videotaped comments as panelist at the Yamamoto Mission Retrospective, Admiral Nimitz Museum, Fredericksburg, TX, April 1988.

293   Would they be on time?: George T. Chandler, videotaped interview with John Mitchell, May 18, 1989, MFP; Mitchell, videotaped comments as panelist at the Yamamoto Mission Retrospective, Admiral Nimitz Museum, Fredericksburg, TX, April 1988.

293   he was right on time: John Mitchell, letter to Burke Davis, January 28, 1968, DA; George T. Chandler, videotaped interview with Mitchell, May 18, 1989, MFP; Mitchell, videotaped comments as panelist at the Yamamoto Mission Retrospective, Admiral Nimitz Museum, Fredericksburg, TX, April 1988; Burke Davis, interview with Jack Jacobson, October 3, 1967, DA.

294   "Bogeys, 10 o'clock high": Note: Nearly all accounts quote Canning as saying "Bogeys, 11 o'clock high." Canning himself told historian Burke Davis in an interview on September 24, 1967, that when he had spotted the Japanese flight, "I just said on the radio, 'Bogeys, 11 o'clock high.'" He said the same time during the 1988 panel at the Nimitz museum. But more recently, during his oral history interview on October 4, 2001, at the National Museum of the Pacific War in Fredericksburg, Texas, he corrected himself. He said, "A lot of books and things that have been written say that I said, 'Bogeys at 11 o'clock.' Really I said, 'Bogies 10 o'clock high.' At 11 o'clock we would have

never been able to get up there in time. We wouldn't have been far enough ahead to intercept them, so it had to have been 10 o'clock high." In addition, Mitch's wingman, Jack Jacobson, provided further backup, telling Burke Davis in their October 3, 1967, interview that someone [Canning] broke radio silence, saying "Target at 10 o'clock."

294   "we didn't even practice this one": Louis Kittel, letter to Carroll V. Glines, November 21, 1988, GA.

294   "This time Yamamoto didn't have surprise on *his* side": Eric Hammel, *Aces Against Japan: The American Aces Speak*, vol. 1, "The Hawk, Besby Holmes,"3–10.

294   Yamamoto was supposed to be in one bomber, that was all: George T. Chandler, videotaped interview with John Mitchell, May 18, 1989, MFP; Burke Davis, interview with Rex Barber, September 29, 1967, DA, during which Barber described the bombers as a glittering silver.

294   The Yamamoto trip had been uneventful: Major Adonis C. Arvanitakis, "Killing a Peacock: A Case Study of the Targeted Killing of Admiral Isoroku Yamamoto," master's thesis, School of Advanced Military Studies, United States Army Command and General Staff College, Fort Leavenworth, KS, 2015, 28.

295   Neither he, Ugaki, nor any of the pilots of the eight Japanese planes was aware: Matome Ugaki, *Fading Victory: The Diary of Admiral Matome Ugaki, 1941–1945* (Pittsburgh: University of Pittsburgh Press, 1991), 353; Hiroyuki Agawa, *The Reluctant Admiral: Yamamoto and the Imperial Navy* (Tokyo: Kodansha International, 1969), 348.

295   they had their target: George T. Chandler, videotaped interview with John Mitchell, May 18, 1989, MFP.

296   "Why the hell don't we go in and get 'em?": Burke Davis, interview with Doug Canning, September 24, 1967, DA.

296   issued the order "Skin 'em": Doug Canning, oral history interview, October 4, 2001, National Museum of the Pacific War; Burke Davis, *Get Yamamoto* (New York: Random House, 1969), 155, quoted Mitchell as saying "Everybody, skin tanks." Other accounts quoting Mitchell use different wording, but consistently they quote Mitchell as using the term "skin" in his order to his men to drop their tanks.

296   "climb like homesick angels": Eric Hammel, *Aces Against Japan: The American Aces Speak*, vol. 1, "The Hawk, Besby Holmes,"3–10.

296   "Tom, he's your meat": George T. Chandler, videotaped interview with John Mitchell, May 18, 1989, MFP; Mitchell, videotaped comments as panelist at the Yamamoto Mission Retrospective, Admiral Nimitz Museum, Fredericksburg, TX, April 1988.

296   Besby Holmes could not shed his belly tanks: George T. Chandler, videotaped interview with John Mitchell, May 18, 1989, MFP; Burke Davis, interview with Rex Barber, September 29, 1967, DA; Eric Hammel, *Aces Against Japan: The American Aces Speak*, vol. 1, "The Hawk, Besby Holmes," 3–10; Besby Holmes, videotaped comments as panelist at the Yamamoto Mission Retrospective, Admiral Nimitz Museum, Fredericksburg, TX, April 1988, during which he said he hadn't thought he had any choice but to first try to get rid of the belly tanks, saying "I had to shake my tanks before I could enter the fight."

297   planner's remorse: John Mitchell, letter to Tom Lanphier, July 9, 1984, GA.

Mitchell's quote in full: "I tell you, for sure, had I not believed, at the time, that the Japanese would most likely have an escort of 50 or so fighters greeting Yamamoto some distance from Kahili, the same as we did for Knox when he came to visit, then you probably would never have fired a shot that day as I could have taken my own flight in for the killer, Mitscher, Viccellio or anyone else notwithstanding, as I was in the front and in perfect position to do just that."

298   Ugaki's pilot "overspeeded": Ugaki, *Fading Victory*, 353–57; interview with bomber pilot Horishi Hayashi, June 1990, in R. Cargill Hall, ed., *Lightning over Bougainville: The Yamamoto Mission Reconsidered* (Washington, DC: Smithsonian Institution Press, 1991), 149–57; interview with Zero pilot Kenji Yanagiya, at the Yamamoto Mission Retrospective, Admiral Nimitz Museum, Fredericksburg, TX, April 1988, in *Lightning over Bougainville*, 111–20.

299   "They were a little bit asleep": George T. Chandler, videotaped interview with Rex Barber, May 18, 1989, MFP.

299   "Get the bombers": Burke Davis, *Get Yamamoto*, 157.

300   the second one penetrated his jaw on the lower left side: Hall, *Lightning over Bougainville*, 45–46; Carroll V. Glines, *Attack on Yamamoto* (Atglen, PA: Schiffer Military History, 1993), 67–69 and 105–06, citing the autopsy report showing the entry points of the two bullets on Yamamoto's left side, meaning that during the attack, Yamamoto had turned to the left in his seat.

301   They were unloading on him: Rex Barber, videotaped comments as panelist at the Yamamoto Mission Retrospective, Admiral Nimitz Museum, Fredericksburg, TX, April 1988; George T. Chandler, videotaped interview with Barber, May 18, 1989, MFP; Burke Davis, interview with Rex Barber, September 29, 1967, DA; John T. Wible, *The Yamamoto Mission: Sunday, April 18, 1943* (Fredericksburg, TX: Admiral Nimitz Foundation, 1988), 19–20.

301   "I got one of the bombers!": Burke Davis, interview with Rex Barber, September 29, 1967, DA; Davis, interview with Besby Holmes, February 14, 1968, DA, during which Holmes said that while he had been struggling to drop his tanks, he had heard Barber's voice over the radio shouting that he'd gotten a bomber.

301   He was dead: Ugaki, *Fading Victory*, 354; Agawa, *The Reluctant Admiral*, 358; Arvanitakis, "Killing a Peacock," 2, citing an interview with the Japanese army officer who was taken to Yamamoto's crash site the next day.

302   it had gone straight in: Besby Holmes, videotaped comments as panelist at the Yamamoto Mission Retrospective, Admiral Nimitz Museum, Fredericksburg, TX, April 1988; Burke Davis, interview with Besby Holmes, February 14, 1968, DA.

302   The action continued nonstop: Rex Barber, videotaped comments as panelist at the Yamamoto Mission Retrospective, Admiral Nimitz Museum, Fredericksburg, TX, April 1988; George T. Chandler, videotaped interview with Barber, May 18, 1989, MFP; Besby Holmes, videotaped comments as panelist at the Yamamoto Mission Retrospective, Admiral Nimitz Museum, Fredericksburg, TX, April 1988.

302   Holmes and Hine went into a steep dive: Note: In later interviews, Barber, Holmes, and Lanphier gave differing accounts of the downing of the bombers. Most notably, Lanphier said that after attacking the Zeros, he had made

a 270-degree left turn, caught up to Yamamoto, and shot down his plane. Holmes and Barber claimed credit for shooting down the bomber at sea. The pilots also said they'd downed various Zeros, although no evidence has ever emerged to support the downing of anything other than two Betty bombers. Distilled here are the parts of their versions that are consistent. For fuller accounts of each man's version, see Wible, *The Yamamoto Mission*; Carroll V. Glines, *Attack on Yamamoto*; Daniel Haulman, *Killing Yamamoto: The American Raid That Avenged Pearl Harbor* (Montgomery, AL: New South Books, 2015), or William Wolf, *13th Fighter Command in World War II: Air Combat over Guadalcanal and the Solomons* (Atglen, PA: Schiffer Publishing, 2004).

302   drifted ten miles out to sea: Burke Davis, interview and correspondence with Doug Canning, September 24, 1967, and after, DA.

303   "Let's get the hell out and go home": Burke Davis, interview with John Mitchell, September 29, 1967, DA.

303   the pilot who had gone down: John Mitchell, letter to John Wible, May 12, 1961, GA; Burke Davis, interview with Mitchell, September 29, 1967, DA; Wible, *The Yamamoto Mission*, 20; Jack Jacobson diary excerpts found online at usmilitariaforum.com in its section on the Yamamoto mission, http://www.usmilitariaforum.com/forums/index.php?/topic/194018-yamamoto-shoot-down-p-38-handwritten-diary/?hl=%2Byamamoto+%2Bmission. See Wolf, *13th Fighter Command in World War II*, 147, which, citing military documents, reported that the search for Hine was called off on May 28, 1943, and that he was last seen by Mitchell at about 9:40 a.m.; another report dated February 9, 1944, stated that Hine "was not seen to crash. No wreckage of the plane was found."

305   Yamamoto's plane was gone: Ugaki, *Fading Victory*, 354–55, 359.

305   The secret cable prompted an emergency gathering: Agawa, *The Reluctant Admiral*, 353–55.

305   "There could be only one Yamamoto": Samuel Eliot Morison, *Breaking the Bismarcks Barrier, 22 July 1942–1 May 1944* (Boston: Little, Brown, 1950), 129, n. 13.

305   "I got the son-of-a-bitch!": First Lieutenant Roger J. Ames, videotaped comments as a panelist at the Yamamoto Mission Retrospective, Admiral Nimitz Museum, Fredericksburg, TX, April 1988; note left at Nimitz Museum reception desk October 6, 1979, by Lieutenant Edward C. Hutcheson, the officer on duty on "Recon," the Guadalcanal fighter director, saying he heard Lanphier yelling those words over the radio as Lanphier approached the airfield; see Glines, *Attack on Yamamoto*, 81. Note: Lanphier did not deny having made a radio call but denied having said these particular words or that he'd said the plane he had shot down had contained Yamamoto. Note: Beginning with a newspaper article he wrote in 1945, Lanphier gave differing accounts over the years. In the 1945 article, he identified the bomber he had gone after as "Yamamoto's bomber. It was skimming the jungle, headed for Kahili," and then, after shooting it down, he wrote in the newspaper article, "That was the end of Admiral Isoroku Yamamoto." The following excerpt is from an article he wrote that appeared in the December 1966 issue of *Reader's Digest*, titled "I Shot Down Yamamoto," in which he described Barber in combat with Zeros while he eludes Zeros and goes after Yamamoto: "I kicked my ship over on

its back and looked for the bomber I had lost in the melee. Sheer panic does wonders for the vision. In one glance I saw Barber tangling with some Zeros even as two other Zeros bored in on me. Then I saw a green shadow streaking across the jungle below—the bomber, skimming just over the trees. I followed it down to treetop level, and began firing a long steady burst. Its right engine and right wing fell off, and the bomber crashed into the jungle."

305　"Lanphier had not cross-checked results": Lou Kittel, letter to Carrol V. Glines, November 21, 1988, GA.

306　the epitome of recklessness: First Lieutenant Roger J. Ames, videotaped comments as a panelist at the Yamamoto Mission Retrospective, Admiral Nimitz Museum, Fredericksburg, TX, April 1988.

306　Lanphier's dramatics: Joseph O. Young, letter to George Chandler, September 22, 1988, GA.

306　"I got him! I got him!": Colonel Bill Harris, letter to Carroll V. Glines, November 29, 1988, GA, reporting that he had heard Lanphier.

306　"I got him, I got him, is what Capt. Lanphier said": Arvanitakis, "Killing a Peacock," 36, quoting a poem from the diary of crew chief Robert Pappake.

306　celebratory reception the returning pilots were receiving at Fighter Two: Burke Davis, interview with John Mitchell, September 29, 1967, DA. Note: John F. Kennedy, a navy lieutenant, happened to be at port in Guadalcanal with his torpedo boat and later described watching the returning P-38s in Robert J. Donovan, *PT 109: John F. Kennedy in World War II* (New York: McGraw-Hill, 1961), 58.

307　gas tanks were nearly empty: Besby Holmes, videotaped comments as panelist at the Yamamoto Mission Retrospective, Admiral Nimitz Museum, Fredericksburg, TX, April 1988.

307　Holmes's plane wasn't the only one damaged: Wible, *The Yamamoto Mission*, 20; Wolf, *13th Fighter Command in World War II*, 147; George T. Chandler, videotaped interview with Rex Barber, May 18, 1989, MFP.

307　"that he had shot the big man down": Lou Kittel, letter to Carroll V. Glines, November 21, 1988, GA.

307　Holmes pressed his case: Besby Holmes, videotaped comments as panelist at the Yamamoto Mission Retrospective, Admiral Nimitz Museum, Fredericksburg, TX, April 1988.

308　"You raised hell with 'em, boys": Burke Davis, *Get Yamamoto*, 186.

308　"a marvel of performance on Mitchell's part": John Condon, oral history interview with the US Air Force, Historical Research Center, Office of Air Force History, March 8, 1989, Washington Navy Yard.

308　there must have been three Betty bombers: Wible, *The Yamamoto Mission*, 29.

308　"I couldn't have cared less who shot him down": George T. Chandler, videotaped interview with John Mitchell, May 18, 1989, MFP.

308　"singing like a bunch of high school kids after a ball game": Mack Morriss, diary, April 18, 1943, *South Pacific Diary, 1942–1943* (Lexington: University Press of Kentucky, 1996).

308　"Congratulations to you and Major Mitchell and his hunters": William Halsey, communication to Pete Mitscher, April 18, 1943, with copies going to Mitchell, Viccellio, Barber, Lanphier, and Holmes, MFP.

309   "We'uns is coming home!!": John Mitchell, letter to Annie Lee, May 3, 1943, MFP.

Epilogue

312   he transmitted official notice of the admiral's death: Hiroyuki Agawa, *The Reluctant Admiral: Yamamoto and the Imperial Navy* (Tokyo: Kodansha International, 1969), 354–58, 366–68; Daniel Haulman, *Killing Yamamoto: The American Raid That Avenged Pearl Harbor* (Montgomery, AL: New South Books, 2015), 18; John T. Wible, *The Yamamoto Mission: Sunday, April 18, 1943* (Fredericksburg, TX: Admiral Nimitz Foundation, 1988), 23–25.

312   a successor was immediately named: Burke Davis, *Get Yamamoto* (New York: Random House, 1969), 200.

313   Chiyoko felt faint: Donald M. Goldstein and Katherine V. Dillon, *The Pearl Harbor Papers: Inside the Japanese Plans* (New York: Brassey's, 1993), 131–34; Agawa, *The Reluctant Admiral*, 383–84.

313   news of Yamamoto's death was finally released on Friday, May 21: "Yamamoto Killed; Boasted He'd Dictate Terms in White House," *Washington Post*, May 22, 1943; "Japanese Admiral Killed in Combat: Commander of Fleet Had Said He Would Dictate Peace Terms in the White House," *New York Times*, May 21, 1943.

313   continuous outpouring of sadness: Agawa, *The Reluctant Admiral*, 386–92.

314   "Admiral Yamamoto's Sweetheart": Goldstein and Dillon, *The Pearl Harbor Papers*, 126–33.

314   "Public enemy number one of the American Navy is dead!": Chet Huntley, CBS radio broadcast, May 22, 1943.

315   every story repeating the jingoistic boast: "Yamamoto Killed; Boasted He'd Dictate Terms in White House," *Washington Post*, May 22, 1943; "Yamamoto Held Among Greatest Japanese Leaders," *Los Angeles Times*, May 22, 1943.

315   "can be forgiven, I hope, if we were pleased": Edwin T. Layton, letter to Burke Davis, July 13, 1968, DA.

316   Finally, FDR spoke: "Gosh!": Ray Brecht, "The President's Press Conference," May 21, 1943, Dispatches from *Time* magazine correspondents: first series, 1942–1955.

316   a faux letter to Yamamoto's widow: Grace Tully (FDR's personal secretary), archives, cited in Adonis C. Arvanitakis, "Killing a Peacock: A Case Study of the Targeted Killing of Admiral Isoroku Yamamoto," master's thesis, School of Advanced Military Studies, United States Army Command and General Staff College, Fort Leavenworth, KS, 2015, 41.

317   "comparable to what the loss of Gen. Douglas MacArthur would be to the United States": "Yamamoto Held Among Greatest Japanese Leaders," *Los Angeles Times*, May 22, 1943.

317   "His stirring fighting spirit lives on": Victor D. Hader, "Decapitation Operations: Criteria for Targeting Enemy Leadership: A Monograph," School of Advanced Military Studies, United States Army Command and General Staff College, Fort Leavenworth, KS, 2004, 34.

317   "the Japanese felt lost at sea": Arvanitakis, "Killing a Peacock," 39, 44. Note: Arvanitakis, on page 42, made an observation about the name commonly

attributed to the mission to get Yamamoto: Operation Vengeance. In his research of air forces archives, he found that the mission had never been given an official name and no mission had ever been named Vengeance.

317   "We had been 99 percent sure": Burke Davis, interview with John Mitchell, September 27–28, 1967, DA.

317   carry on, business as usual: R. Cargill Hall, ed., *Lightning over Bougainville: The Yamamoto Mission Reconsidered* (Washington, DC: Smithsonian Institution Press, 1991), 26.

318   they knew too much: Burke Davis, interview with Doug Canning, September 24, 1967, DA; Canning, letter to John T. Wible, November 9, 1962, GA.

318   "This operation was excellently conceived and splendidly executed": Lieutenant General M. F. Harmon, US Army Air Forces, South Pacific Area, letter to Major John Mitchell, May 26, 1943, MFP; Harmon, letter to Lieutenant General H. H. Arnold, commanding general of the Army Air Forces, Washington, DC, May 26, 1943, MFP.

319   paperwork was under way for everyone to get honors: Lieutenant General M. F. Harmon, U.S. Army Air Forces South Pacific Area, letter to Lieutenant General H. H. Arnold, commanding general of the Army Air Forces, Washington, DC, May 1, 1943, MFP.

319   "I'm tickled to death": John Mitchell, letter to Annie Lee, May 3, 1943, MFP.

320   "You're not worthy": Rex Barber, letter to Carroll V. Glines, January 5, 1989, GA.

320   the prowess of "our baby, the P-38": Office of Air Staff Intelligence, copy of confidential interview with Mitchell and Lanphier on June 15, 1943, and prepared on July 30, 1943, during which the pilots discuss the strengths and weaknesses of the P-38 Lightning, MFP; War Department press release, June 17, 1943, Washington, DC; transcript of Mitchell and Lanphier appearance on the "Army Hour," June 15, 1943, Washington, DC.

321   the Tom and Mitch show: "Lanphier Stars as 'Hot' Pilot in Pacific War," *Washington Post*, June 19, 1943; "P-38s Without Bombs Destroy Two Jap Ships," *Los Angeles Times*, July 24, 1943.

321   "the art of knocking Japs out of the sky": "Pilot Will Demonstrate Air Tactics to Cadets," *Los Angeles Times*, August 1, 1943; John Mitchell, diary number two, July 31, 1943, MFP.

322   "Stewart was really tickled": John Mitchell, letters to Annie Lee, July 20, 22, and 27, 1943; Mitchell, diary, July 25, 1943, MFP.

322   Mitch sneaked off to see Annie Lee: John Mitchell, diary, September 1 and 10, 1943, MFP.

322   he and Lanphier had been presented with Navy Crosses: Public Relations Officer, Fourth Air Force, background material on Mitchell and Lanphier for the July 16, 1943, award of the Navy Cross at Fourth Air Force Headquarters in San Francisco, MFP; John Mitchell, letters to Annie Lee, July 15 and November 3, 1943, MFP.

323   he was assigned to take command of the 412th Fighter Group at Muroc Army Airfield: John Mitchell, letter to Annie Lee, November 5, 1943, MFP; Mitchell, diary, December 29, 1943, MFP.

324   "It is entirely lawful": Eric H. Holder, Jr., "Attorney General Eric Holder Speaks

at Northwestern University School of Law," March 5, 2012, https://www
.justice.gov/opa/speech/attorney-general-eric-holder-speaks-northwestern
-university-school-law.

324 "Decapitation does not increase the likelihood of organizational collapse":
Jenna Jordan, "When Heads Roll: Assessing the Effectiveness of Leadership
Decapitation," *Security Studies* 18, no. 4 (December 2, 2009): 791–95; David
Ignatius, "Killing Top Terrorists is Not Enough," *Washington Post*, March 5,
2015; James A. Warren, "The Hit and Miss Record of U.S. Targeted Killing
Programs," Daily Beast, May 25, 2018.

324 "targeted killing needs to be more than eliminating the next person": Arvanitakis, "Killing a Peacock," 43–44.

325 "my feelings and thoughts were varied": John Mitchell, diary, April 21, 1945,
MFP.

325 he liked Iwo Jima better than Guadalcanal: Ibid.

326 "I got one Zeke and should have got another": John Mitchell, diary, June 27,
1945, MFP.

326 "will the Japs elect to fight it out": John Mitchell, diary, May 9, 1945, MFP.

326 "It's almost unbelievable": John Mitchell, diary, August 8, 1945, MFP.

327 Lanphier, the report said, had "nosed over, and went down to the tree tops":
"Fighter Command Debriefing," April 18, 1943, quoted in Carroll V. Glines,
Attack on Yamamoto (Atglen, PA: Schiffer Military History, 1993), 161.

328 "the man who shot down the plane": "Yamamoto's Killer Identified by Army:
Lieut. Col. T. G. Lanphier Jr., Son of Army Officer, Shot Down Admiral's
Plane in Trap," *New York Times*, September 11, 1945.

328 "He really hogs the show": John Mitchell, letters to Annie Lee, September 16
and 27, 1945, MFP.

328 a status he exploited: Neil Sheehan, *A Fiery Peace in a Cold War: Bernard Schriever
and the Ultimate Weapon* (New York: Random House, 2009), 253–61.

329 the two pilots had shot at the same plane: Glines, *Attack on Yamamoto*, 148.

329 Lanphier never budged: Tom Lanphier, "At All Costs Reach and Destroy," autobiography, GA. Note: For a thorough, succinct, and up-to-date account of
the credit dispute, see Haulman, *Killing Yamamoto*, 18–23.

330 Today, the island's downtown is a bustling port: Author visit to Guadalcanal,
May 2019.

331 "None of the accounts of the Yamamoto Mission give full credit to the superb
leadership": Besby F. Holmes, "Who Really Shot Down Yamamoto?," *Popular
Aviation*, March–April 1967, 64.

331 "the greatest fighter pilot that ever lived on this earth, is John Mitchell": Doug
Canning, videotape as speaker at the Yamamoto Mission Retrospective, Admiral Nimitz Museum, Fredericksburg, TX, April 1988.

331 "I have felt badly over this ever since": Rex Barber, letter to Carroll V. Glines,
January 5, 1989, GA.

SELECTED BIBLIOGRAPHY

Papers and Archives

Burke Davis Papers (DA), 1920–1987, #4569, Southern Historical Collection, The Wilson Library at the University of North Carolina at Chapel Hill.

C. V. Glines, Jr., Papers (GA), Special Collections and Archives Division, History of Aviation Collection, University of Texas Library, Dallas.

DENSHO Digital Archives, https://www.densho.org/archives/.

Dispatches from *Time* magazine correspondents, first series, 1942–1955, Harvard College Library Special Collections, Houghton Library.

Mitchell Family Papers (MFP): the letters of John Mitchell, Annie Lee Mitchell, and Noah Mitchell, along with diaries, military records, and newspaper clippings.

Nimitz Education and Research Center at the National Museum of the Pacific War, Fredericksburg, Texas: oral histories, 1988 Yamamoto Mission Retrospective.

Rohwer Reconstructed Archive, https://risingabove.cast.uark.edu/archive.

Rosalie Santine Gould–Mabel Jamison Vogel Collection, the Butler Center for Arkansas Studies, Little Rock.

Books, Articles, and Other Materials

Agawa, Hiroyuki. *The Reluctant Admiral: Yamamoto and the Imperial Navy.* Translated by John Bester. Tokyo: Kodansha International, 1969.

Arvanitakis, Adonis C. "Killing a Peacock: A Case Study of the Targeted Killing of Admiral Isoroku Yamamoto." Master's thesis, School of Advanced Military Studies, United States Army Command and General Staff College, Fort Leavenworth, KS, 2015.

Bradt, Hale. *Wilber's War: An American Family's Journal Through World War II.* Salem, MA: Van Dorn Books, 2015.

Brieger, James F. *Hometown Mississippi.* Jackson, MS: Town Square Books, 1997.

Carlson, Elliot. *Joe Rochefort's War: The Odyssey of the Codebreaker Who Outwitted Yamamoto at Midway.* Annapolis, MD: Naval Institute Press, 2011.

Cary, Otis, ed. *From a Ruined Empire: Letters—Japan, China, Korea, 1945–46.* Tokyo: Kodansha International, 1984.

Collie, Craig. *Code Breakers: Inside the Shadow World of Signals Intelligence in Australia's Two Bletchley Parks.* Sydney, Australia: Allen & Unwin, 2017.

Cook, Haruko Taya, and Theodore F. Cook. *Japan at War: An Oral History*. New York: New Press, 1992.

Davis, Burke. *Get Yamamoto*. New York: Random House, 1969.

Davis, Donald. *Lightning Strike: The Secret Mission to Kill Admiral Yamamoto and Avenge Pearl Harbor*. New York: St. Martin's Press, 2005.

Field, James A., Jr. "Admiral Yamamoto." *United States Naval Institute Proceedings 75*, no. 10 (October 1949): 1105–13.

Gamble, Bruce. *Fortress Rabaul: The Battle for the Southwest Pacific, January 1942–April 1943*. Minneapolis: Zenith Press, 2010.

Glines, Carroll V. *Attack on Yamamoto*. Atglen, PA: Schiffer Military History, 1993.

Goldstein, Donald M., and Katherine V. Dillon. *The Pearl Harbor Papers: Inside the Japanese Plans*. New York: Brassey's, 1993.

Grapes, Bryan J., ed. *Japanese American Internment Camps*. San Diego: Greenhaven Press, 2001.

Hader, Victor D. "Decapitation Operations: Criteria for Targeting Enemy Leadership: A Monograph." School of Advanced Military Studies, US Army Command and General Staff College, Fort Leavenworth, KS, 2004.

Hall, R. Cargill, ed. *Lightning over Bougainville: The Yamamoto Mission Reconsidered*. Washington, DC: Smithsonian Institution Press, 1991.

Hammel, Eric. *Aces Against Japan II: The American Aces Speak*, vol. 3. Pacifica, CA: Pacifica Press, 1996.

Haufler, Hervie. *Codebreakers' Victory: How the Allied Cryptographers Won World War II*. New York: New American Library, 2003.

Haulman, Daniel. *Killing Yamamoto: The American Raid That Avenged Pearl Harbor*. Montgomery, AL: NewSouth Books, 2015.

Heppenheimer, T. A. "Yamamoto and the Hijackers." *Defense World* 1, no. 2 (1989): 46–50.

Hersey, John. *Into the Valley: A Skirmish of the Marines*. New York: Schocken Books, 1942.

Hess, William. *Ace Profile* (American Fighter Pilot Series). Vol. 1, no. 2: *Col. Rex T. Barber*. Tucson, AZ: Mustang International, 1993.

Holmes, W. J. *Double-Edged Secrets: U.S. Naval Intelligence Operations in the Pacific During World War II*. Annapolis, MD: Naval Institute Press, 1979.

Honan, William H. *Visions of Infamy: The Untold Story of How Journalist Hector C. Bywater Devised the Plans That Led to Pearl Harbor*. New York: St. Martin's Press, 1991.

Horne, Alistair. *Hubris: The Tragedy of War in the Twentieth Century*. New York: HarperCollins, 2015.

Hoyt, Edwin P. *Yamamoto: The Man Who Planned the Attack on Pearl Harbor*. Guilford, CT: Lyons Press, 1990.

Hull, Michael D. "Japan's Naval War Leader," *WWII History* 1, no. 7 (Fall 2012): 26–31.

Kahn, David. *The Codebreakers: The Comprehensive History of Secret Communications from Ancient Times to the Internet*. New York: Scribner, 1996.

Lanphier, Thomas G., Jr. "At All Costs Reach and Destroy." Unpublished manuscript, US Army Military History Institute, 1984–85.

Lasswell, Alvin B. Interviewed by Benis M. Frank, historical division unit chief, Oral History Program, Historical Division, United States Marine Corps, April 1, 1968, at Lasswell's home in Rancho Santa Fe, CA.

Layton, Edwin T., Roger Pineau, and John Costello. "And I Was There": Pearl Harbor and Midway—Breaking the Secrets. New York: William Morrow, 1985.

McWilliams, Bill. Sunday in Hell: Pearl Harbor Minute by Minute. New York: E-Rights/ E-Reads, 2011.

Miller, John, Jr. Guadalcanal: The First Offensive. Washington, DC: Center of Military History, United States Army, 1949.

Morison, Samuel Eliot. History of United States Naval Operations in World War II. Vol. 5: The Struggle for Guadalcanal, August 1942–February 1943. Boston: Little, Brown, 1949.

——. History of United States Naval Operations in World War II. Vol. 6: Breaking the Bismarcks Barrier, 22 July 1942–1 May 1944. Boston: Little, Brown, 1950.

——. "Six Minutes That Changed the World," American Heritage 14, no. 2 (February 1963): 50–56.

Morriss, Mack. South Pacific Diary, 1942–1943. Edited by Ronnie Day. Lexington: University Press of Kentucky, 1996.

Morton, Louis. "Japan's Decision for War." United States Naval Institute Proceedings 80 (December 1954): 1325–34.

Okumiya, Masatake, Jiro Horikoshi, and Martin Caidin. Zero! New York: E. P. Dutton, 1956.

Potter, E. B. Bull Halsey. Annapolis, MD: Naval Institute Press, 1985.

Prados, John. Combined Fleet Decoded: The Secret History of American Intelligence and the Japanese Navy in World War II. New York: Random House, 1995.

——. Islands of Destiny: The Solomons Campaign and the Eclipse of the Rising Sun. New York: NAL Caliber, 2012.

Sledge, E. B. With the Old Breed: At Peleliu and Okinawa. New York: Ballantine, 2010.

Smith, Michael. The Emperor's Code: Breaking Japan's Secret Ciphers. New York: Arcade, 2000.

Stanaway, John. P-38 Lightning Aces of the Pacific and CBI. Oxford, England: Osprey, 1997.

Taylor, Blaine. "Ambush in Hostile Skies." Military History 5, no. 1 (1988): 42–49.

Taylor, Theodore. The Magnificent Mitscher. Annapolis, MD: Naval Institute Press, 1954.

Toland, John. The Rising Sun: The Decline and Fall of the Japanese Empire, 1936–1945, vol. 2. New York: Random House, 1970.

Toll, Ian W. The Conquering Tide: War in the Pacific Islands, 1942–1944. New York: W. W. Norton, 2015.

——. Pacific Crucible: War at Sea in the Pacific, 1941–1942. New York: W. W. Norton, 2012.

Twomey, Steve. *Countdown to Pearl Harbor: The Twelve Days to the Attack*. New York: Simon & Schuster, 2016.

Ugaki, Matome. *Fading Victory: The Diary of Admiral Matome Ugaki, 1941–1945*. Translated by Masataka Chihaya. Pittsburgh: University of Pittsburgh Press, 1991.

Wels, Susan. *Pearl Harbor: America's Darkest Day*. New York: Time-Life Books, 2001.

Wey, Adam Leong Kok. *Killing the Enemy: Assassination Operations During World War II*. London: I. B. Taurus, 2015.

Wible, John T. "The Yamamoto Mission." *AAHS Journal* no. 3 (Fall 1967): 159–68.

———. *The Yamamoto Mission: Sunday, April 18, 1943*. Fredericksburg, TX: Admiral Nimitz Foundation, 1988.

Wings at War Series. No. 3: *Pacific Counterblow*. Washington, DC: Headquarters, Army Air Forces, 1943. New imprint by the Center for Air Force History, Washington, DC, 1992.

Wolf, William. *13th Fighter Command in World War II: Air Combat over Guadalcanal and the Solomons*. Atglen, PA: Schiffer Publishing, 2004.

Yamashita, Samuel Hideo. *Daily Life in Wartime Japan, 1940–1945*. Lawrence: University Press of Kansas, 2015.

Zacharias, Ellis M. *Secret Missions: The Story of an Intelligence Officer*. New York: G. P. Putnam's Sons, 1946.

INDEX

Page references in *italics* refer to illustrations.

Aaron Ward, 251

Achilles, 139

acrobatics (planes), 56–57, 103, 140

aircraft carriers: *Akagi*, 47, 112–14, 173–74; in the Aleutians, 172; *Argus*, 42; at Battle of the Coral Sea, 158–59; bombers, 80, 83; and Bywater, 72; and Doolittle, 144, 158, 269; *Enterprise*, 154; *Hiryu*, 174; *Hornet*, 269; *Kaga*, 173–74; *Langley*, 132, 152; *Lexington*, 154, 159; Midway, 157, 169–70, 173–74, 203; Operation I-Go, 238; Pearl Harbor, 83, 110, 113, 152, 154; *Saratoga*, 154; Scout Bomber Douglas (SBD), 203; *Shoho*, 159; *Shokaku*, 159; *Soryu*, 174; Yamamoto's vision of, 41, 45–47, 77–78, 80; *Yorktown*, 159, 173, 175

Akagi, 47, 173–74, 112–14

Aleutians, 172, 176

American war effort, 155–56

Ames, Roger J., 225, 283, 306

Anglin, Everett H., 283

Argus, 42

Arizona, 2, 263

Army Hour, 320

Arnold, Henry "Hap," 182

Arvanitakis, Adonis C., 324

atabrines, 191, 222

Atlanta Constitution, 20

"Authorization for Use of Military Force," 323

B-17 Flying Fortress, 139; at Bougainville, 229; on Guadalcanal, 219; and Lanphier, 183; at Munda, 226–27; *My Lovin' Dove*, 285; on New Caledonia, 190

B-24 Liberators, 139; at Nadi, 144; on New Caledonia, 190; and Jimmy Stewart, 322

"Banzai," 214

Barber, Rex T., 118–20, *179, 209*, 226; death of, 329; on Faisi, 248–50; first kill, 225; leak, *319*, 331; on Lanphier, 183, 227, 328; Operation I-Go, 251; P-38s, 181, 229, 260; target fixation, 260; Yamamoto mission, aftermath, 318–20, 328–29; Yamamoto mission, attack, 299–301, 378n302; Yamamoto mission, briefing, 283–84; Yamamoto mission, return, 305, 307; Yamamoto mission, takeoff, 287

"barnstormers," 22, 26, 116

Bataan Death March, 137, 145

Battle of Bloody Ridge, 190

Battle of Santa Cruz, 220

Battle of the Coral Sea, 158–59, 162

Battle of the Java Sea, 152

battleship admirals, 47, 76, 77–78

"Battleship March," 152

battleships: *Arizona*, 2–3, 263; at Battle of Santa Cruz, 219–220; and Bywater, 45; *California*, 2;

battleships (*continued*): German, 139; Guadalcanal, 210–11, 219; *Haruna*, 210; and Henderson Field, 210–11; *Kongo*, 210; at Midway, 169, 175; *Mikasa*, 39; *Musashi*, 76, 236, 252; *Mutsu*, 169; *Nagato*, 80, 82, 86–87, 109–10, 112, 125, 127, 169; *Nevada*, 2; *Oklahoma*, 2; at Pearl Harbor, 2–3, 7, 83, 110, 113, 124, 125, 151, 154; in *Popeye*, 138; in Russo-Japanese War, 39; Three Nights to Remember, 210–11; *West Virginia*, 2; Yamamoto, 77–78, 80; *Yamato*, 76, 80, 167, 169–70, 174, 199, 234, 236

Bell Buckle, 15

Berengaria, 68

Biddle, Francis, 149

"Bigfoot Wallace," 141

Bilirakis, Michael, 331

bin Laden, Osama, 8, 323

Birth of a Nation, The, 20

Blitz, 89, 91, 95

Boone, Daniel, 14

Borneo, 137

Boshin War, 37

Bowen, Ben, 63

Bougainville, 229, 238. *See also* Yamamoto Mission

Brown, Ben, 63

Brown, G. O., 147–48

Buck, A. J., 225, 227

Bush, George W., 323

Buttons. *See* Espiritu Santo

Bywater, Hector C., 45–46, 72–73, 83

Cactus. *See* Guadalcanal

California, 2

Canning, Doug, 133; in Australia, 221–22; foxhole, 205; on Guadalcanal, 216; Independence Day party, 185; John Hersey on, 193; and Mitchell, 331; on Nadi,

140; on New Caledonia, 189–93; with Seabees, 212; skip bomb, 261; with 339th Fighter Squad, 224; Yamamoto Mission, aftermath, 318; Yamamoto Mission, attack, 302; Yamamoto Mission, briefing, 283–84, 375n283; Yamamoto Mission, first legs, 292; Yamamoto Mission, return, 305; Yamamoto Mission, sighting, 294, 376n294; Yamamoto Mission, take off, 286–87

Carson, Kit, 14

Carson, Jane, 14

CAST, 164

Chatfield, Ernle, 71

Chicago Tribune, 176, 218, 318, 359–60n185

Clark, Chase, 149

Classen, Thomas J., 285

Clausewitz, Carl von, 264

Coast Artillery Corps, 24, 59, 124

code breakers, 164, 330; Battle of the Coral Sea, 161–62; *Chicago Tribune*, 176, 318; coastal watchers, 186; Guadalcanal, 250; island code names, 193, 221, 240; Japanese code, 165–66; Japanese declaration of war, 85, 111–12; at Midway, 167–69, 175; outposts, 164; at Pearl Harbor, 8; and Yamamoto itinerary, 255–59, 263–65, 372n275. *See also* JN-25, Lasswell, Layton, Rochefort, *and* Station Hypo

Condon, John P., 228, 276, 281, 308, 373n277

Cromie, Robert, 218, 220, 247, 368n247

cruisers: *Achilles*, 139; Battle of the Java Sea, 152; convoy, first wartime to Pacific, 134; 5:5:3 formula, 47; Guadalcanal, 202, 219; *Isuzu*, 47; Midway, 169, 173, 175;

Mikuma, *161*, 175; *Nisshin*, 39; Pearl
Harbor, 83, 110, 113; *Phoenix*, 134;
in Russo-Japanese War, 39; Suva,
139; upper tonnage limit, 71
Czech pilots, 99

Daily Telegraph, 45, 72
Davis, Burke, 373n277
Davis, Norman H., 73
Dean, Jeanne, 184, 359–60n185
destroyers: *Aaron Ward*, 251;
Akagi, 174; American loaning
of, 121; Battle of the Java Sea,
152; and Bywater, 45; convoy,
first to Pacific, 134; *Greer*, 105;
at Guadalcanal, 188, 199, 202,
210, 218–19, 222, 238, 251; guns,
47; at Midway, 159, 169, 173–75;
Mitchell, 57; Pearl Harbor, 2, 83,
110, 113, 154; in Russo-Japanese
War, 39; Suva, 139; upper tonnage
limit, 71
Dickinson, Lillian Florence, 16, 20
Dinn, Wallace L., 141; Australia, 221–
22; death of, 230; on Guadalcanal,
212–19; Independence Day party,
180; on New Caledonia, 189; P-38s,
229; at Rekata Bay, 220–21; with
339th Fighter Squad, 224
Doolittle, Jimmy, 144–45, 158, 263,
269, 355n144
Douglas DC-3, 26
Dugout Sunday, 218
Dungeon. *See* Station Hypo
Dutch East Indies, 137

Earl Abel's, 30
Eisenhower, Dwight D., 122
Elliott, A. J., 150
Enterprise, 154
Espiritu Santo, 212, 239–40
Esquire, 184–85
Excambion, 101

F4F Wildcats: exercises, 181; Fighter
Two, 240, 242; Guadalcanal
dogfight, 217–18; handling of, 242;
at Henderson Field, 191, 211; in
Yamamoto Mission, 265, 269
Fantan. *See* Fiji
Fighter One, 191, 200, 211, 219, 240
Fighter Two, *197*, 204, 220, 286;
and crosswinds, 242; ground
crew, 285; new structure, 240;
operations dugout, 271; Three
Nights to Remember, 210; P-38s,
280; present day, 330; Washing
Machine Charlie, 231; Yamamoto
Mission, 283
Fiji, 134–44; bigwigs, 182, 240;
Chief, 142; evacuation plans, 137;
Independence Day party, 180,
185; Japanese plans, 144, 156; and
Lautoka, 140–41; and Midway, 177;
movies shown on, 138; and Nadi,
139–46; National Guard, 184; P-39s,
136–38; pilots, 152; rain, 136; Suva,
139; training, 140; Viti Levu, 136, 139
Finnegan, Joseph, 162–69, 256
5:5:3 agreement: Japanese
concessions, 47; Japanese hawks,
71; London Naval Treaty, 68; Pearl
Harbor, 152; Washington Naval
Treaty, 43; Yamamoto, 70
Fleet Radio Unit Pacific Fleet
(FRUPAC). *See* Station Hypo
Fleming, Ian, 94
Forrest, Nathan Bedford, 15
Fraser, Phyllis, 121, 122
Fuchida, Mitsuo, 86, 115, 157
Fukudome, Shigeru, 83, 305

Genda, Minoru, 83, 86
go, 35–36
Goerke, Delton C., 225; and skip
bombs, 261; Yamamoto Mission,
283, 287, 292, 305

Goettge, Frank B., 187
Gone with the Wind, 30, 66
Goshu Maru, 167
Graebner, Lawrence A., 283
Great Pacific War, The (Bywater), 45, 72–73, 83
Greer, 105
Griffith, D. W., 20
Gross, Ellery, 90–105
Guadalcanal, 186–94; "drum method" resupply, 223; "Pistol Pete," 204, 210; "Starvation Island," 222; "Washing Machine Charlie," 204–5; diseases, 3–4; Flanders Field, 247, 368n247; Japanese offensives, 175, 209–20, 222; Japanese retreat, 234–36; Kokumbona, 220; present day, 329–30; record player, 246, 368n246; Seabees, 212; "1,000 yard stare," 206–7; Three Nights to Remember, 210; trophy taking, 217
Guthrie, Woody, 32

Hague Convention of 1907, 259
Halsey, William F. "Bull," Jr., 214–15, 233, 239–41; and December 7 anniversary, 224; Guadalcanal, 236; inspection tour, 240; Yamamoto mission, aftermath, 320; Yamamoto mission, congratulations, 308; Yamamoto mission, itinerary, 266
Hamzawa, Nobuko, 199
Harmon, Millard Fillmore "Miff," 318; on Fiji, 182–83; on Guadalcanal, 189, 193; and Mitchell, 31
Harper's Magazine, 36–37, 41, 151
Harris, Field, 276
Haruna, 210
Hayashi, Hiroshi, 372n275, 376n289, 378n298

Henderson, Lofton R., 188–89
Henderson Field, 188, 194; Battle of Bloody Ridge, 190; bombing of, 199; Christmas dinner on, 227; corpses, 206, 220; final Japanese attack, 222; new aerial defense, 229; Operation I-Go, 238, 250–51; Operation KE, 235–26; Opium Den, 273; present day, 330; reorganization, 240; Third Battle of Matanikau, 202, 206, 213, 215–18; Three Nights to Remember, 210–11; Washing Machine Charlie, 230
Hero, Andrew, Jr., 25
Hersey, John, 193, 201–2
Hie-maru, 67
Hine, Raymond K., 283; Yamamoto Mission, aftermath, 307, 319; Yamamoto Mission, attack, 284, 287, 297, 299, 301–2; death, 303–4
Hirohito, 7, 86, 87, 110, 234, 264
Hiryu, 174
Holder, Eric H., 324
Holmes, Besby Frank, 5, 179; on Bougainville, 229; on Guadalcanal, 200–201; jungle rot, 242; and Mitchell, 331; at Pearl Harbor, 115–18, 200; with 339th Fighter Squad, 225; Yamamoto Mission, aftermath, 318–20, 321; Yamamoto Mission, attack, 287, 294, 378n301, 302, 378n302; Yamamoto Mission, belly tanks, 296–97, 301–2; Yamamoto Mission, briefing, 283–84; Yamamoto Mission, first legs, 292; Yamamoto Mission, return, 305, 307
Holmes, J. "Jasper," 168–69, 257–59, 262
Honolulu Advertiser, 25
Hoover, Herbert, 22

Hornet, 269
Hull, Cordell, 105–6, 111
Huntley, Chet, 314

Imamura, Hitoshi, 268, 275
Inoue, Shigeyoshi, 162
Isuzu, 47

Jacobson, Julius "Jack," 189; in
    Australia, 221–22; on Guadalcanal,
    216, 219; and Mitchell, 192; at
    Rekata Bay, 220–21; with 339th
    Fighter Squad, 224; Yamamoto
    Mission, 283, 286,305, 318, 322
Japan, declaration of war, 87
Japan, military establishment, 74
Japanese Americans, internment of,
    8–9, 145, 149–50, 198, 361n199
Japanese naval code. *See* JN-25
Jenkins, Jack, 55
JN-4, 121
JN-25, 8, 165–66, 181; at Midway, 169,
    176–77; in Yamamoto Mission,
    256–57
Johnson, Lyndon Baines, 182, 184
Johnston, Stanley, 176
Jordan, Jenna, 324
Joshima, Takaji, 268, 275–76, 372n275

Kaga, 173–74
Kates, 115
Kawai, Chiyoko, 6, 50–51; in
    Ginza District, 77, 87; at home,
    80; Itsukushima, 86–87; in
    Kamiyacho, 237; Kure, 170–71;
    Osaka, departure for, 79;
    pleurisy, 170–71; troika, 77; and
    Yamamoto's death, 312–14
Kawai, Chiyoko, Yamamoto letters
    to: code breaking, 176; dream, 75;
    exploitation, feelings of, 71; fan
    mail, 126, 153; hair, 239; his health,
    238–39; *kwu kwu*, 153; longest

letter, 238–39; and Midway, 171,
    238; Operation I-Go, 239; Pearl
    Harbor, departure for, 110; poem,
    239; staff, 238; Tokyo, hatred of,
    75; war and love, 171
Kenney, George C., 182
Kido Butai, 169
King, Ernest J., 176
Kittel, Louis R., 283, 294, 305
Knox, Frank, 4, *233*; Guadalcanal,
    215; inspection tour, 240; Los
    Angeles, 148; targeted killing,
    373n277; and Yamamoto's
    itinerary, 264
Kobayashi, Mary, 198
Koga, Mineichi, 312–13
Kokumin, 45
Kongo, 210
Konoe, Fumimaro, 81
Kopecky family, 26–29
Krock, Arthur, 155
Ku Klux Klan, 15, 20
Kusaka, Ryunosuke, 113, 114–15
kwu kwu, 153

Langley, 132, 152
Lanphier, Thomas G., Jr., 5, 120–23,
    179; ambitions, 183, 227, 249–50,
    359n183; death of, 329; on Faisi,
    248–49; and Jimmy Doolittle, 144;
    leave, 248; on Munda airfield, 226;
    in Operation I-Go, 251; P-38s, 229,
    260–61, 320–21; and skip bomb,
    261; and targeted killing, 373n277;
    with 339th Fighter Squad, 225–26;
    Yamamoto Mission, aftermath,
    318–23; Yamamoto Mission,
    attack, 287, 299, 303, 377n297,
    378n302; Yamamoto Mission,
    briefing, 283–84, 374n278;
    Yamamoto Mission, claims,
    305–7, 327–29, 379n305, 383n329;
    Yamamoto Mission, report, 308

Lasswell, Alva B. "Red," 162–63, 257–60, 370n258; code altering, 169; as cryptanalyst, 164, 257; Dungeon, 166; in Japan, 164; and Joe Finnegan, 163; in language school, 164; and Midway, 165–66, 172; and Nimitz, 259–60; at Pearl Harbor, 165; at Port Moresby, 162; at Rochefort, 256; at Station Hypo, 164; Yamamoto Mission, 168, 256, 257–58

Layton, Edwin T., 48; and Clausewitz, 264; and targeted killing, 259, 263–64, 373n277; and Yamamoto, 48, 262; and Yamamoto's death, 315; Yamamoto's itinerary, 259, 263–66

Lee, Gypsy Rose, 184

Lexington, 154, 159

Life, 191, 202

Lindbergh, Charles, 46, 92, 122

Lodge, J. Norman, 319

London, bombing of, 91

London Naval Conference, 47, 50

London Naval Treaty, 68, 70

Long, Albert R., 283

Los Angeles Times, 147, 215, 321

Lott, Evelyn, 20–21

MacArthur, Douglas, 153, 317

malaria, 3–4, 191, 222, 227

Malaysia, 137

Mariposa, 134

Marshall, George C., 122

Maruyama, Masao, 213, 216, 220, 252

Massey, Eunice, 21–22

McGuigan, Joseph E., 282, 307, 308

McLanahan, Jim, 278, 283–84, 287

Midway Island, 156–59, 161–77; *Akagi*, 174; American press, 175–76, 318; American stakes, 157; damage, 175; Japanese armada, 169; Japanese attack, 173;

Japanese bombers, 170; Japanese cover-up, 175, 224; JN-25 code, 165–69, 176, 256–58, 265; *Mikuma*, 161; Nagumo, 173; Nimitz, 175; operation end, 174; Point Luck, 173; Roosevelt and, 234; SBDs, 203; Yamamoto, effect on, 200; Yamamoto's plan, 156–59, 170, 172; *Yorktown*, 173

Mihashi, Reiko, 6, 41

Mikasa, 39

Mikuma, 161, 175

Mitchell, Annie Lee (née Miller), 26–29, *131*; anniversary of, 224; Aunt Ludma, 30, 91, 100–101, 215; birthday, 64; childhood, 26–28; Christmas, 65, 133; correspondence, 62, 63, 142–43, 145, 244–45; dating, 30–31; handwriting, 97; Joan, 326; meeting Mitchell, 29–30; Noah, 91, 96; open relationship, 64; in San Anselmo, 326–27; in San Antonio, 320; support letter, 262; telegram, 222–23; and Theresa Ann, 325; university, 28–29; Waco, 102; wedding of, 66, 106–7, *131*, 132

Miller, Armin W., 26–28

Mitchell, Billy, 122

Mitchell, Eunice Massey, 21, 22, 91, 96, 322

Mitchell, Joan, 326

Mitchell, John William, 14, 17–19; ace pilot, 241; acrobatics, 56–57; after the war, 326–27; and Annie Lee, 29–30, 62–64, 224; Auckland, 260; Australia, 221–22; with "barnstormers," 22; basketball, 61; Ben Bowen, 63; Billy, 326; birthday of, 185; on Bougainville, 229–30; and the British, 96; daughter, 325; death of, 331; dives, 57; "drum method" resupply, 223; Edith, 107;

education, 23–24; England, 93–102, 350n100; and Eunice, 22; 15th Fighter Group, 325–26; 51st Fighter Wing, 326; on Fiji, 136–46, 180, 184; first kill, 203, 206; 412th Fighter Group, 324; in Georgia, 25; at Golden Gate Bridge, 135; at Grand Canyon, 32; on Guadalcanal, 193–94, 216–17; on Hamilton Field, 53; Harmon, 318; Harris tweed, 97; in Hawaii, 24–25; Independence Day, 179, 185–86; Iwo Jima, 325–26; Joan, 326; Korea, 326; *Life*, 191–92; malaria, 222; in Massachusetts, 123–24; at Matanikau, Third Battle of, 202; at Midway, 177; mock dogfights, 57–58; on Moffett Field, 32–33; moon, 33, 62, 94, 124, 244; morale, 191–92; movies, 63; at Muroc Dry Lake, 58–59; on New Caledonia, 189–93; P-36 Hawk, 53; P-38, 190, 228–29; P-39, 203; P-40, 54, 58; P-51, 326; P-59, 325; P-80, 325; Pacific, trip to, 134; and Pearl Harbor attack, 124; pilot wings, 31; *President Monroe*, 135; proposal, 66; and queen of England, 95–96; RAF, 90; Rekata Bay, 220–21; "rugged individualism" of, 22; the slot, 202–3; Spitfire, 98–99; strip maps, 282; and Strother, 230; telegram, 223; with 339th Fighter Squadron, 200, 224; training exercise, 60; trip home, 320; Vandegrift, 200; war games, 102; Washing Machine Charlie, 230–32, 366n231; wedding, 106–7, *131*, 132; Wendell Willkie, 92; on Wright Field, 91; wristwatch, 243; Yamamoto Mission, aftermath, 318–23; Yamamoto Mission, attack, 296, 377n297; Yamamoto Mission, briefing, 283–84, 375n283; Yamamoto Mission, final leg, 293; Yamamoto Mission, first legs, 291–93; Yamamoto Mission, Lanphier, 303; Yamamoto Mission, operation meeting, 276–81, 374n278; Yamamoto Mission, return, 305–7; Yamamoto Mission, takeoff, 286; Yamamoto Mission, two bombers, 295

Mitchell, Lillian Florence Dickinson, 16,

Mitchell, Noah Boothe, *13*, 15–23, 198, 243, 322; and Annie Lee, 91, 96–97; poem, 13–14, 340n14; racism, 31

Mitchell, Theresa Ann, 325

Mitchell, William "Billy," 69

Mitchell, William Carson, 14, 18

Mitscher, Marc "Pete," 241; and Doolittle, 269; on Guadalcanal, 251; Yamamoto Mission, 268–79, 286, 308, 373n277, 374n278; Yamamoto Mission, aftermath, 317–19, 330

Mitsubishi, 7, 47, 76, 265

Montgomery, Monty, 90

Moore, Boyd, 103

Moore, Joseph, 225, 278, 283–84, 287

Morrison, William, 282, 307

Moore, Boyd, 103

Morriss, Mack, 231, 236, 308, 250–51

Musashi, 76, 236, 252

Mutsu, 169

My Lovin' Dove, 285

Nagano, Osami, 85–86

Nagato, 80, 82, 86–87, 109–10, 112, 125, 127, *169*

Nagumo, Chuichi, 86, 110–13, 126, 167–74

navigational instrumentation, 46

"Negro Mammy," 21

Nevada, 2
New Guinea: and Guadalcanal, 238; and Hirohito, 234; and Johnson, 182; and New Britain, 5; and Operation I-Go, 251; and Operation MO, 158; and Port Moresby, 144, 158; and Yamamoto, 5
New York Times, 45, 155, 327–28
Nimitz, Chester W., 176, *233*; and Battle of the Coral Sea, 161–62; and Guadalcanal, 214, 217, 239–40; and Midway, 168–69, 175–76; Midway, aftermath, 180–81; and targeted killing, 263–64, 371n264, 373n277; Yamamoto Mission, 259–60, 263–66, 269, 318
Nisshin, 39

Obama, Barack, 323
ocean liners: *Berengaria*, 68; *Hiemaru*, 67; *Mariposa*, 134; *Phoenix*, 134; *President Coolidge*, 134; *President Monroe*, 132, 133–36
Ohnishi, Takijiro, 83
Oikawa, Koshiro, 82
Oklahoma, 2
Olson, Mae, 242
On War (Clausewitz), 264
OP-20-G, 164, 167, 256
Operation Flipper, 105
Operation I-Go, 238–39, 250–52; American loses, 251, 273, 369n251; and Chiyoko Kawai, 239; and Faisi, 250; and Milne Bay, 267
Operation KE, 235
Operation MO, 158
Operation Watchtower, 187
Operation Z, 83
"Opium Den, the," 273
"Outline of Proposed Basis for Agreement Between the United States and Japan," 106

P-38 Lightning, 93, *271*; "fork-tailed devil," 248; cockpit bubble, 287; compass, 278; cone of fire, 228, 285, 296; Fighter Two, 240; high-altitude capability, 298; kinks, 103–5, 190; showcasing, 320–23; transition to, 227–29, 242; Yamamoto's itinerary, 269, 275
P-39 Airacobra, 103, 134, 140; gun, 203; handling, 137–38, 218; weight, 181
P-40 Warhawk, 54–58, 103, 117, 152, 211
P-59 Airacomet, 325
P-80 Shooting Star, 325
Pacific Affairs, 72
paranoia, 123, 148–49
Pearl Harbor, 46–47, 151; aircraft carriers, 41, 47; American fleet, 57; American losses, 151; attack on, 1–3, 87, 111–18; "Combined Fleet Top Secret Operational Order Number 1," 86; dredging, 46; and Fuchida, 115; and Holmes, 115–18; Japanese pilot captured, 149, 355n149; Japanese training for, 85; Kusaka, 113, 115; Operation Z, 83; and Roosevelt, 3; tactical shortcomings, 154; and Taylor, 117; and Welch, 117; Yamamoto's plan, 7, 9, 46, 72, 83–86. *See also* Station Hypo
Petit, Robert, 225
Philippines: CAST, 164; "Combined Fleet Top Secret Operational Order Number 1," 3, 86, 125, 137; Bataan Death March, 145; and Bywater, 45, 72; and Nagano, 85
Phoenix, 134
"Pistol Pete," 204, 210, 212, 218
Popeye, 138
Popular Aviation, 331
Poppy. *See* Caledonia

Point Cruz, 187
Port Moresby, 158–59, 161–62
Preliminary Naval Limitation
    Conference, 70–74
President Coolidge, 134
President Garfield, 118–20, 122
President Monroe, 132, 133–36
Price, Willard, 36–37, 40–41, 151
Pugh, Edwin L., 276
Purnell, Fred, 219

Rabaul, 6; and Guadalcanal, 222,
    238, 239, 250, 251; invasion of, 137;
    party, 266; Seventeenth Army, 188.
    See also Yamamoto Mission
racism: Griffith, 20; Hawaii, 149;
    Japanese submarine, 149; Kyser,
    246–47; Lott, 20–21; Mitchell,
    142, 146, 180, 215, 221, 223, 269;
    Mitchell family, 20–21; Price,
    41; *Time*, 150; Vardaman, 20–21.
    See also Japanese Americans,
    internment of
Read, William A., 276
Reader's Digest, 329
Rice Bird, 28
Rickenbacker, Eddie, 120
Ring, Stanhope C., 276
Ringbolt. *See* Tulagi
Rochefort, Joseph, 48–49, 162–69,
    256
Rohwer War Relocation Center,
    198–99
Rommel, Erwin, 105, 259
Roosevelt, Franklin Delano: "arsenal
    of democracy," 32; infamy speech,
    3; Japanese Americans, internment,
    145, 149, 198; Japanese declaration
    of war, 111–12; neutrality, 105;
    State of the Union, 233–34; and
    targeted killing, 364, 373n277; War
    Relocation Authority, 198; War
    Shipping Administration, 134;

wartime footing, 182; Yamamoto's
    death, 315–16
Roosevelt, Nicholas, 45
Rozhestvensky, Zinovy, 39–40
Russell, Jane, 184
Russo-Japanese War, 38–39

Saratoga, 154
Sasakawa, Ryoichi, 84, 125–26
Seabees, 191, 210, 212, 219
Second Yamamoto Mission
    Association, 331
sharks, 148, 188, 213, 292
Shepley, Jim, 261
Shibata, Takeo, 199
Shoho, 159
Shokaku, 159
Siboney, 94
skip bombs, 261
Smith, William E., 283
Solomon Islands, 188, 238; and
    Australia, 175; coastal watchers,
    186; Florida, 187; Honiara, 330;
    Nagano, 85; New Britain Island, 5,
    137, 175, 238; Slot, the, 199; Tulagi,
    186–87, 193, 219; and Yamamoto,
    5, 238, 265. *See also* Bougainville,
    Guadalcanal, *and* Rabaul
Soryu, 174
Spitfire, 98–100, 273
Standley, William H., 71, 73
Station Hypo, 162–69, 256–59
Stevens, John Paul, 256, 369n256
Stewart, Jimmy, 321–22
Stimson, Henry, 148
Stratton, Eldon E., 283
strip maps, 282, 286
Strother, D. C. "Doc," 183, 189, 192,
    230, 240
subchaser, 258, 275, 277–78
submarine, Japanese, 147–49
Sugita, Shoichi, 275
Suwa-maru, 41

Takahashi, Sankichi, 126
Takano family, 37, 40
targeted killing: ACLU, 323–24;
    aerial interception, 309;
    Arvanitakis, 324; bin Laden, 8,
    323, 340n8; Hague Convention
    of 1907, 259; Holder, 324; Jenna
    Jordan, 324; Lasswell, 258–59;
    Layton, 263–64; Nimitz, 263–64;
    precedent, 323–324; "ramming
    Yamamoto's plane," 277; Rommel,
    105, 259; Roosevelt, 264, 371n264,
    373n277; Stevens, 369n256
Taylor, Kenneth M., 117
TBF Avenger, 191
Third Battle of Matanikau, 202
339th Fighter Squadron: Harmon,
    189; Mitchell, 5, 224; pilots,
    200–201; Viccellio, 242; Yamamoto
    Mission, 278
Three Nights to Remember, 210
Time, 149, 150, 261, 315–16
Togo, Heihachiro, 39–40, 45, 112,
    151, 314
Tojo, Hideki, 85–86, 125, 313
Topoll, George, 225
tri-power talks, 68
Tripartite Pact, 55, 78, 81
Tsurushima, Masako, 40–41, 77, 110
Tulagi, 186–87, 193, 219

Ugaki, Matome: on Guadalcanal,
    189, 238, 250–51, 273; Musashi, 236;
    on New Guinea, 238; in Operation
    I-Go, 250–51, 273; Yamamoto's
    inspection, 239, 268, 274–75,
    288–90, 294–95, 298–99, 304–5, 312,
    376n289; Yamamoto's mood, 237
Uryu, Sotokichi, 36, 40

Vals, 114–15
Vandegrift, Alexander, 188, 200, 213
Vardaman, James K., 20–21

Vargas, Alberto, 184–85, 359–60n185
Viccellio, Henry "Vic," 59; on Fiji,
    137, 140, 152, 181; on Guadalcanal,
    240, 242; and Harmon, 183; and
    Lanphier, 183; and Mitchell,
    270; and Pearl Harbor, 122–24;
    promotion, 140; skip bombs,
    261; war games, 101; Yamamoto
    Mission, 270–73, 280, 284
"victory disease," 153
Virden, Ralph, 104
Vultee BT-13, 32

war bonds, 156, 224
War Relocation Authority, 198
Washing Machine Charlie, 205, 215,
    229–32, 330, 366n231
Washington Naval Treaty, 43, 73
Washington Post: Guadalcanal, 236;
    Lanphier, 249; Pearl Harbor, 224;
    Yamamoto, 150, 315, 321
Watanabe, Yasuji: Yamamoto's
    death, 305; Yamamoto's funeral,
    313; Yamamoto itinerary, 268,
    275, 280, 372n275; Yamamoto's
    remains, 311–12
Webb School, 15–16, 23
Weekly Asahi, 314
Welch, George S., 117
West Virginia, 2
Wheeler Field, 1, 115–18
"When Heads Roll: Assessing
    the Effectiveness of Leadership
    Decapitation" (Jordan), 324
Whitaker, Gordon, 283
Willkie, Wendell, 58, 92
Wilson, Woodrow, 19

Yamamoto Mission, 286–87, 291–94,
    291, 296, 380n306; cockpit bubble,
    287; cover-up flights, 317–18; name
    of, 381n317; operation meeting,
    271–84; pilots, 283–84, 291, 317

Yamamoto, Isoroku, 5–6, 37–38, 40; Aeronautics Department, 47, 75–76; aircraft carriers, 41–47, 75; America, tours of, 41–46, 343n37; battleships, 77–78; Bywater, 45–46, 72; children, 49; China, 49, 78; Chiyoko, 170, 238–39; Combined Fleet, 79–80; "Combined Fleet Top Secret Operational Order Number 1," 86; death, 300–1, 314–16, 327–29; declaration of war, 111, 126–27; fatalism, 200, 224, 234; First Carrier Division, 50; football, 68; funeral, 311, 313–14; go, 35–36; gambling, 43, 44; Guadalcanal, 199, 235; health, 172, 237, 239; Hirohito, 87, 110; Hitler, 73; inspection tour, 7, 274–76, 288–90, 294–98, 372n275, 376n289; Japanese hawks, 49–50, 74, 78, 82–83; Kasumigaura Air Corps, 43–44; and Lincoln, 42; and Lindbergh, 46; London Naval Conferences, 47, 50, 70; London Naval Treaty, 68, 70; marriage of, 49; Midway Island, 156–58, 169–76; Nagaoka, 37–38, 75; and Napoleon, 200; as national hero, 151; Naval Affairs Bureau, 74; Naval Cadet School, 38; and Naval General Staff, 50; navigational instrumentation, 46; Navy Torpedo School, 46–47; New York, 68–69; Oikawa, letter to, 82–83, 349n83, 349n84; oil, 43; Operation I-Go, 238, 250–52, 267, 273; Operation Z, 83; "Opinions on War Preparations," 82–83, 349n83, 349n84; peace, desire for, 154–55, 157; and Pearl Harbor, 84–85, 87, 110–14, 125, 224; Preliminary Naval Limitation Conference, 70–74; propaganda poster, 147; Rabaul base party, 266–67; Reiko, 6, 41, 43, 49; Russo-Japanese War, 38–39; Sasakawa letter, 125–26, 155, 315; Tripartite Pact, 68, 81; Tsurushima, Masako, 40, 77, 110; vice minister of navy, 76; war games, 80; Washington DC, 44; Washington Naval Treaty, 43; Willard Price, 36–37, 41

Yamamoto, Reiko, 6, 41, 49, 312, 313

Yamato, 76, 80

Yanagiya, Kenji, 372n275, 376n289, 378n298

Yank, 231, 236, 250, 308

Yonai, Mitsumasa, 76, 78–79

Yorktown, 159, 173, 175

Yoshida, Zengo, 51

Young, Joseph, 306

Yoyogi parade grounds, 153

Zacharias, Ellis M., 44, 46

Zero (fighter plane), 47, 75, 114

ABOUT THE AUTHOR

Dick Lehr is a professor of journalism at Boston University. He previously wrote for the *Boston Globe,* where he was a member of the *Globe*'s Spotlight Team, a special-projects reporter, and a magazine writer. While at the *Globe,* he was a Pulitzer Prize finalist in investigative reporting and the winner of numerous national awards. Lehr is the author of seven previous works of nonfiction and fiction, including the Edgar Award winner and *New York Times* bestseller *Black Mass: Whitey Bulger, the FBI, and a Devil's Deal,* which became the basis of the Warner Bros. feature of the same name. *The Birth of a Movement: How* Birth of a Nation *Ignited the Battle for Civil Rights* became the basis for a PBS/Independent Lens documentary. Lehr lives outside Boston with his family.